Antique Dresser Sets

1890s to 1950s

Roseann Ettinger

4880 Lower Valley Road, Atglen, PA 19310 USA

Dedication

To my father, Vito Rodino,
who was always a pillar of strength,
and to my mother and best friend, Marie Rodino,
who are both so very dearly missed.

Acknowledgments

Sincere thanks go to my wonderful husband, Terry Ettinger, for his continuing patience and support throughout the research and writing stages of this book. Also to my younger daughters, Lexie and Sabrina, for trying to understand that, at times, I needed peace and quiet! And again, thanks to Lexie, for teaching me a thing or two on the computer. I am especially grateful to Bruce Waters, at Schiffer Publishing, for his photography expertise. And last, but certainly not least, thanks to Peter and Nancy Schiffer, for their continuing faith in my work.

Library of Congress Control Number: 2005929656

Designed by Mark David Bowyer
Type set in ShelleyAllegro BT / Souvenir Lt BT

ISBN: 0-7643-2237-0
Printed in China
1 2 3 4

Published by Schiffer Publishing Ltd.
4880 Lower Valley Road
Atglen, PA 19310
Phone: (610) 593-1777; Fax: (610) 593-2002
E-mail: Info@schifferbooks.com

For the largest selection of fine reference books on this and related subjects, please visit our web site at
www.schifferbooks.com
We are always looking for people to write books on new and related subjects. If you have an idea for a book please contact us at the above address.

This book may be purchased from the publisher.
Include $3.95 for shipping.
Please try your bookstore first.
You may write for a free catalog.

In Europe, Schiffer books are distributed by
Bushwood Books
6 Marksbury Ave.
Kew Gardens
Surrey TW9 4JF England
Phone: 44 (0) 20 8392-8585; Fax: 44 (0) 20 8392-9876
E-mail: info@bushwoodbooks.co.uk
Free postage in the U.K., Europe; air mail at cost.

Contents

Introduction

Anything related to a lady's dressing table has fascinated me since I was a little girl. Maybe, I was captivated by them while watching old movies, when a 1930s or 1940s star gazed into a mirror on her dressing table. Reflections from within the mirror displayed crystal perfume bottles, cut crystal powder jars, and wonderful sterling silver dresser items. Perhaps I was struck by the ivory-colored celluloid set on my grandmother's dressing table. It consisted of twenty or more matching pieces, everything necessary to apply make-up and give a manicure. And most likely, I was struck by the fancy, gold-plated filigree and enameled perfume bottles and matching dresser accessories that decorated my mother's bedroom dresser. In any case, dresser sets and matching boudoir accessories have always provided me with a fond memory, which in turn created impetus for my collecting craze over the last twenty-five years. It seems that I was always drawn to the colorful early plastic sets, especially those still perched within the folds of luxurious fabric in their original presentation boxes. They are so glamorous, having been cherished and preserved for many years. This book attempts to investigate the early uses of Celluloid, French Ivory, Pyralin and Lucite in the dressing room.

The varieties that exist in dresser accessories are endless. This volume only touches the surface of its topic, and hopefully entices readers to continue the journey that this book initiates. The words "dresser set," "toilet set" and "boudoir accessory" are used interchangeably here, to refer to the common brush, mirror and comb sets and other accoutrements used for dressing well. While conducting the research for this work, it became obvious that during the 19th century, the word "toiletware" was consistently used to refer to women's grooming items and other dressing furnishings. As the decades progressed, terminology changed: by the 1920s the word "boudoir" seems to have been the common term, and by the late1940s "dresser set" defined the basic brush, mirror, and comb set.

While many varieties of porcelain, cut crystal, and glassware sets are presented briefly here, the collecting possibilities are endless. I am focused on the diverse metals and early plastics that were used to create "necessary items" for women from 1880 until 1960, with a high concentration on plastics from the 1920s and 1930s.

Art Nouveau hair brush coupled with a striking clothes brush both of quadruple plate and heavy embossing. The cloth brush has been monogrammed with fancy script lettering. $150-200 each.

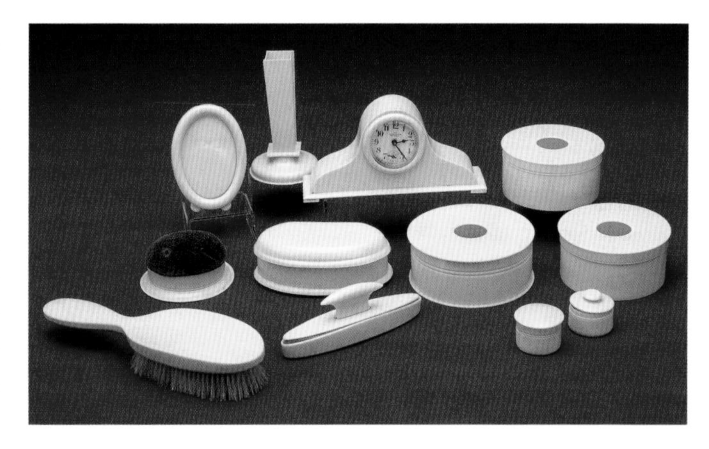

Wonderful assortment of French Ivory boudoir accessories to include a dresser clock, flower vase, picture frame, hair receivers, jewel box, pincushion, hair brush, nail buffer and two salve jars. Of all twelve pieces, only the two hair receivers are stamped *Palmer's French Ivory*. The clock movement was made by the New Haven Clock company. Clock $125-150, Nail buffer $30-40, Vase$ 40-50, jars $25-35, hair receivers $50-75, pincushion $50-70, picture frame $55-85 and brush $30-40.

Chapter One

Metal Dresser Sets & Accessories

Sterling Silver & Silver Plate

In the 1890s, solid silver ware, silver-plated hollow ware, and sterling silver novelties were abundant and offered for sale through mail-order and trade catalogs. The consumer was able to choose from a wide spectrum of goods in various price ranges. Ladies' vanity items were commonplace, well made, and extremely beautiful. Toilet items, including brush, mirror, and comb sets; manicure sets; or even men's military brush sets were made of sterling silver with heavily embossed or engraved decorations. An inexpensive alternative would have been a similar set made of quadruple plated silver, German silver, or Silveroin.

In 1895, wonderful silver-plated hollowware sets were manufactured by the Meriden Britannia Company. Hand mirrors, hairbrushes, and combs with heavily embossed tops sold wholesale for up to $11.65 each. Gorgeous embossed brushes came in all shapes and sizes. There were regular hairbrushes as well as bonnet brushes, infant brushes, nailbrushes, and men's military brushes. Many techniques were employed to create the right toilet accessory. Embossing, engraving, and chasing were three of the most popular late Victorian ornamental techniques used, and the more ornate the better. Other items were hand hammered and engine turned, or made with a combination of engraved and engine-turned designs. Still others were plain and simple, yet highly polished. These techniques were popular in the Edwardian era and first two decades of the 20th century. Some, today, might even consider these decorative styles to being gaudy. Reproducing them would be cost prohibitive now. Therefore, people collect the wonderful originals.

Lacquer finishes were sometimes applied to sterling silver and quadruple plated silver toilet sets and dresser accessories, which would keep the silver from tarnishing. Sterling silver and silver plated toilet goods were occasionally satin engraved, and some even had "French Gray" finishes.

Novelty items and boudoir accessories, including hairpin boxes, pin holders, jewel trees, glove and handkerchief boxes, trinket trays, toothbrush holders, picture frames, and jewel caskets, were common items elaborately crafted in sterling silver and quadruple plate. Atomizers, perfume bottles, and vinaigrettes were usually heavily embossed or engraved. Brushes were designed not only for hair, but for brushing bonnets or hats, nails, and mustaches. The silver could be polished, burnished, satin-finished, embossed, chased, or engraved. Mirrors were fitted with beveled-edge glass, brushes were fitted with natural bristles, and combs had tortoiseshell teeth. Mail order and trade catalogs were wish books of their day, filled with wonderful sterling silver and quadruple plated boudoir accessories, toilet sets, and novelty items.

Round mirror and matching oval hair brush decorated with embossed feathers and flowers with a smooth area in the center for monogramming. Oblong cloth brush similarly embossed with flowers and feathers and a space for initials. All three pieces are quadruple plated in silver. Mirror and brush set $195-245; cloth brush $95–135.

A garden of flowers was exquisitely embossed on this late 19th century hair brush made of sterling silver. $300-400.

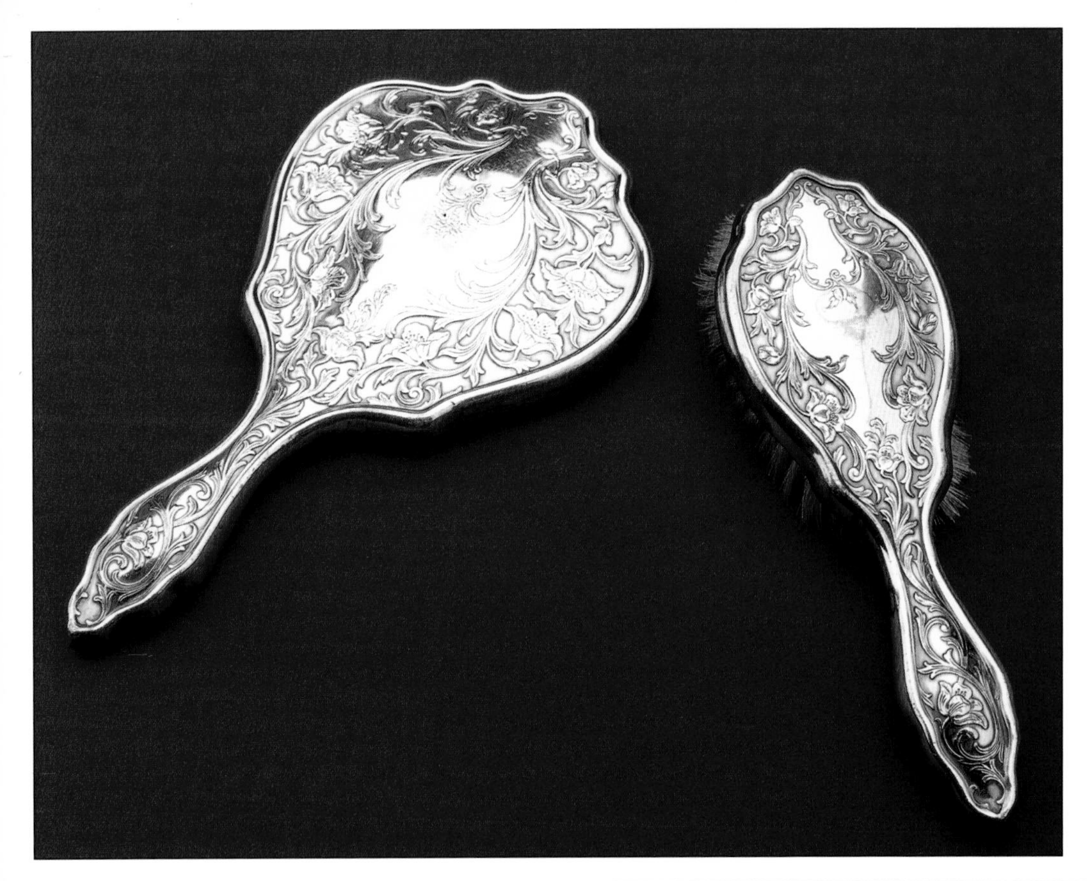

An early 20th century dresser set consisting of beveled glass mirror and hair brush made of quadruple silver-plate. Beautiful embossed floral designs and scrolling vines embellish each piece. $140-195 set.

Beautiful matched set consisting of a hand held beveled glass mirror and hair brush with heavily embossed rose designs made of quadruple plated silver. You can see some residue of a sticky tag on the back of the mirror and also some spots where the plating has worn completely off. $125-145 set. If perfect, $200-250 set.

Three-piece military brush set consisting of two military hair brushes and one cloth brush. These sets were popular for men in the late 19th and early 20th centuries. All three pieces have silver-plated backs with heavy embossed floral decorations. All are stamped *Quadruple Plate*. $225-275 set.

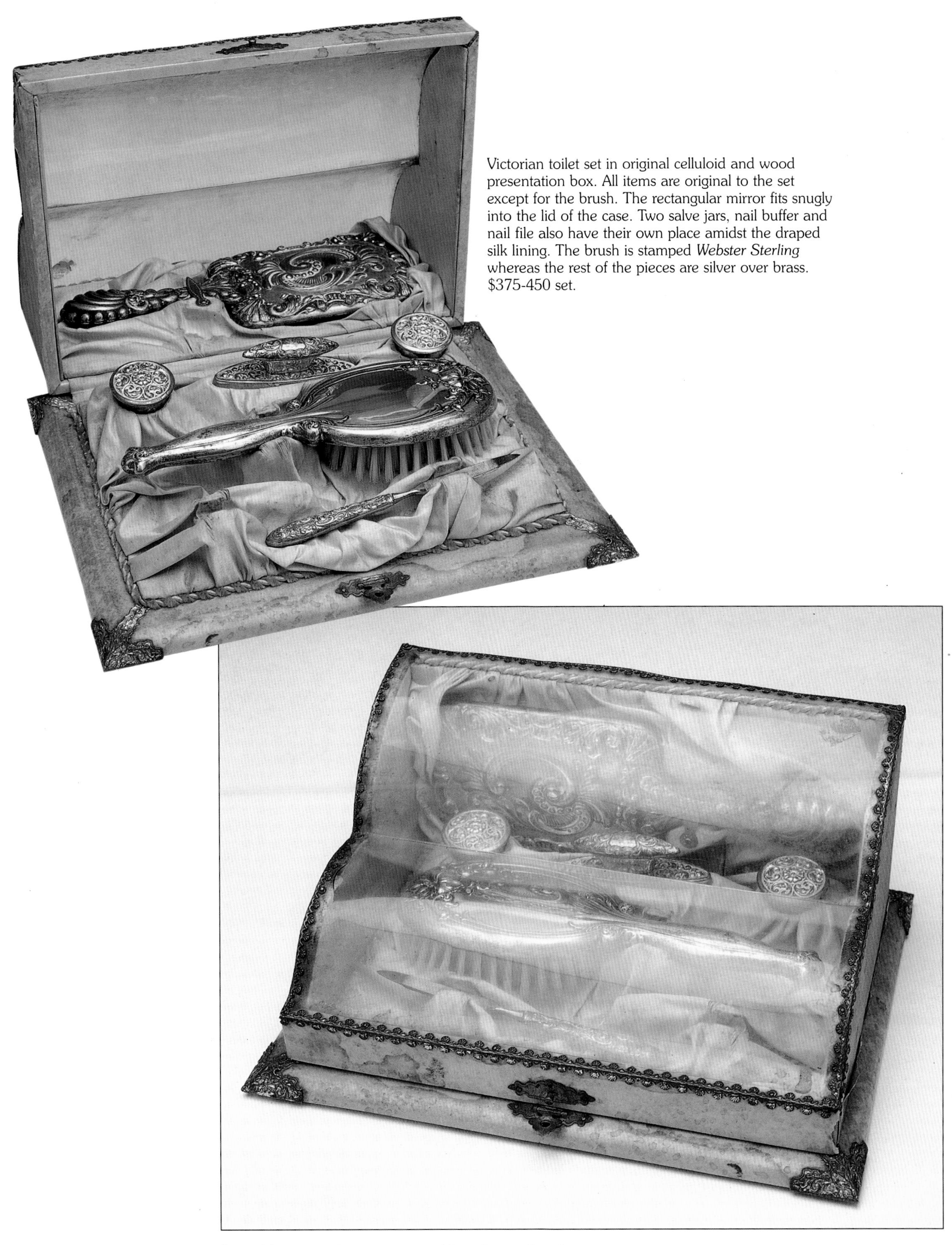

Victorian toilet set in original celluloid and wood presentation box. All items are original to the set except for the brush. The rectangular mirror fits snugly into the lid of the case. Two salve jars, nail buffer and nail file also have their own place amidst the draped silk lining. The brush is stamped *Webster Sterling* whereas the rest of the pieces are silver over brass. $375-450 set.

Same Victorian toilet set in original box shown closed.

STERLING SILVER $\frac{925}{1000}$ FINE TOILET ARTICLES.

Illustrations below are One-half Size.

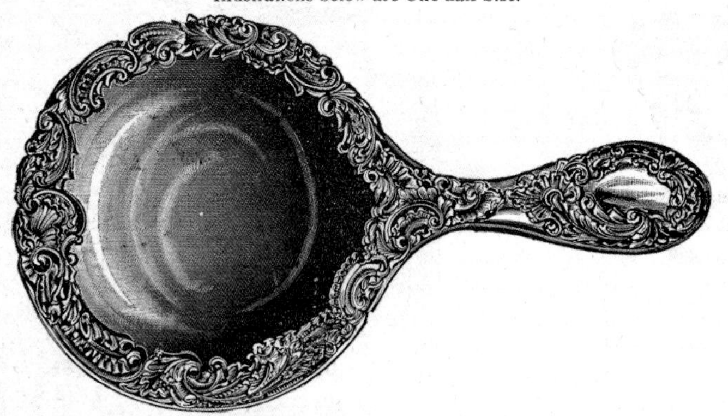

No. 7024. MIRROR, 8½ x 4½ inches........................[SUMPTER] **$27 00**
No. 7025. Plain Satin....................................[SUMPTUARY] 21 00

Illustration Two-thirds Size. Length, 4½ inches.

No. 7036. POLISHER, 4½ inches long........**$10 00**
[SUNDER.]
No. 7037. Polisher, Plain Satin.......[SUPERB] 7 50

No. 7026. CLOTH BRUSH, Embossed.........[SUPERFINE] **$13 00**
No. 7027. " " Plain Satin.....[SUPERHUMAN] 10 75
No. 7028. Hat Band Brush, Embossed, 6x¾ in...[SUPERNAL] 6 00

Illustration Half-size.

No. 7038. HAT BRUSH, Embossed.......[SUPERSEDE] **$9 50**
No. 7039. " " 6 x 1½ inches.....[SUPERVISE] 7 00
No. 7040. " " Like Cut, Satin.[SUPERVISION] 7 50

No. 7029. HAIR BRUSH, 8¾ inches long ....[SUPINE] **$14 50**
No. 7030. " " 8¼ " " ....[SUPPLANT] 13 25
No. 7031. " " Plain Satin, 8¾ inches.............[SUPPLICATE] 12 00
No. 7032. " " " " 8¼ " ....[SUPPLEMENT] 10 50

Width of Box, 3¾ inches.

No. 7041. PUFF BOX, Repousse.........**$27 00**
[SUPPLETIVE.]
No. 7042. Puff Box, Plain Satin....[SUPPLY] 21 50

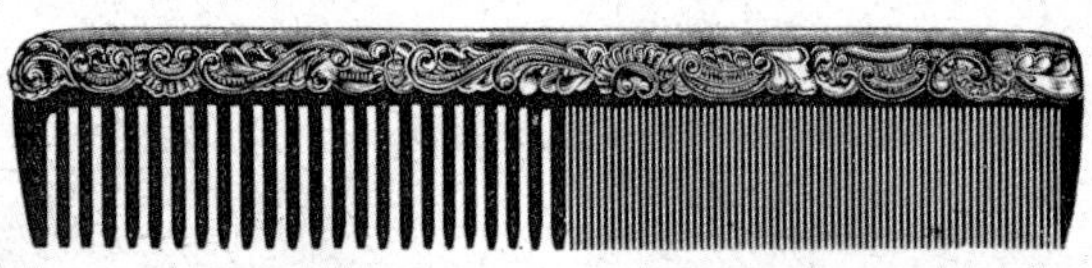

No. 7033. LADIES' COMB, 7½ inches..........................[SUPPORT] **$10 00**
No. 7034. Gents' " 7 " ....[SUPPOSABLE] 9 50
No. 7035. Ladies' " Satin, 7½ inches....................[SUPPRESS] 8 50
Real Tortoise Shell Combs.

Illustration Full Size.

No. 7043. SALVE OR POWDER BOX.....$7 00
[SUPREMACY.]

224

Fine embossed sterling silver toilet articles offered for sale in 1895 from the BHA Illustrated Catalog.

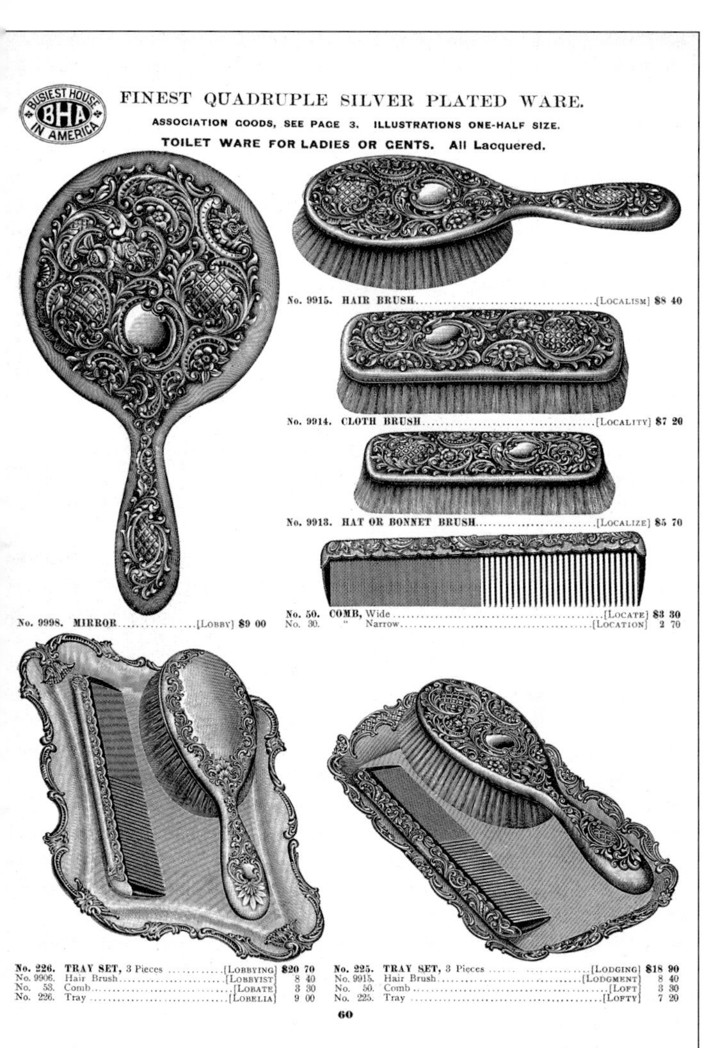

BUSIEST HOUSE IN AMERICA (BHA)

FINEST QUADRUPLE SILVER PLATED WARE.

ASSOCIATION GOODS, SEE PAGE 3. ILLUSTRATIONS ONE-HALF SIZE.

TOILET WARE FOR LADIES OR GENTS. All Lacquered.

No. 9915. HAIR BRUSH....................[Localism] $8 40

No. 9914. CLOTH BRUSH....................[Locality] $7 20

No. 9913. HAT OR BONNET BRUSH....................[Localize] $5 70

No. 50. COMB, Wide....................[Locate] $3 30
No. 30. " Narrow....................[Location] 2 70

No. 9998. MIRROR....................[Lobby] $9 00

No. 226. TRAY SET, 3 Pieces....................[Lobbying] $20 70
No. 9906. Hair Brush....................[Lobbyist] 8 40
No. 53. Comb....................[Lobate] 3 30
No. 226. Tray....................[Lobelia] 9 00

No. 225. TRAY SET, 3 Pieces....................[Lodging] $18 90
No. 9915. Hair Brush....................[Lodgment] 8 40
No. 50. Comb....................[Loft] 3 30
No. 225. Tray....................[Lofty] 7 20

60

Toiletware made of the finest quadruple plate fashionable in 1895. The silver plate was lacquered to prevent tarnishing.

FINEST QUADRUPLE SILVER PLATED WARE.

BUSIEST HOUSE IN AMERICA (BHA)

ASSOCIATION GOODS, SEE PAGE 3. ILLUSTRATIONS ONE-HALF SIZE.

TOILET WARE FOR LADIES OR GENTS. All Lacquered.

No. 8900. HAIR BRUSH....................[Loiter] $9 30
No. 8950. " " smaller....................[Loitering] 7 80

No. 9000. HAIR BRUSH....................[Lonely] $8 40

No. 9977. MIRROR, 5½ inch...[Loftiness] $9 30
No. 9977½. " 4½ " ......[Logic] 7 20
Above Measurements are of the Glass. Beveled Plate Glass Mirrors.

No. 8800. CLOTH BRUSH....................[Longing] $9 30

No. 8700. HAT OR BONNET BRUSH....................[Loricate] $6 60

No. 34. BONNET BRUSH.......[Logical] $4 20
Bristles in all these Brushes are of the Finest Quality.

No. 50. COMB....................[Lossable] $3 30

61

Fine quality bristle brushes were designed in all shapes and sizes and for many purposes.

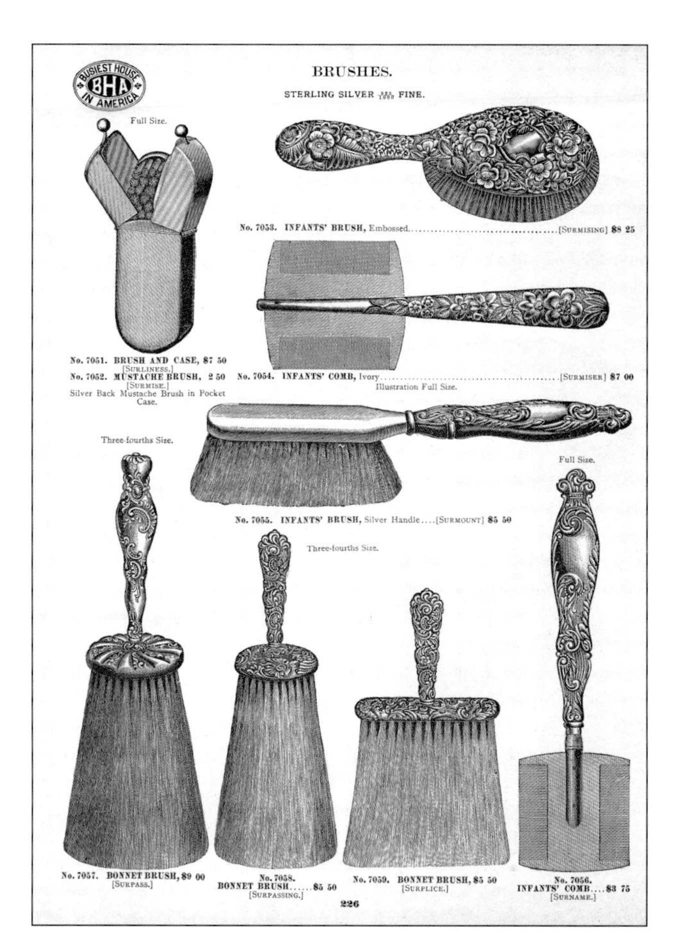

BUSIEST HOUSE IN AMERICA (BHA)

BRUSHES.

STERLING SILVER 925/1000 FINE.

Full Size.

No. 7053. INFANTS' BRUSH, Embossed....................[Surmising] $8 25

No. 7051. BRUSH AND CASE, $7 50
[Surliness.]
No. 7052. MUSTACHE BRUSH, 2 50
[Surmise.]
Silver Back Mustache Brush in Pocket Case.

No. 7054. INFANTS' COMB, Ivory....................[Surmiser] $7 00
Illustration Full Size.

Three-fourths Size.

No. 7055. INFANTS' BRUSH, Silver Handle....[Surmount] $5 50

Full Size.

Three-fourths Size.

No. 7057. BONNET BRUSH, $9 00
[Surpass.]

No. 7058. BONNET BRUSH...... $5 50
[Surpassing.]

No. 7059. BONNET BRUSH, $5 50
[Surplice.]

No. 7056. INFANTS' COMB.... $3 75
[Surname.]

226

Bonnet brushes, infant brushes and mustache brushes were made of heavily embossed sterling silver and offered for sale in 1895.

126 PUFF BOXES, TOILET BOTTLES AND ATOMIZERS.

QUADRUPLE PLATE.

No. B19. Puff Box.
Bright Cut.................$2 50
Bright Cut, Gold Lined...... 3 00
Height, 2¾ inches.

No. M31. Puff Box and Tray.
Gold Lined....................................$8 10

No. B21. Puff Box.
Bright Repousse..............$3 50
Bright Repousse, Gold Lined... 4 00
Height, 3¼ inches.

No. B2002. Cologne.
Bright Embossed.....$3 50
Height, 6¼ inches.

No. M2800. Toilet Bottle.
Engraved, as shown......$6 60
Plain Satin............. 5 40

No. B2003. Cologne, Glass Lining.
Satin.................$4 50
Bright Cut........... 5 00
Height, 7¾ inches.

No. B50. Atomizer.
Satin................................$4 00
Satin, Bright Cut 5 00
Height, 7¼ inches.

No. S911. Puff Box.
Satin Finish, Gold Lined, with Puff Ball........................... $3 85

No. P2357. Atomizer.
Bright Cut.................$4 00

No. T30. Puff Box and Puff.
Satin, Bright Cut, Gold Lined......$4 10

No. P2363. Atomizer.
Burnished Silver..............$5 50

Puff boxes, toilet bottles and atomizers made of quadruple plate with fancy embossed and repoussé decorations.

TOILET WARE AND PHOTOGRAPH FRAMES. 137

QUADRUPLE PLATE.

No. R32. Military Brush.
Satin Engraved......$6 65
Cut One-third Size.

No. M45. Photo Frame.
Gold Finish......................$4 50
Cut One-half Size.

No. R32. Comb.
Satin Engraved......$3 35
Cut One-third Size.

No. R146. Mirror.
Satin Engraved..........$9 15
Cut One-third Size

No. R32. Mirror.
Satin Engraved......$8 90
Cut One-third Size.

No. R32. Hair Brush.
Satin Engraved.....$7 80
Cut One-third Size.

No. R32. Cloth Brush.
Satin Engraved........$6 65
Cut One-third Size.

No. R43. Cabinet Photograph Frame.
Satin Finish, Rococo Border...........................$8 00
Cut One-third Size.

No. R32. Hat Brush.
Satin Engraved...$4 90
Cut One-third Size.

For Sterling Silver Toilet Ware see pages 187 to 190 inclusive.

Fancy satin engraved toiletware for a lady's boudoir made of quadruple plate and offered for sale in 1896 from Marshall Field & Company.

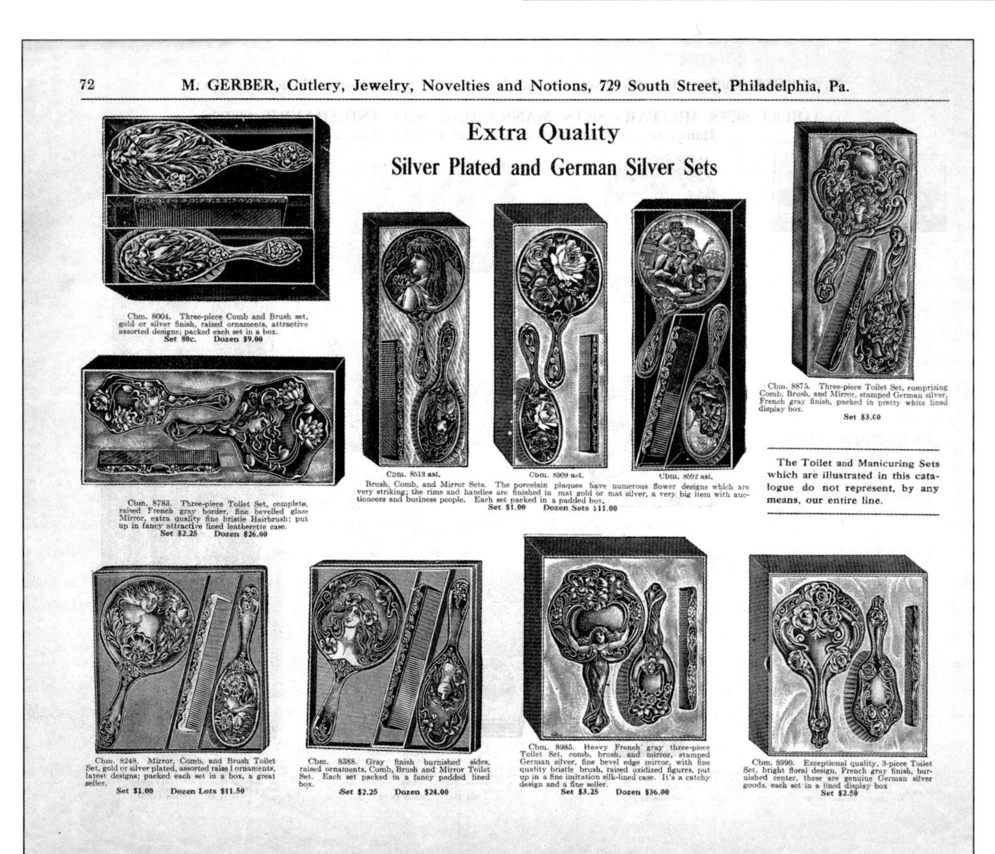

72 M. GERBER, Cutlery, Jewelry, Novelties and Notions, 729 South Street, Philadelphia, Pa.

Extra Quality

Silver Plated and German Silver Sets

Cbm. 8004. Three-piece Comb and Brush set, gold or silver finish, raised ornaments, attractive assorted designs; packed each set in a box.
Set 80c. Dozen $9.00

Cbm. 8783. Three-piece Toilet Set, complete, raised French gray border, fine bevelled glass Mirror, extra quality fine bristle Hairbrush; put up in fancy attractive lined leatherette case.
Set $2.25 Dozen $26.00

Cbm. 8518 ast. Cbm. 8509 ast. Cbm. 8592 ast.
Brush, Comb, and Mirror Sets. The porcelain plaques have numerous flower designs which are very striking; the rims and handles are finished in mat gold or mat silver, a very big item with auctioneers and business people. Each set packed in a padded box.
Set $1.00 Dozen Sets $11.00

Cbm. 8875. Three-piece Toilet Set, comprising Comb, Brush, and Mirror, stamped German silver, French gray finish, packed in pretty white lined display box.
Set $3.00

The Toilet and Manicuring Sets which are illustrated in this catalogue do not represent, by any means, our entire line.

Cbm. 8248. Mirror, Comb, and Brush Toilet Set, gold or silver plated, assorted raised ornaments, latest designs; packed each set in a box, a great seller.
Set $1.00 Dozen Lots $11.50

Cbm. 8388. Gray finish burnished sides, raised ornaments, Comb, Brush and Mirror Toilet Set. Each set packed in a fancy padded lined box.
Set $2.25 Dozen $24.00

Cbm. 8985. Heavy French gray three-piece Toilet Set, comb, brush, and mirror, stamped German silver, fine bevel edge mirror, with fine quality bristle brush, raised oxidized figures, put up in a fine imitation silk-lined case. It's a catchy design and a fine seller.
Set $3.25 Dozen $36.00

Cbm. 8990. Exceptional quality, 3-piece Toilet Set, bright floral design, French gray finish, burnished center, these are genuine German silver goods, each set in a lined display box
Set $2.50

The M.Gerber Company of Philadelphia, Pa. offered wonderful silver plate and German silver dresser sets in 1899 with gorgeous embossed Art Nouveau designs. The three sets pictured in the center of the page have hand painted porcelain plaques on the backs of the mirrors and brushes.

ADAMS, QUINCY, FRANKLIN, FIFTH AVE., CHICAGO. 257

STERLING SILVER MOUNTED TOILET GOODS.

French Gray Finish.

PRICES EACH.

No. 809. Pumice Stone..................$3 60

No. 809. Nail Polisher..................$3 60

No. 809. Military Brush..............................$8 50

No. 809. Cloth Brush..$9 00

For Combs to Match this Set, See Page 257. Ebony Toilet Ware, See Pages 296 to 301.

Pumice stone, nail polisher, military brush and cloth brush with floral embossed sterling silver tops offered from Marshall Field in 1900.

256 MARSHALL FIELD & CO.,

STERLING SILVER MOUNTED TOILET GOODS.

French Gray Finish. Illustrations are Three-fourths Size.

PRICES EACH.

No. 809. Velvet Brush........................$3 10

No. 809. Hair Brush........$12 00

No. 809. Handle Mirror....$17 50

For Combs to Match this Set, See Page 257. Ebony Toilet Ware, See Pages 296 to 301.

Sterling silver mounted toilet goods with French gray finishes and embossed floral decorations offered for sale in 1900 from the Marshall Field Company.

IMITATION TORTOISE SHELL DRESSING COMBS.

Sterling Silver Mounted. Illustrations Actual Size.

PRICES EACH.

No. 806 .... $1 00

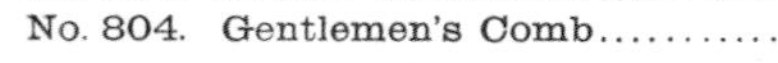

No. 804. Ladies' Comb $2 50
No. 804. Gentlemen's Comb 2 50

No. 803 .... $1 30

No. 809 .... $2 50

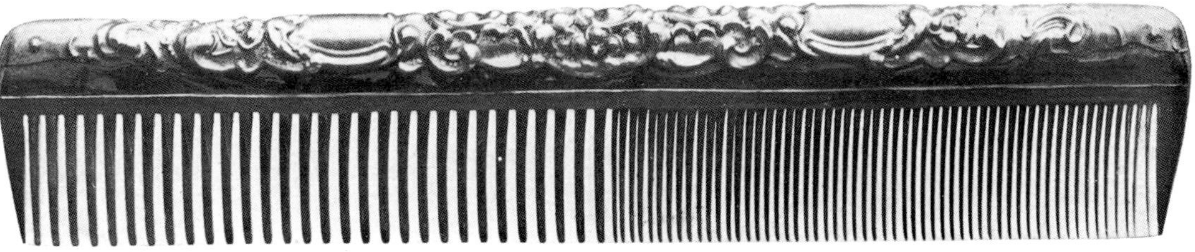

No. 805 .... $1 00

Imitation tortoiseshell dressing combs with embossed sterling silver mounts popular in 1900.

GENUINE EBONY TOILET GOODS.

Sterling Silver Trimmings.

These Goods are Solid One-Piece Ebony, Finely Finished and First Quality Bristles.

PRICES EACH.

Genuine ebony toilet goods with sterling silver trimmings were fashionable in 1900. They were expertly crafted of one solid piece of ebony and fitted with the finest quality bristles.

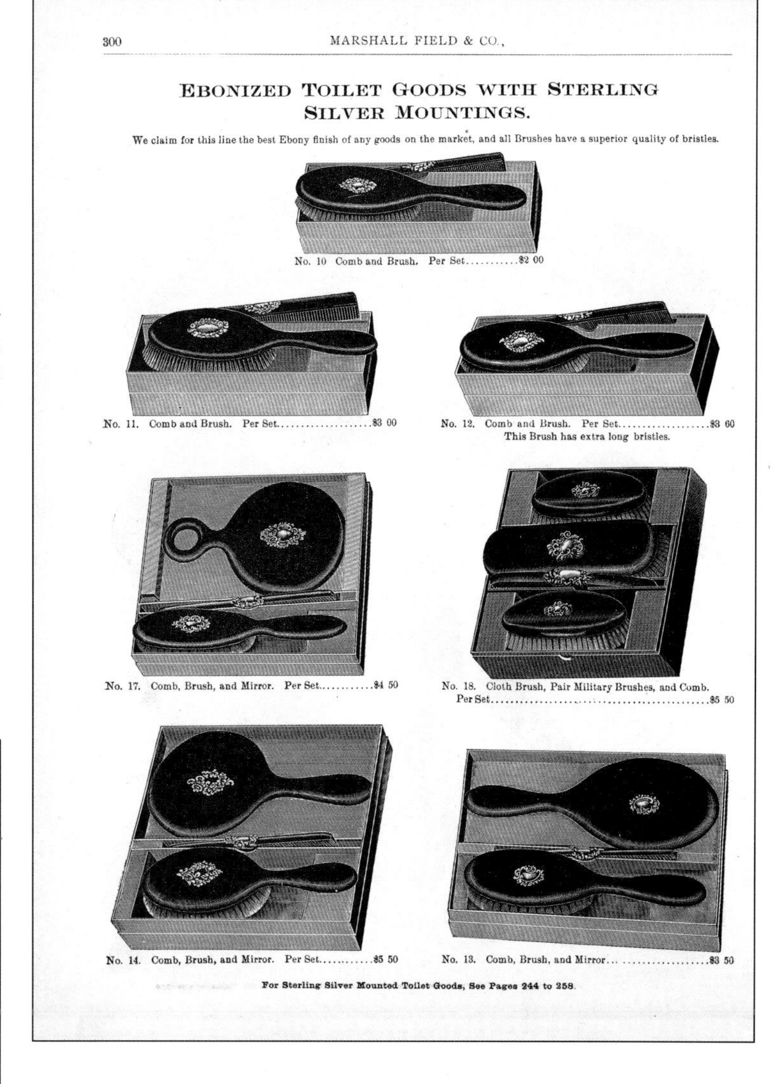

300 MARSHALL FIELD & CO.,

EBONIZED TOILET GOODS WITH STERLING SILVER MOUNTINGS.

We claim for this line the best Ebony finish of any goods on the market, and all Brushes have a superior quality of bristles.

No. 10 Comb and Brush. Per Set..........$2 00

No. 11. Comb and Brush. Per Set..................$3 00

No. 12. Comb and Brush. Per Set..................$3 60
This Brush has extra long bristles.

No. 17. Comb, Brush, and Mirror. Per Set...........$4 50

No. 18. Cloth Brush, Pair Military Brushes, and Comb.
Per Set...$5 50

No. 14. Comb, Brush, and Mirror. Per Set...........$5 50

No. 13. Comb, Brush, and Mirror... ....$3 50

For Sterling Silver Mounted Toilet Goods, See Pages 244 to 258.

Ebonized toilet sets with sterling silver trimmings were also offered for sale in 1900. They looked exactly like the genuine ebony pieces but cost slightly less.

296 MARSHALL FIELD & CO.,

EBONIZED TOILET SETS.

Sterling Silver Mountings.
PRICES PER SET.

No. 36. Hair Brush and Comb..$4 00

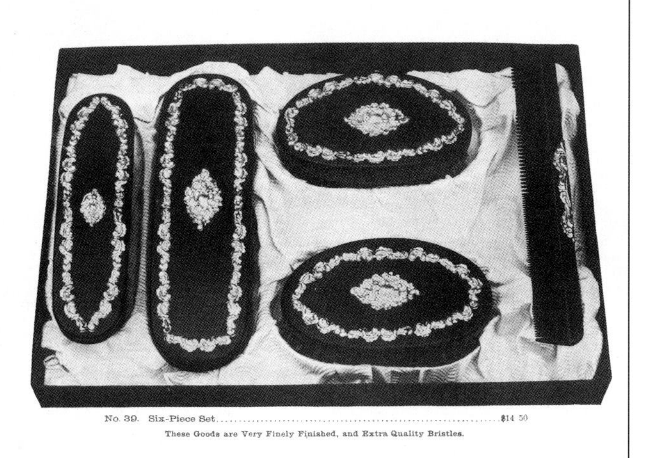

No. 39. Six-Piece Set..$14 50
These Goods are Very Finely Finished, and Extra Quality Bristles.

Fancy ebonized toilet sets from 1900 with ornate sterling silver trimmings, fine finishes and extra quality bristles. A six-piece set in a silk-lined box sold for $14.50.

In 1899, mail order and wholesale trade catalogs offered numerous toilet sets with embossed Art Nouveau motifs. Their common themes include ladies with flowing hair, sinuous vines, and floral designs. German silver was also used as well as Silveroin, which resembled sterling or quadruple plate. In 1910, silver-plated sets were advertised as "Empire Art Silver." "Victor silver" was another imitation for sterling silver.

By 1920, there were so many variations in sterling silver toilet ware that each pattern was given a specific name. For example, a handsome pattern with a gray finish and a border design was called the *Claremont* pattern. The *Georgian* pattern was slightly fancier with more embossing and the center was left plain so it could be monogrammed. An even fancier set was the *Colonial* pattern, which was heavily hand-engraved. The *Victoria* pattern was plain with a beveled edge and a monogrammed center while the *Duchess* pattern incorporated engine turned and engraved designs.

In 1923, toilet ware patterns made of quadruple plate were offered from the Ft. Dearborn Watch and Clock Company in patterns reminiscent of the Arts and Crafts styles popular three decades earlier. Beautiful hammered patterns combined with rich burnished and French gray finishes were common choices. Three-piece sets for women sold for $14.75, while a four-piece men's military brush set was $19.75. The same catalog offered a sterling silver pattern called *Fleur de Champ,* which was advertised as *"positively the richest, most magnificent and elaborate pattern in Sterling Silver Toiletware ever offered for sale."* The mirror alone cost $51.35, which was an exorbitant sum to pay at the time.

Two more examples of exquisite Art Nouveau embossing techniques. This time, a cloth brush stamped *Silveroin* and a hair brush made of quadruple plate. Both brushes exemplify the strong influence of the typical woman with flowing hair surrounded by sinuous lines and floral designs. Silveroin and quadruple plate were used in abundance to imitate the more expensive sterling silver items. Hair brush $150-225; cloth brush $125-175.

Military hair brush, hat or bonnet brush and nail brush all made with long bristles and silver-plated backs. The smaller nail brush has an embossed lady design in the center surrounded by floral designs; the other two brushes display embossed floral decorations. $65-120 each.

Heavily embossed Art Nouveau beveled glass hand mirror depicting a woman with flowing hair and holding a hand mirror and surrounded by sinuous lines and floral designs. The mirror is stamped *Sterling.* $350-400.

Exquisite silver-plated brush and beveled glass mirror set with engine-turned and engraved designs, made in the Edwardian tradition of the early 20th century. The center shield has been monogrammed with fancy script lettering. $145-185 set.

Six-inch cloth brush with intricate engraving and monogrammed shield in the center made of quadruple plate; small child's hand mirror with beveled glass and silver-plated embossed and engraved back and handle. You can see where the plating has worn off the handle and the brass is showing through. $85-145 each.

Two six-inch cloth brushes both decorated with embossed and intricate engraved designs. The brushes have long bristles and both are silver-plated. $75-100 each.

Delicate hand held oval-shaped mirror with beveled glass and embossed flowers and leaves around the borders of the mirror back and on both sides of the handle. A smooth center could have been used for a monogram. The mirror is stamped *G. Silver* for German silver which was used to imitate the finer sterling silver items. $95-135.

Cloth brush made of silver-plate with embossed, engraved and hammered designs; heavily embossed cloth brush with elaborate floral and scroll decoration and a center space for monogramming. The later is marked *Sterling.* Silver-plated brush $75-100; Sterling brush $150-200.

268

THE HOME OF TRUE ECONOMY | BOSTON STORE CHICAGO ILL. | YOUR MONEY BACK IF YOU ARE DISSATISFIED

BRUSH, COMB AND MIRROR SETS—MANICURE, ETC.

THREE PIECE TOILET SET

No. J26801. Three piece toilet set consisting of comb, brush and mirror. China back painted with fancy flower design, white metal frame.
Special price $1.89

GERMAN SILVER SET

No. J26802. Three piece German silver set consisting of comb, brush and mirror, heavily embossed polished edges, good quality bristles, fancy shape mirror and brush to match, in white lined box. Price complete $3.48

OUR $3.98 TOILET SET

No. J26803. Three piece toilet set consisting of comb, brush and mirror of large size, made of Victor silver, looks just like Sterling silver. Heavy embossed figure on brush and mirror, bright finish, good quality bristle. Put up in neat lined box. Special price complete..... $3.98

STERLING SILVER SET

No. J26804. Sterling silver comb and brush set, heavy embossed rose pattern, extra good quality bristles. Each piece stamped "Sterling Silver." Put up in neat white lined box.
Extra special price complete..... $2.98

Great Value at 48c

GLASS CIGAR JAR

No. J26805. Heavy glass cigar jar, embossed oxidized metal cover. Special price,
48c

DRESDEN CHINA TWO PIECE TOILET SET

No. J26806. Two piece toilet set, comb and brush, neatly embossed pattern, hand painted China back. This set can also be had with a China back. Makes an attractive and acceptable gift. Price complete............ 79c

SPECIAL $2.48 SET

No. J26807. Two piece toilet set consisting of comb and brush, made of Victor silver, neat embossed figure pattern, good quality bristles. Something any person will appreciate. In a neat lined box. A new and handsome pattern.
Special price complete $2.48

DRESDEN CHINA TOILET SET

No. J26808. Dresden China toilet set consisting of comb and brush with extra good quality bristles and large mirror, painted with fancy flower design and neatly etched white metal frame. In white lined box.
A bargain at $2.48

BEAUTIFUL TOILET SET

No. J26809. Three piece metal set consisting of comb, brush and mirror, heavily embossed cupid design and brush and mirror oxidized finish with burnished edges.
Special price complete $2.69

MANICURE SCISSORS

ONLY 48c

No. J26810

No. J26810. Sterling silver manicure scissors, heavy embossed patterns, curved points.
Our special price 48c

MANICURE SET

No. J26811. Sterling silver manicure set consisting of salve jar, nail file, tooth brush, manicure scissors, buttonhook and a chamois covered buffer.
Our special price.. $2.48

MATCH SAFE

No. J26812. Regulation size German silver match safe. Sterling silver plated. Price.. 24c

By 1910, three-piece sets consisting of a brush, comb and hand mirror were offered for sale in fancy fabric-lined boxes. Sterling silver sets were available as well as German silver and an imitation called Victor silver. China backed sets made of Dresden china sold for $2.48 set.

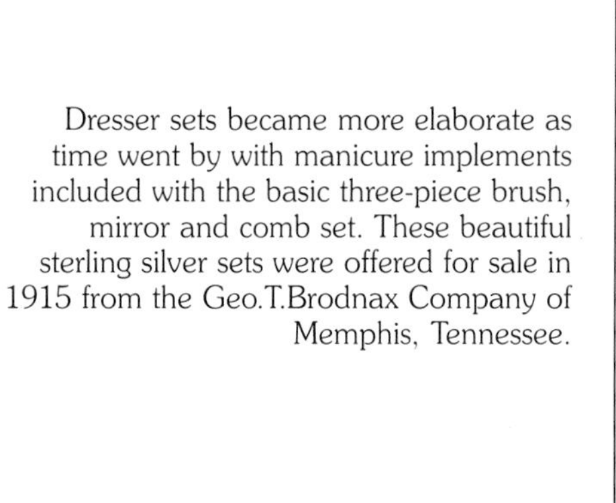

Dresser sets became more elaborate as time went by with manicure implements included with the basic three-piece brush, mirror and comb set. These beautiful sterling silver sets were offered for sale in 1915 from the Geo.T.Brodnax Company of Memphis, Tennessee.

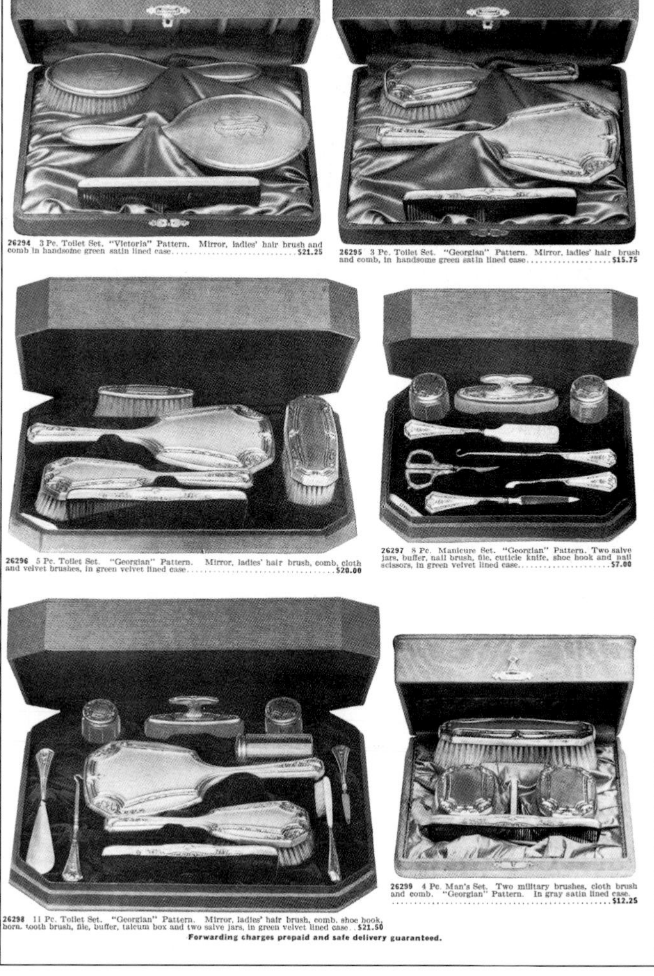

GEO. T. BRODNAX, INC., GOLD AND SILVERSMITHS, MEMPHIS, TENN. 105

26294 3 Pc. Toilet Set. "Victoria" Pattern. Mirror, ladies' hair brush and comb in handsome green satin lined case............................$21.25

26295 3 Pc. Toilet Set. "Georgian" Pattern. Mirror, ladies' hair brush and comb, in handsome green satin lined case....................$15.75

26296 5 Pc. Toilet Set. "Georgian" Pattern. Mirror, ladies' hair brush, comb, cloth and velvet brushes, in green velvet lined case..................................$20.00

26297 8 Pc. Manicure Set. "Georgian" Pattern. Two salve jars, buffer, nail brush, file, cuticle knife, shoe hook and nail scissors, in green velvet lined case......................$7.00

26298 11 Pc. Toilet Set. "Georgian" Pattern. Mirror, ladies' hair brush, comb, shoe hook, horn, tooth brush, file, buffer, talcum box and two salve jars, in green velvet lined case..$21.50

26299 4 Pc. Man's Set. Two military brushes, cloth brush and comb. "Georgian" Pattern. In gray satin lined case. $12.25

Forwarding charges prepaid and safe delivery guaranteed.

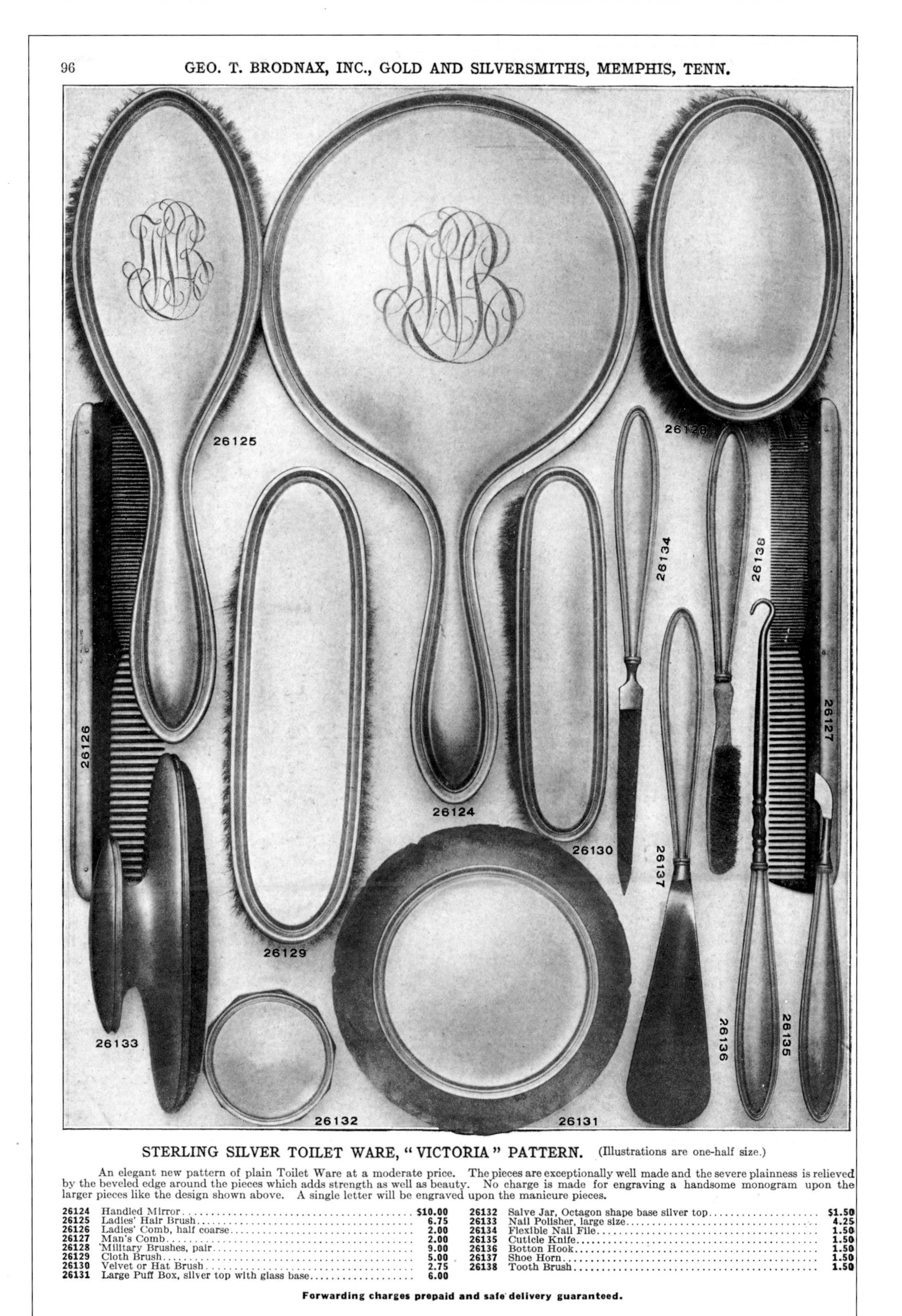

96 GEO. T. BRODNAX, INC., GOLD AND SILVERSMITHS, MEMPHIS, TENN.

STERLING SILVER TOILET WARE, "VICTORIA" PATTERN. (Illustrations are one-half size.)

An elegant new pattern of plain Toilet Ware at a moderate price. The pieces are exceptionally well made and the severe plainness is relieved by the beveled edge around the pieces which adds strength as well as beauty. No charge is made for engraving a handsome monogram upon the larger pieces like the design shown above. A single letter will be engraved upon the manicure pieces.

26124	Handled Mirror	$10.00	**26132**	Salve Jar, Octagon shape base silver top	$1.50
26125	Ladies' Hair Brush	6.75	**26133**	Nail Polisher, large size	4.25
26126	Ladies' Comb, half coarse	2.00	**26134**	Flexible Nail File	1.50
26127	Man's Comb	2.00	**26135**	Cuticle Knife	1.50
26128	Military Brushes, pair	9.00	**26136**	Botton Hook	1.50
26129	Cloth Brush	5.00	**26137**	Shoe Horn	1.50
26130	Velvet or Hat Brush	2.75	**26138**	Tooth Brush	1.50
26131	Large Puff Box, silver top with glass base	6.00			

Forwarding charges prepaid and safe delivery guaranteed.

Plain toiletware made of sterling silver was accented with beveled edges around the ends of each piece. This pattern was called *Victoria*. This was the perfect pattern to have a personal monogram engraved on each piece for a fashionable look in 1915.

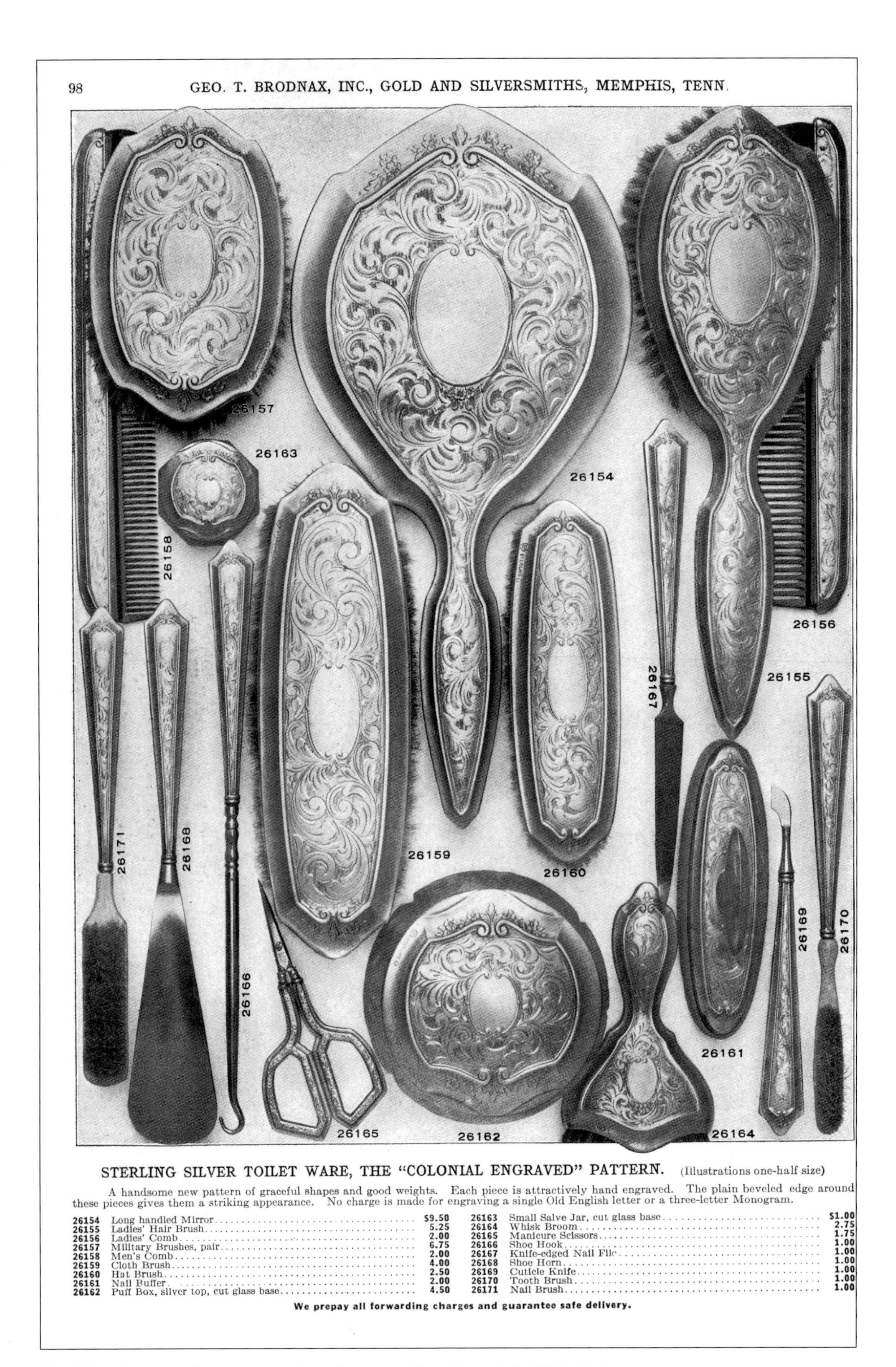

98 GEO. T. BRODNAX, INC., GOLD AND SILVERSMITHS, MEMPHIS, TENN.

STERLING SILVER TOILET WARE, THE "COLONIAL ENGRAVED" PATTERN. (Illustrations one-half size)

A handsome new pattern of graceful shapes and good weights. Each piece is attractively hand engraved. The plain beveled edge around these pieces gives them a striking appearance. No charge is made for engraving a single Old English letter or a three-letter Monogram.

26154	Long handled Mirror	$9.50	26163	Small Salve Jar, cut glass base	$1.00
26155	Ladies' Hair Brush	5.25	26164	Whisk Broom	2.75
26156	Ladies' Comb	2.00	26165	Manicure Scissors	1.75
26157	Military Brushes, pair	6.75	26166	Shoe Hook	1.00
26158	Men's Comb	2.00	26167	Knife-edged Nail File	1.00
26159	Cloth Brush	4.00	26168	Shoe Horn	1.00
26160	Hat Brush	2.50	26169	Cuticle Knife	1.00
26161	Nail Buffer	2.00	26170	Tooth Brush	1.00
26162	Puff Box, silver top, cut glass base	4.50	26171	Nail Brush	1.00

We prepay all forwarding charges and guarantee safe delivery.

The *Colonial* engraved pattern in sterling silver was offered for sale in 1915. Each piece was *attractively hand-engraved* with a center shield left plain for monogramming.

GEO. T. BRODNAX, INC., GOLD AND SILVERSMITHS, MEMPHIS, TENN. 171

FINE QUADRUPLE PLATED TOILET SETS. (Illustrations one-half size.)

These are two of the most attractive patterns ever offered in plated ware. The designs are similar to the much more expensive goods in Sterling silver. The goods come in lined boxes, in sets of three pieces for ladies, comb, brush and mirror, and for men, one pair of military brushes and a comb. We will engrave, without charge, a single Old English or Script letter. If additional engraving is wanted an extra charge will be made.

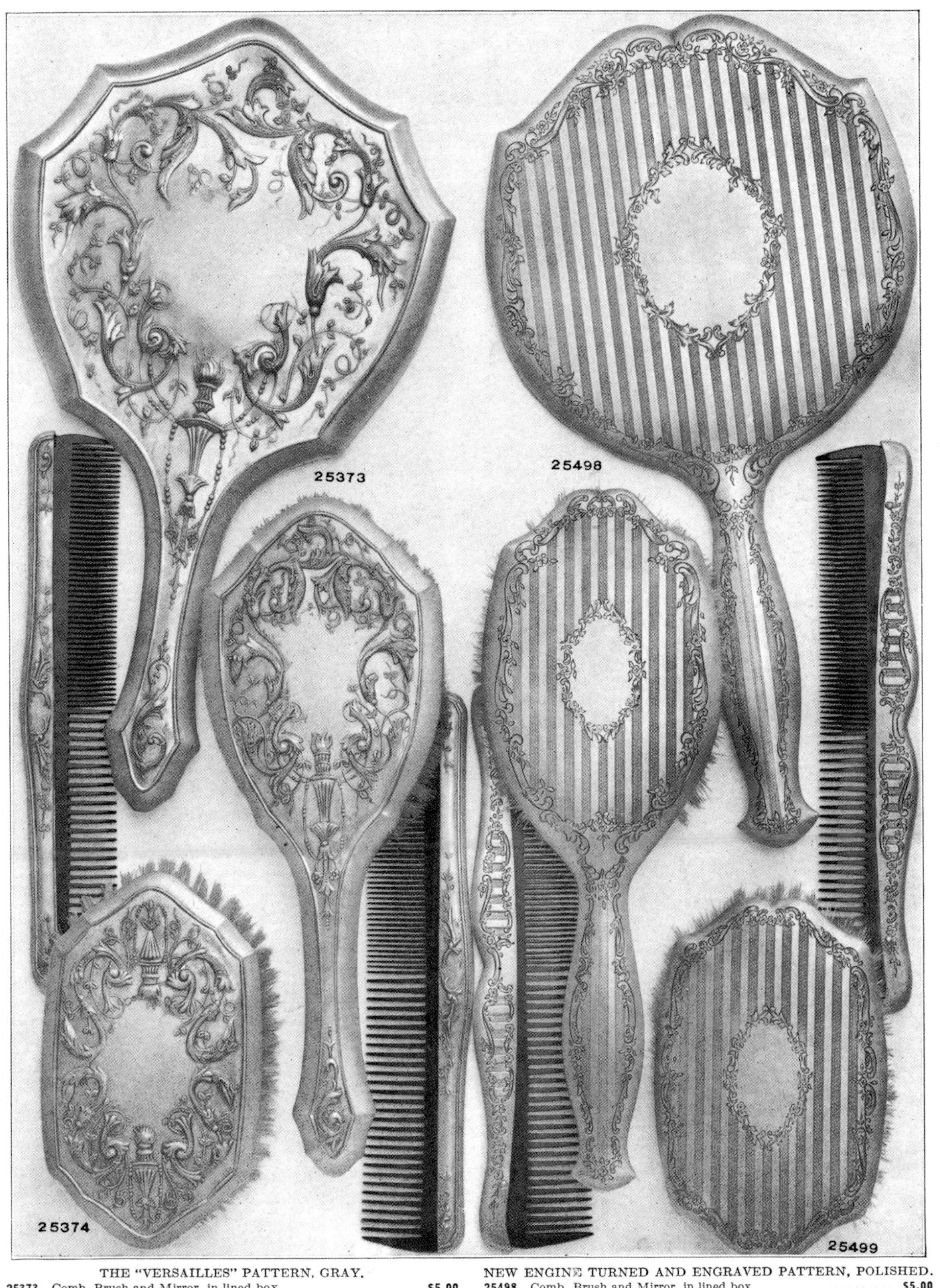

THE "VERSAILLES" PATTERN, GRAY.

25373 Comb, Brush and Mirror, in lined box $5.00
25374 2 Military Brushes and Comb in lined box 4.50

NEW ENGINE TURNED AND ENGRAVED PATTERN, POLISHED.

25498 Comb, Brush and Mirror, in lined box $5.00
25499 Military Brush Set, 2 brushes and comb, in lined box 4.50

Express charges prepaid as far West as Denver and as far East as New York. To points beyond we ship express collect but allow 10% off the catalogue price.

Advertised as two of the most attractive styles in fine quadruple plate, the *Versailles* pattern with a fancy embossed French gray finish and the engine-turned and engraved pattern with a polished finish were desirable styles in 1915. Their designs were compared to the finer sets made of sterling silver.

In 1930, the May and Malone Company advertised sterling silver dresser sets that had a little something extra: beautiful, hard French enamel backs with wonderful hand painted flower decorations in a pattern called the *Tourville*. These items, made by skilled artisans, were expensive. A single mirror could cost up the $86. A three-piece set cost nearly $170 in 1930, an extravagant sum at that time.

The 1932 silver patterns for dresser sets from the N. Shure Company were given the pattern names *Lady Wynne, Lady Louise, Patricia, and Isabeau*. The Lady Wynne pattern was plain and simple with a gray finish, wheras the Isabeau pattern incorporated striped and brocade engine-turned and hand-engraved designs. In 1933, sterling silver toilet articles made in the new *Anne* pattern were described as being satin finished with polished bars and shield. This was "*a new creation by master silversmiths in solid sterling silver. Simple, yet distinguished by its graceful shape and outline.*"

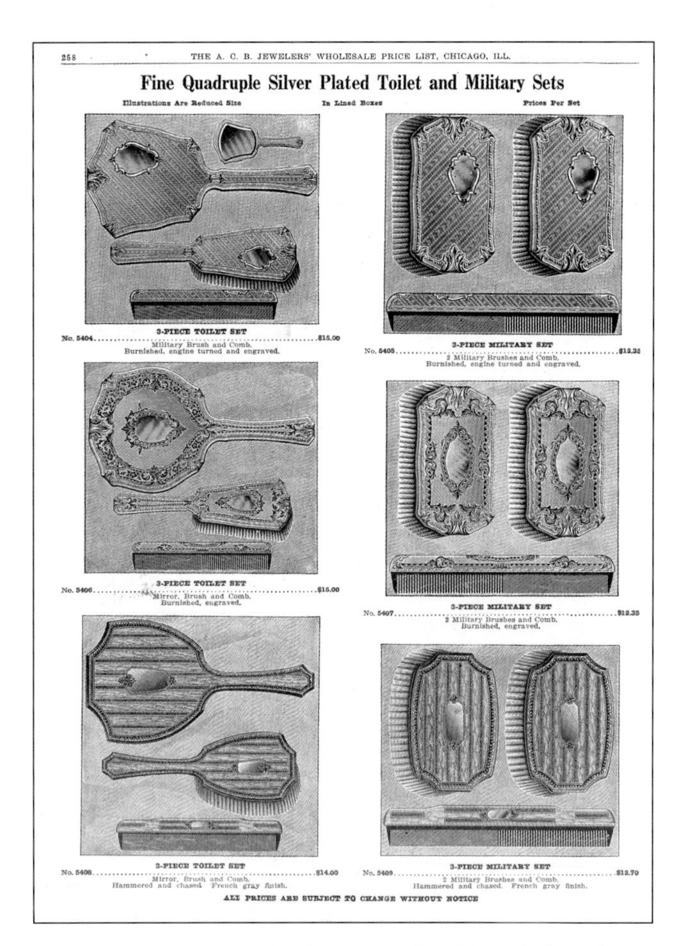

258 THE A. C. B. JEWELERS' WHOLESALE PRICE LIST, CHICAGO, ILL.

Fine Quadruple Silver Plated Toilet and Military Sets

Illustrations Are Reduced Size — In Lined Boxes — Prices Per Set

3-PIECE TOILET SET
No. 5404 .... $15.00
Military Brush and Comb.
Burnished, engine turned and engraved.

3-PIECE MILITARY SET
No. 5405 .... $12.35
2 Military Brushes and Comb.
Burnished, engine turned and engraved.

3-PIECE TOILET SET
No. 5406 .... $15.00
Mirror, Brush and Comb.
Burnished, engraved.

3-PIECE MILITARY SET
No. 5407 .... $12.35
2 Military Brushes and Comb.
Burnished, engraved.

3-PIECE TOILET SET
No. 5408 .... $14.00
Mirror, Brush and Comb.
Hammered and chased. French gray finish.

3-PIECE MILITARY SET
No. 5409 .... $12.70
2 Military Brushes and Comb.
Hammered and chased. French gray finish.

ALL PRICES ARE SUBJECT TO CHANGE WITHOUT NOTICE

Three-piece toilet sets and military sets offered for sale from the A.C.B.Jewelers' Wholesale price list from Chicago, Illinois in 1920.

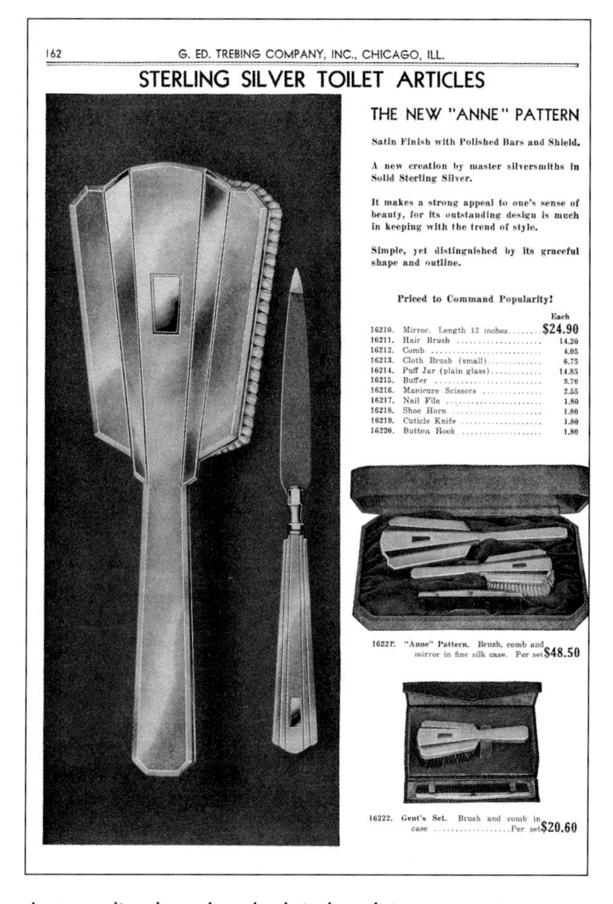

162 G. ED. TREBING COMPANY, INC., CHICAGO, ILL.

STERLING SILVER TOILET ARTICLES

THE NEW "ANNE" PATTERN

Satin Finish with Polished Bars and Shield.

A new creation by master silversmiths in Solid Sterling Silver.

It makes a strong appeal to one's sense of beauty, for its outstanding design is much in keeping with the trend of style.

Simple, yet distinguished by its graceful shape and outline.

Priced to Command Popularity!

		Each
16210.	Mirror. Length 13 inches	$24.90
16211.	Hair Brush	14.20
16212.	Comb	4.05
16213.	Cloth Brush (small)	6.75
16214.	Puff Jar (plain glass)	14.85
16215.	Buffer	3.70
16216.	Manicure Scissors	2.55
16217.	Nail File	1.80
16218.	Shoe Horn	1.80
16219.	Cuticle Knife	1.80
16220.	Button Hook	1.80

16221. "Anne" Pattern. Brush, comb and mirror in fine silk case. Per set $48.50

16222. Gent's Set. Brush and comb in case .... Per set $20.60

A streamlined modern look in boudoir accessories became evident in the 1930s when the Art Deco movement was in full bloom. The new *Anne* pattern in sterling silver was designed with a satin finish and polished bars and shield. A three-piece set listed for $48.50 in 1933.

89

Sterling Silver Toilet Ware

John V. Farwell Company CHICAGO

LADY AVALON PATTERN; PLATINUM FINISH; FLORAL DECORATIONS
GUARANTEED 925/1000 FINE

Illustrations are about ⅞ actual size.

J33584—Mirror; length 10 inches	Each 20.50
J33585—Hair Brush; length 8¾ inches	Each 13.15
J33586—Ladies' Comb; length 7½ inches	Each 4.75
J33587—Hair Brush, Comb and Mirror in silk case	Set 44.65
J33588—Military Brush; length 4½ inches	Each 9.50
J33589—Men's Comb; length 7½ inches	Each 4.75
J33590—2 Military Brushes and Comb in silk case	Set 28.15
J33591—Cloth Brush; length 6¾ inches	Each 12.50
J33592—Hat Brush; length 4¼ inches	Each 5.90
J33593—Hair Receiver	Each 14.60
J33594—Puff Box; medium	Each 14.60
J33595—Cold Cream Jar; small	Each 2.90
J33596—Cold Cream Jar; medium	Each 5.15
J33597—Button Hook	Each 3.25
J33598—Corn Knife	Each 3.25
J33599—Cuticle Knife	Each 3.25
J33600—Nail File	Each 3.25
J33601—Shoe Horn	Each 3.25
J33602—Nail Polisher	Each 6.25
J33603—Nail Scissors	Each 4.40
J33604—Cuticle Scissors	Each 5.15

The *Lady Avalon* pattern was made in sterling silver with a platinum finish and dainty floral decorations. It was popular in 1920.

STERLING
SILVER
Other matching pieces are also available.
98-92
77-44
98-94
96-22
77-44 "BAROQUE" 3-piece Ladies' Set. $50.00
96-22 "ECSTASY" 3-piece Ladies' Set. $35.00
98-92 "KIRSTEN" 3-piece Ladies' Set. $35.00
98-94 "SONJA" 3-piece Ladies' Set. $30.00
119

Four different styles in sterling silver dresserware available from W.H.Sims in 1940. The patterns were called *Baroque, Ectasy, Kirsten* and *Sonja.*

The International Silver Company manufactured lovely sterling silver dresser sets in 1935 that were offered for sale from the Benjamin Allen Company. The *Maid of Orleans* was a plain pattern with hand-engraved shields and a lustrous finish. Brushes, mirrors and combs were available in addition to cologne bottles, pin trays, hair receivers and manicure implements. A three-piece set in a satin-lined presentation box listed for $66.00. The less expensive *Lady Louise* pattern was $37.80 for a three-piece set. Evans Case Company also manufactured sterling silver toilet sets with old world characteristics of engraved and engine-turned designs. A ten-piece set listed for $66.00.

In 1936, a New York wholesale house called L & C Mayers Company offered a wonderful line of sterling silver toilet ware which was extremely elegant. The *Antoinette* pattern was designed with a decorative oval shield and an embossed border. The *Countess* pattern had broad and narrow engine-turned stripes and an oval shield for engraving. The *La Reine* pattern was made with a hammered design and a French gray finish and an embossed border. The *Imperial* pattern was rather plain with a shield for engraving in a soft gray satin finish. The *Majestic* pattern was polished sterling silver with black enamel back and a shield for engraving a monogram. The *Avon* and *Christina* patterns were similar in design both displaying broad and narrow engine turning and center shields for monogramming. The only difference was their overall shape. Complete sets of up to fourteen pieces were attractively packaged in silk-lined boxes. Dresser sets became wonderful gifts for women and the manufacturers were adding plenty of new lines each season.

The 1940s brought more styles, unusual pattern names and all price ranges. Sterling silver was still extremely common, more so than other metals because of the war effort. The W. H. Sims Company offered a wonderful assortment of sterling silver dresser sets in the early

I purchased this lovely sterling silver dresser set about 10 years ago from a woman who received it as a wedding present in the 1940s. It was a three-piece set consisting of a mirror, brush and comb in this satin-lined presentation box. Over the years, the comb, having celluloid teeth, began to deteriorate the way some early plastics unfortunately do and through the process, the fabric in the presentation box changed colors and the silver began to tarnish on the sides closest to where the comb had been resting. When I opened the box a few years later, the comb had completely decomposed. I immediately removed it from the box so that no further deterioration would occur. Thank goodness I caught it in time or the rest of the set would have been ruined. It is stamped *International Silver.* $145-195.

1940s. Pattern names were *Baroque, Ecstasy, Kirsten, Landsdowne, Christine, Helga, and Cambridge*. By 1950, patterns were becoming simpler and the majority of the old world charm was beginning to wane. A departure from Victorian, Edwardian and Art Nouveau styles was taking place. An occasional revival piece would surface, but it was short-lived. For instance, in 1953, the Bennett Blue Book catalog offered dresser sets crafted of sterling silver in the *Chantille* pattern, "reminiscent of a romantic era;" the *Colonial Maid* pattern, "classically simple;" and the *Christina* pattern, engine-turned with a shield for engraving. A few years later, these sets disappeared from the catalog altogether. Room was made for modern designs made with modern materials.

Previously, sterling silver and quadruple silver-plated sets were the height of fashion. Because of the popularity of the metal in the previous decades, imitations were also common. Dresser sets and matching dresser accessories were also made of German silver, Silveroin and Victor silver to name a few. Sterling and quadruple plated dresser sets had been extremely popular since the last quarter of the 19th century. They continued to be desirable throughout the early 20th century. They were always viewed as something that had lasting value. They were classics and classics never go out of style. Many American silver companies continued to manufacture sterling and silver-plated dresser sets in large numbers until the 1960s.

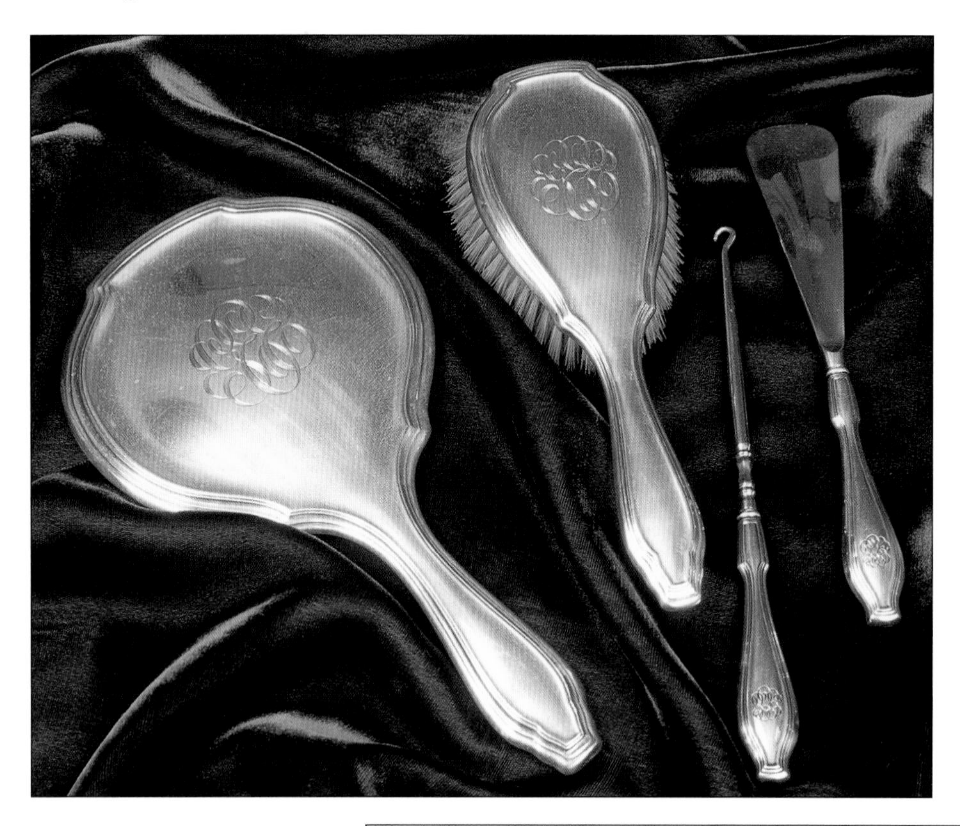

Four-piece dresser set consisting of a beveled glass mirror, hair brush, shoe buttoner and shoe horn made of sterling silver with a satin finish. All four pieces are monogrammed. The set is marked *Sterling Pat. Applied for.* Other hallmarks are visible but barely legible. $395-450 set.

Small Victorian sterling silver hand mirror with heavy embossed designs depicting flowers and scroll work bordering the mirror with a small polished left plain in the center $85-135. Victorian bonnet brush with sterling silver handle and back, richly embossed with ornate designs. $75-125.Dressing comb made with an embossed sterling silver back and coarse and fine tortoise shell patterned teeth $50-75.

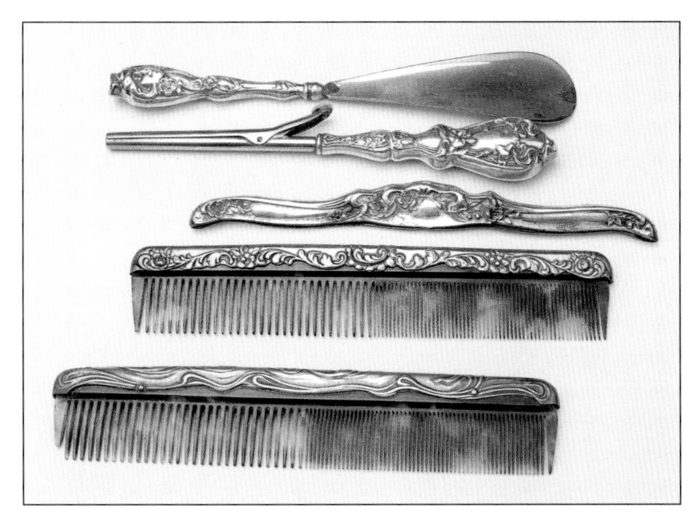

Shoe horn, curling iron, two tortoise combs and a single comb back all made of richly embossed sterling silver dating to the late 19th and early 20th centuries. Shoe horn $85-115; curling iron $100-150; tortoise combs $75-100; comb back $40-50.

This early 20th century ruby to clear cut glass jar has a jeweled domed lid. Around the top of the jar is an ormolu rim encrusted with bezel-set paste stones. The ormolu filigree lid is also accented with glass stones in many shapes, sizes and colors. The hand-cut diamond and star patterns in the glass looks similar to early Baccarat pieces made in the first two decades of the 20th century. Definitely a piece fit for a queen. $400-500.

Brass & Ormolu

By the end of the 19th century, the popularity of sterling silver, quadruple plate, German silver, and even aluminum toilet items, including dresser sets and other accessories, was secure, but brass, gilded, and jeweled sets also were desirable.

Possessions made of solid gold have been coveted since ancient times. Metal that looked like gold but sold for a fraction of the cost was now what the consumer wanted. Ormolu was extremely common at this time and had been used in the manufacture of many decorative pieces for the home. Ormolu was a type of brass, which was made to resemble gold by altering the metals' composition by using less zinc and more copper. Sometimes a lacquered finish was applied to the completed item to create an even more gold hue. Acid was also used for the same reason. Dresser sets, jewel boxes, atomizers, powder jars, trays, picture frames, and many other boudoir items were made of ormolu from the late 19th century to the 1930s. Earlier ormolu *objets de art,* from the 18th and early 19th centuries, were made from gilded bronze.

Wonderful display of seven dresser pieces, four of which are a matched set, stamped *Empire Art Gold.* Set $375-425; dresser box $125-175; pin tray $50-75; glass tray $85-115.

Exquisite three-piece boudoir set consisting of a large velvet-lined footed jewel box with hinged lid and matching round mirror and hair brush. All three pieces are gold-plated with beautiful embossed designs and set with faux emerald stones. The jewel box is stamped *Apollo*. $500-750 set.

Powder and bath crystal jars made with clear glass bottoms and jewel encrusted lids. The bath crystal jar on the right still retains a paper label on the bottom which reads: *Bath Crystals Dermay Paris, New York*. A paper insert still remains tucked in the lid. $125-175 each.

Four square and rectangular metal jewel boxes or caskets, as they were sometimes referred to in the late 19th and early 20th centuries. The boxes are accented with chased and engraved designed, applied filigree decorations and set with paste stones. The large rectangular box is stamped *Apollo*. Two of the boxes are wooden-lined. $200-300 each.

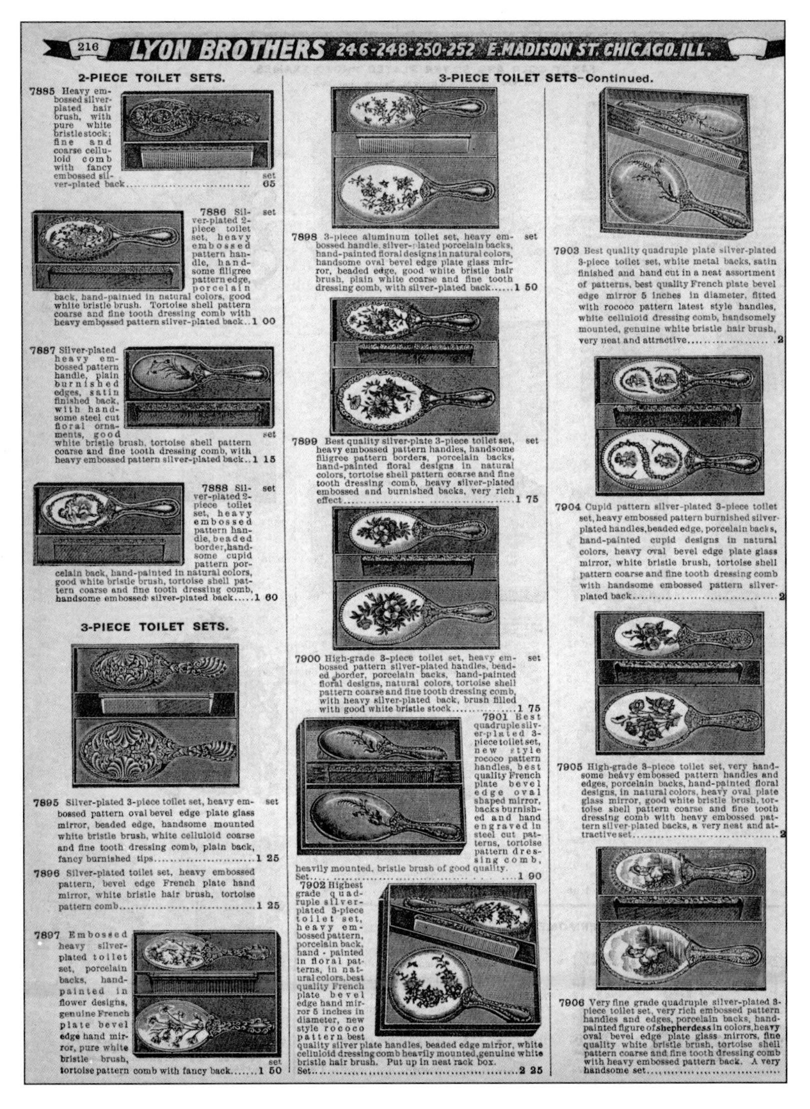

216 LYON BROTHERS 246-248-250-252 E. MADISON ST. CHICAGO. ILL.

2-PIECE TOILET SETS.

7885 Heavy embossed silver-plated hair brush, with pure white bristle stock; fine and coarse celluloid comb with fancy embossed silver-plated back.......... set 65

7886 Silver-plated 2-piece toilet set, heavy embossed pattern handle, handsome filigree pattern edge, porcelain back, hand-painted in natural colors, good white bristle brush. Tortoise shell pattern coarse and fine tooth dressing comb with heavy embossed pattern silver-plated back.. set 1 00

7887 Silver-plated heavy embossed pattern handle, plain burnished edges, satin finished back, with handsome steel cut floral ornaments, good white bristle brush, tortoise shell pattern coarse and fine tooth dressing comb, with heavy embossed pattern silver-plated back.. set 1 15

7888 Silver-plated 2-piece toilet set, heavy embossed pattern handle, beaded border, handsome cupid pattern porcelain back, hand-painted in natural colors, good white bristle brush, tortoise shell pattern coarse and fine tooth dressing comb, handsome embossed silver-plated back..... set 1 60

3-PIECE TOILET SETS.

7895 Silver-plated 3-piece toilet set, heavy embossed pattern oval bevel edge plate glass mirror, beaded edge, handsome mounted white bristle brush, white celluloid coarse and fine tooth dressing comb, plain back, fancy burnished tips.......... set 1 25

7896 Silver-plated toilet set, heavy embossed pattern, bevel edge French plate hand mirror, white bristle hair brush, tortoise pattern comb.......... 1 25

7897 Embossed heavy silver-plated toilet set, porcelain backs, hand-painted in flower designs, genuine French plate bevel edge hand mirror, pure white bristle brush, tortoise pattern comb with fancy back....... set 1 50

3-PIECE TOILET SETS–Continued.

7898 3-piece aluminum toilet set, heavy embossed handle, silver-plated porcelain backs, hand-painted floral designs in natural colors, handsome oval bevel edge plate glass mirror, beaded edge, good white bristle hair brush, plain white coarse and fine tooth dressing comb, with silver-plated back...... set 1 50

7899 Best quality silver-plate 3-piece toilet set, heavy embossed pattern handles, handsome filigree pattern borders, porcelain backs, hand-painted floral designs in natural colors, tortoise shell pattern coarse and fine tooth dressing comb, heavy silver-plated embossed and burnished backs, very rich effect.......... set 1 75

7900 High-grade 3-piece toilet set, heavy embossed pattern silver-plated handles, beaded border, porcelain backs, hand-painted floral designs, natural colors, tortoise shell pattern coarse and fine tooth dressing comb, with heavy silver-plated back, brush filled with good white bristle stock.......... set 1 75

7901 Best quadruple silver-plated 3-piece toilet set, new style rococo pattern handles, best quality French plate bevel edge oval shaped mirror, backs burnished and hand engraved in steel cut patterns, tortoise pattern dressing comb, heavily mounted, bristle brush of good quality. Set.......... 1 90

7902 Highest grade quadruple silver-plated 3-piece toilet set, heavy embossed pattern, porcelain back, hand-painted in floral patterns, in natural colors, best quality French plate bevel edge hand mirror 5 inches in diameter, new style rococo pattern best quality silver plate handles, beaded edge mirror, white celluloid dressing comb heavily mounted, genuine white bristle hair brush. Put up in neat rack box. Set.......... 2 25

7903 Best quality quadruple plate silver-plated 3-piece toilet set, white metal backs, satin finished and hand cut in a neat assortment of patterns, best quality French plate bevel edge mirror 5 inches in diameter, fitted with rococo pattern latest style handles, white celluloid dressing comb, handsomely mounted, genuine white bristle hair brush, very neat and attractive.......... 2

7904 Cupid pattern silver-plated 3-piece toilet set, heavy embossed pattern burnished silver-plated handles, beaded edge, porcelain backs, hand-painted cupid designs in natural colors, heavy oval bevel edge plate glass mirror, white bristle brush, tortoise shell pattern coarse and fine tooth dressing comb with handsome embossed pattern silver-plated back.......... 2

7905 High-grade 3-piece toilet set, very handsome heavy embossed pattern handles and edges, porcelain backs, hand-painted floral designs, in natural colors, heavy oval plate glass mirror, good white bristle brush, tortoise shell pattern coarse and fine tooth dressing comb with heavy embossed pattern silver-plated backs, a very neat and attractive set.......... 2

7906 Very fine grade quadruple silver-plated 3-piece toilet set, very rich embossed pattern handles and edges, porcelain backs, hand-painted figure of **shepherdess** in colors, heavy oval bevel edge plate glass mirrors, fine quality white bristle brush, tortoise shell pattern coarse and fine tooth dressing comb with heavy embossed pattern back. A very handsome set..........

Porcelain backed dresser sets offered for sale in 1899.

24Kt gold-plated toilet ware made by *Apollo* accented with imported painted miniatures under glass and exquisite embossed ornamentation. These sets were offered for sale in 1930 from May & Malone.

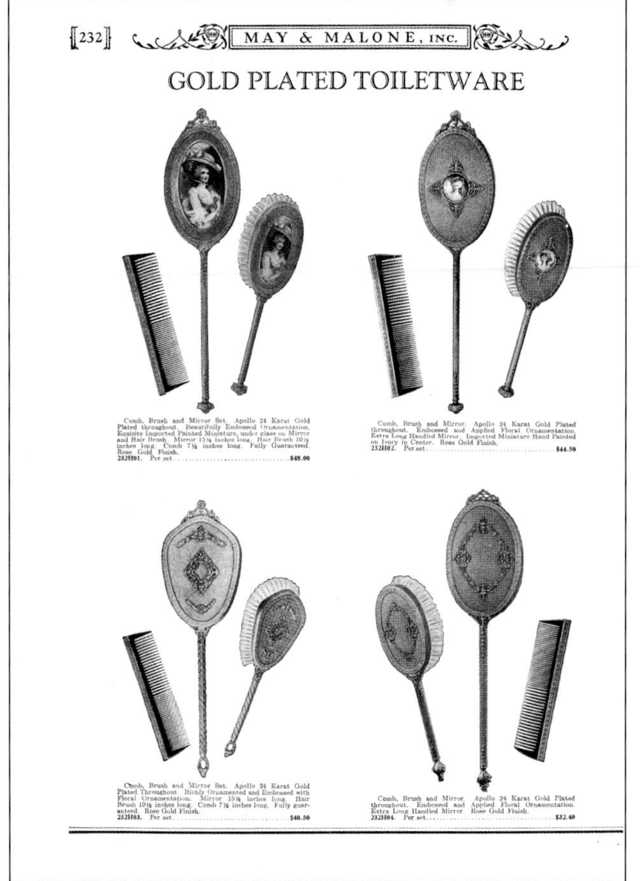

[232] MAY & MALONE, INC.

GOLD PLATED TOILETWARE

Comb, Brush and Mirror Set. Apollo 24 Karat Gold Plated throughout. Beautifully Embossed Ornamentation. Equisite Imported Painted Miniature, under glass on Mirror and Hair Brush. Mirror 15¾ inches long. Hair Brush 10½ inches long. Comb 7½ inches long. Fully Guaranteed. Rose Gold Finish.
232H01. Per set.......... $48.00

Comb, Brush and Mirror. Apollo 24 Karat Gold Plated throughout. Embossed and Applied Floral Ornamentation. Extra Long Handled Mirror. Imported Miniature Hand Painted on Ivory in Center. Rose Gold Finish.
232H02. Per set.......... $44.50

Comb, Brush and Mirror Set. Apollo 24 Karat Gold Plated Throughout. Richly Ornamented and Embossed with Floral Ornamentation. Mirror 15½ inches long. Hair Brush 10½ inches long. Comb 7½ inches long. Fully guaranteed. Rose Gold Finish.
232H03. Per set.......... $40.50

Comb, Brush and Mirror. Apollo 24 Karat Gold Plated throughout. Embossed and Applied Floral Ornamentation. Extra Long Handled Mirror. Rose Gold Finish.
232H04. Per set.......... $32.40

Ormolu dresser items often were designed with ornate filigree, and most were also jeweled with multi-color paste stones from Czechoslovakia. Different size glass stones made in various shapes were usually bezel-set on the fancy gilded or ormolu filigree, creating pieces fit for the boudoir of a queen. Items of this style manufactured by Empire Art were common, as well as those by Apollo Studios of New York.

Jeweled

Jeweled and ormolu filigree dresser sets and matching boudoir accessories from the early 20th century are fantastic. Elegant brush, hand mirror, and comb sets, in addition to perfume bottles and atomizers, powder jars and jewel boxes, were encrusted with bezel-set faux gemstones or foil-backed pastes. Many items of this nature were manufactured by the Apollo Studios of New York.

Apollo Studios opened in 1909 and manufactured many wonderful jeweled dresser sets and accessories for over a decade, until its doors closed in 1922. Powder jars, trinket boxes, dresser clocks, jewel caskets, perfume atomizers and even picture frames were wonderfully crafted with openwork filigree designs and carefully studded with bezel-set foil-backed pastes. Thin layers of intricate filigree would be applied to a cut crystal perfume bottle and then jeweled with multi-colored glass stones. The filigree resembled lace wrapped around a glass bottle. These items created an opulent look for the early 20th century dressing table. They are coveted by collectors in the 21st century. Because of their limited production years, these items are scarce today and they command high prices.

Two rectangular picture frames and two dresser jars made of gold-plated metal and enhanced with glass stones. The two dresser jars are stamped *Apollo, Ovington, New York.* Frames $250-325 pair; jars $200-300 each.

This long-handled ormolu mirror is unusual in the fact that it was designed with elements of Victorian, Edwardian and Art Nouveau styles combined into one piece. $150-195.

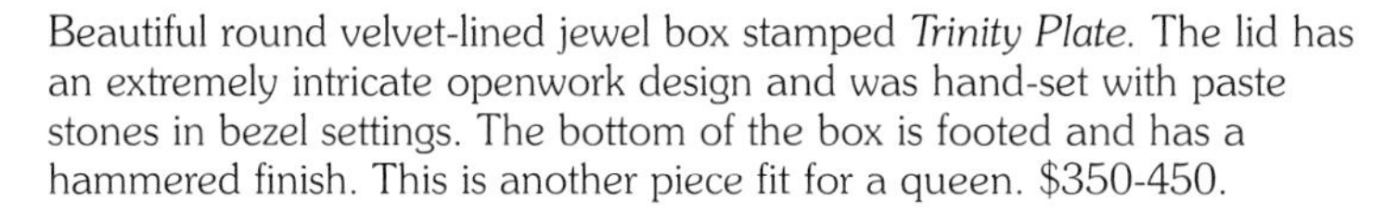

Beautiful round velvet-lined jewel box stamped *Trinity Plate.* The lid has an extremely intricate openwork design and was hand-set with paste stones in bezel settings. The bottom of the box is footed and has a hammered finish. This is another piece fit for a queen. $350-450.

Four-piece dresser set made up of two scent bottles, powder jar and tray. Filigree *fleur de les* decorate the metal holders for the glass scent bottles. All of the pieces are further enhanced with paste stones. $400-500 set.

This lovely embossed dresser tray accented with openwork designs and colored glass stones is stamped *Mfg. by Art Metal Works, Newark, N.J.* The jeweled atomizer, perfume bottle and scent bottle holders are stamped *Empire Art Gold.* Tray $115-155; atomizer $225-325; perfume bottle $195-275; letter opener $85-115.

Round metal powder box and lid. The base of the powder box is fitted with a blue glass insert. The outside is accented with filigree and set with colored glass stones. Three different match holders made of fancy gold plate and accented with embossing and paste stones. Powder box $145-195; match box holders $70-95.

Matched set of round hinged dresser boxes with embossed designs in addition to openwork decoration and jeweled enhancements. $275-325 pair.

Two powder jars with amber glass bottoms and embossed lids with applied filigree decorations. Both lids are further enhanced with paste stones. The larger jar is stamped *24 Kt Gold Plated.* $145-185 each.

In the 1940s, jeweled dresser sets were still in vogue and offered for sale through many venues, however they were made a little differently: the jewel was much larger. Instead of a wide variety of small glass stones, these dresser sets had one large stone mounted in the centers of the mirror and brush backs. The large faceted jewel was surrounded by applied floral decorations and filigree designs.

Fancy embossed dresser sets made of gilded brass sometimes were decorated with porcelain plaques. The embossed handles and borders provided the framework for lovely and delicate porcelain. In 1910, Chicago's Boston Store advertised China Back Sets with metal frames of brass or gilt as being Dresden china. The beautiful plaques were sometimes floral and other times figural. Hand mirrors were either round or oval.

Extremely ornate 24 Kt gold-plated dresser set consisting of round mirror, brush and two perfume bottles all set with large amber glass stones. The hammered jewel box with the filigree ornamentation is set with a topaz glass stone. It is not part of the set and was made earlier. $400-495 set; jewel box $150-200.

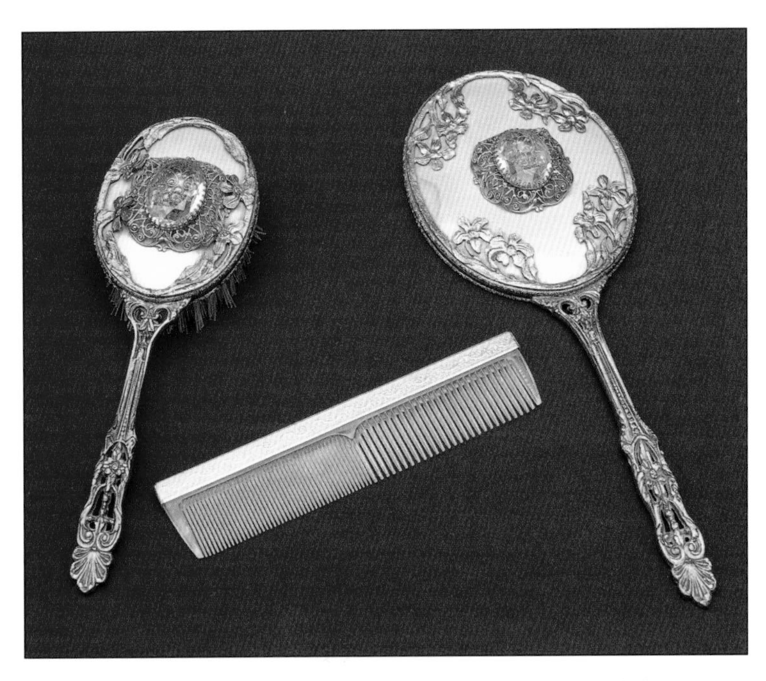

Three-piece ensemble made up of mirror with beveled glass, brush and comb, ornately decorated with floral designs around the rims and large pink glass stones mounted on fancy filigree medallions. This was a very popular style in the 1950s. $145-195.

Jeweled electric dresser clock made during the Victorian revival period of the 1930s. The ormolu filigree is further embellished with colored glass stones. The clock movement was made by *Sessions*. $250-325.

Attractive rhinestone and faux pearl studded dresser clock from the 1950s. The clock still retains the original paper label which reads *Fabulous Boutique Originals by Robert*. The clock movement was made by *RENSIE and Made in Germany*. $135-185.

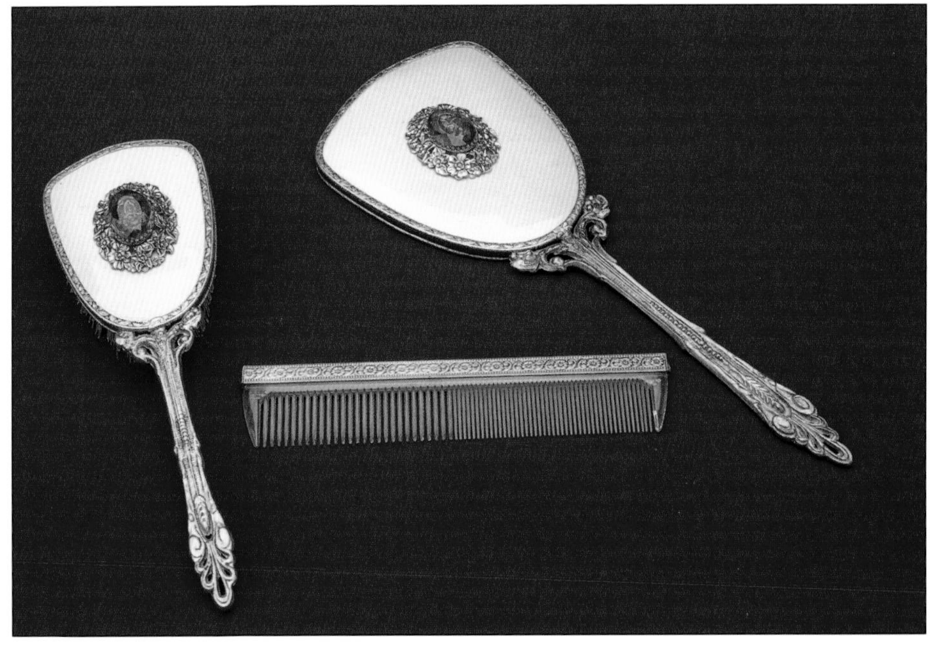

Dresser set made of 24 Kt gold plate manufactured by the Globe Silver Company. The backs of the mirror and brush are also accented with a large faceted amber glass stone surrounded by an embossed floral wreath and fancy rim and handle. The comb has clear Lucite teeth. This was another popular style in the late 1950s and continuing into the early 1960s. $125-150 set.

Two ormolu hair brushes with embossed pattern handles and delicate porcelain plaques accented with hand painted floral and cupid designs. $115-145 each.

Beautiful hand mirror made of brass and fitted with an oval hand-painted porcelain plaque. The handle is richly embossed. This was a very popular style at the turn of the 20th century. $115-145.

A heavily embossed pattern handle and frame is further accented with a porcelain back depicting a French courting scene. This type of dresser accessory was extremely fashionable in the late 19th and early 20th centuries. Some of these mirror and brushes were actually fitted with Dresden china. $125-175.

Three piece ormolu dresser set consisting of round beveled edge mirror, mirror-backed brush and tortoise hair comb. The mirror and brush handles are richly embossed with floral and cupid patterns. $165-225 set.

Gold Plate

New creations for the dressing table were constantly being manufactured. Gold plated items in 22kt and 24kt gold proved tasteful. They were advertised as "elegance at a reasonable cost." Glass-covered miniatures were popular in gold-plated dresser ware. In the 1930s and 1940s, toilet sets and boudoir accessories appeared with antique gold finishes. Again, filigree was common and sometimes enhanced with cloisonné medallions or applied floral decorations. Elegant, extra-long handles on mirrors were common in this period.

In the 1940s and 1950s, Jeweler's Bronze was another popular finish used for metal dresser sets. Two-toned jeweler's bronze was popular in what was called a peacock finish in addition to a rose and green finish in a quilted design. This medium was perfect for engraved and chased designs. Ribbons, scrolls, classic stripes and hearts and flowers were popular themes in jeweler's bronze.

In the 1950s and 1960s, Astorloid dresser sets and accessories were the height of fashion. Gold finished metal with filigree designs were coupled with silk brocade backs in many patterns, making exquisite modern and affordable creations for the dressing table.

Three-piece dresser set made of gold-plated metal with portrait miniatures under thin sheets of clear celluloid. The brush and mirror have long fluted handles. The smaller comb belongs to the set. $95-135.

Two glass-lined embossed metal powder jars with portrait lids. The metal is covered with a lacquer coating. $95-140 pair.

GOLD PLATED DRESSER SETS

Three outstanding designs. The beauty of these well styled sets will add grace to any lady's dressing table. They are plated with 24 Kt. pure gold and guaranteed not to tarnish. Packed in fine gift cases. Unusual value.

PRICES SUBJECT TO CATALOG DISCOUNTS. SEE PAGE 1.

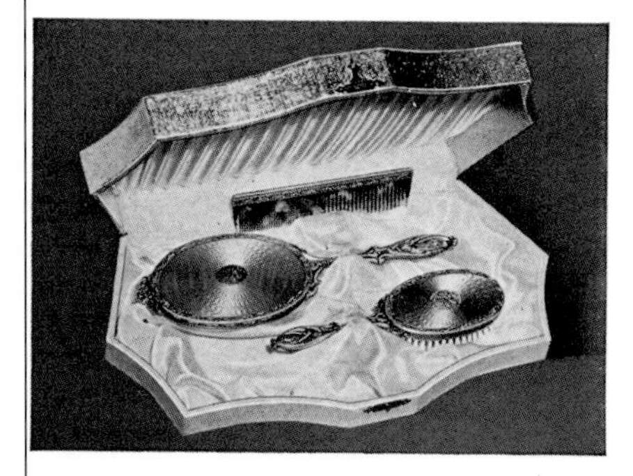

AS5003 Dresser Set, 3 pieces $18.75
Jade Green colored cloisonne effect with filigree metal ornamentation. Oval shaped. Mirror has bevelled glass. Length 13 inches. Brush has fine quality white bristles. Length 9 inches. Comb has half fine and half coarse teeth. Length 7½ inches.

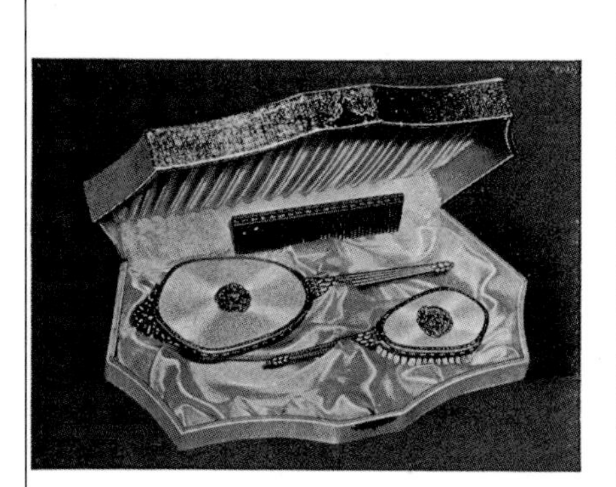

AS5004 Dresser Set, 3 pieces.................... $21
Ivory colored cloisonne effect with filigree metal ornamentation. Pear shaped. Mirror has bevelled glass. Length 13½ inches. Brush has fine quality white bristles. Length 9¼ inches. Comb has half fine and half coarse teeth. Length 7½ inches.

AS5005 Dresser Set, 3 pieces.................. $28.50
Beautiful chased design with genuine cloisonne center. Maize color, floral decoration. Distinctive filigree handle. Mirror has bevelled glass. Length 13½ inches. Brush has fine white bristles. Length 10 inches. Comb has half fine and half coarse teeth. Length 7½ inches.

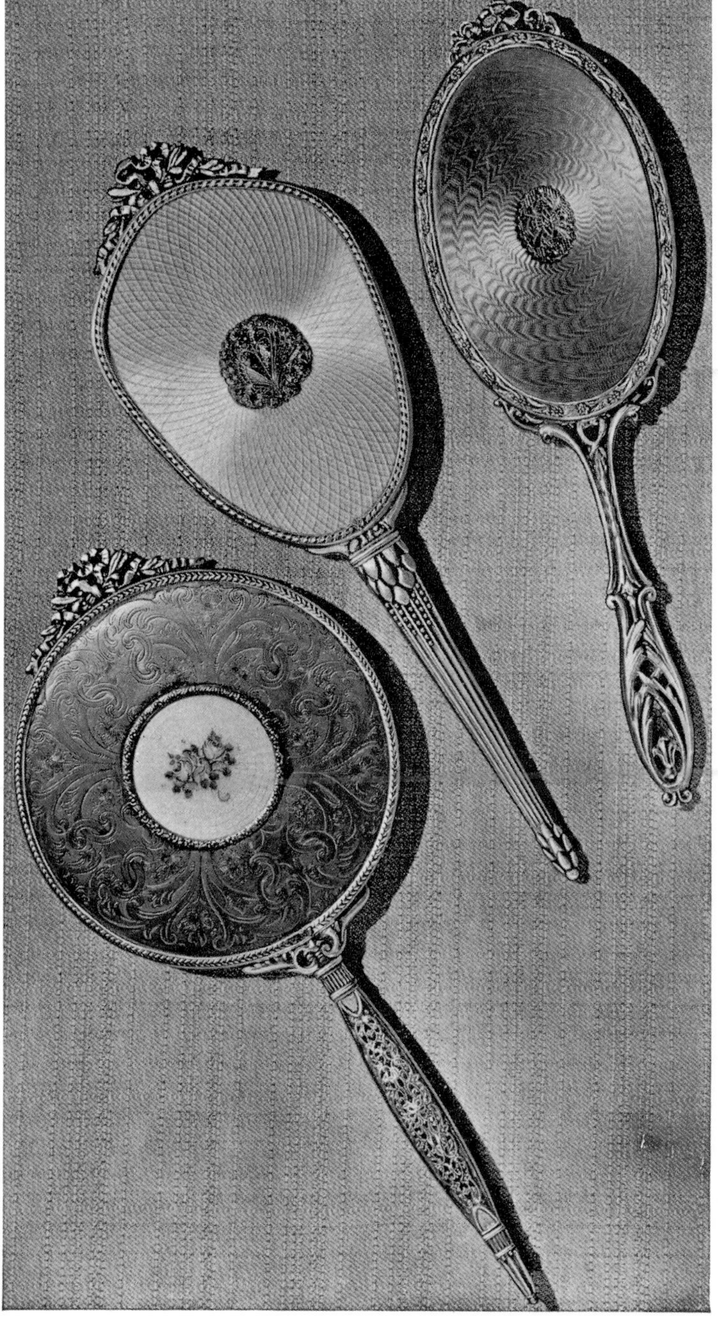

L. & C. MAYERS CO. FIFTH AVE., NEW YORK

227

Gold-plated dresser sets offered for sale in 1939. Oval, round and pear-shaped mirrors were common at this time.

Beauty Accessories For Milady

IVORY ROSE 5-PC. DRESSER SET. Tasteful elegance with matte goldtone finish and raised ivory ornament. 13¾" mirror, brush, comb, powder jar and atomizer. Packed in satin lined gift box.
1506 .... Retail $30.00
3-pc. set: comb, brush and mirror. In satin lined gift box.
1505 .... Retail $20.00

IMPERIAL 5-PC. DRESSER SET. Beautiful chased design in contrasting matte and bright silver. Mirror and brush 1-pc. construction. 13¼" mirror, 10¼"x16¼" tray with filigree border, comb, brush, powder jar.
1508 In satin lined case .... Retail $48.00
3-PC. SET: comb, brush and mirror. In satin lined gift box.
1507 .... Retail $25.00

ASTORLOID DRESSER SETS

- *All Hand Mirrors Have Bevel Edge Glass*
- *Nylon Bristles Snap Out of Frame for Easy Cleaning*
- *Mirrors Are Protected With Copper Backs*

GOLDEN JADE ENSEMBLE. Recaptures all the luxurious elegance of the Renaissance. Filigree rims and handles finished in 24 Kt. antique gold. Moire back is decorated with filigree mounted jade ornament. In gift box.
(A) 3-pc. set: Comb, brush and 14" long hand mirror.
1509 .... Retail $30.00
(B) Matching mirror tray with jade ornaments on handles. Size 11$\frac{5}{16}$x20$\frac{5}{16}$"
1510 .... Retail $28.50
(C) Matching 30-hour alarm clock with jade ornament.
1511 .... Retail $24.00
plus tax
(D) Matching jewel box with hinged bevelled glass cover and jade ornament. Velvet lined. 6¼x5⅛x3".
1512 .... Retail $20.00

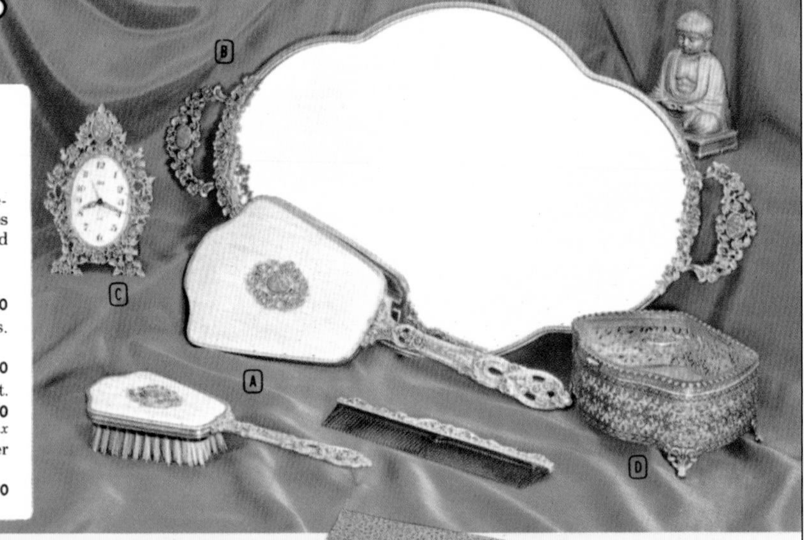

BOUTIQUE 4-PC. DRESSER SET. Beauty that will win its way into the heart of every woman. Decorated with 3-dimensional rosebuds on white silk background. Includes comb, brush, 13¼" hand mirror and matching perfume tray. Satin lined gift box.
1502 .... Retail $24.00
3-pc. set: comb, brush, hand mirror. Satin lined gift box.
1501 .... Retail $10.00

SWEETHEART 5-PC. DRESSER SET. Beautifully decorated with a raised cupid design in gold and matte silver background. 13¼" mirror, comb, brush, powder jar and glass perfume tray. In satin lined case.
1504 .... Retail $30.00
3-pc. set: comb, brush and hand mirror. In satin lined gift box.
1503 .... Retail $14.00

Stylish *Astorloid* dresser sets offered for sale in 1967.

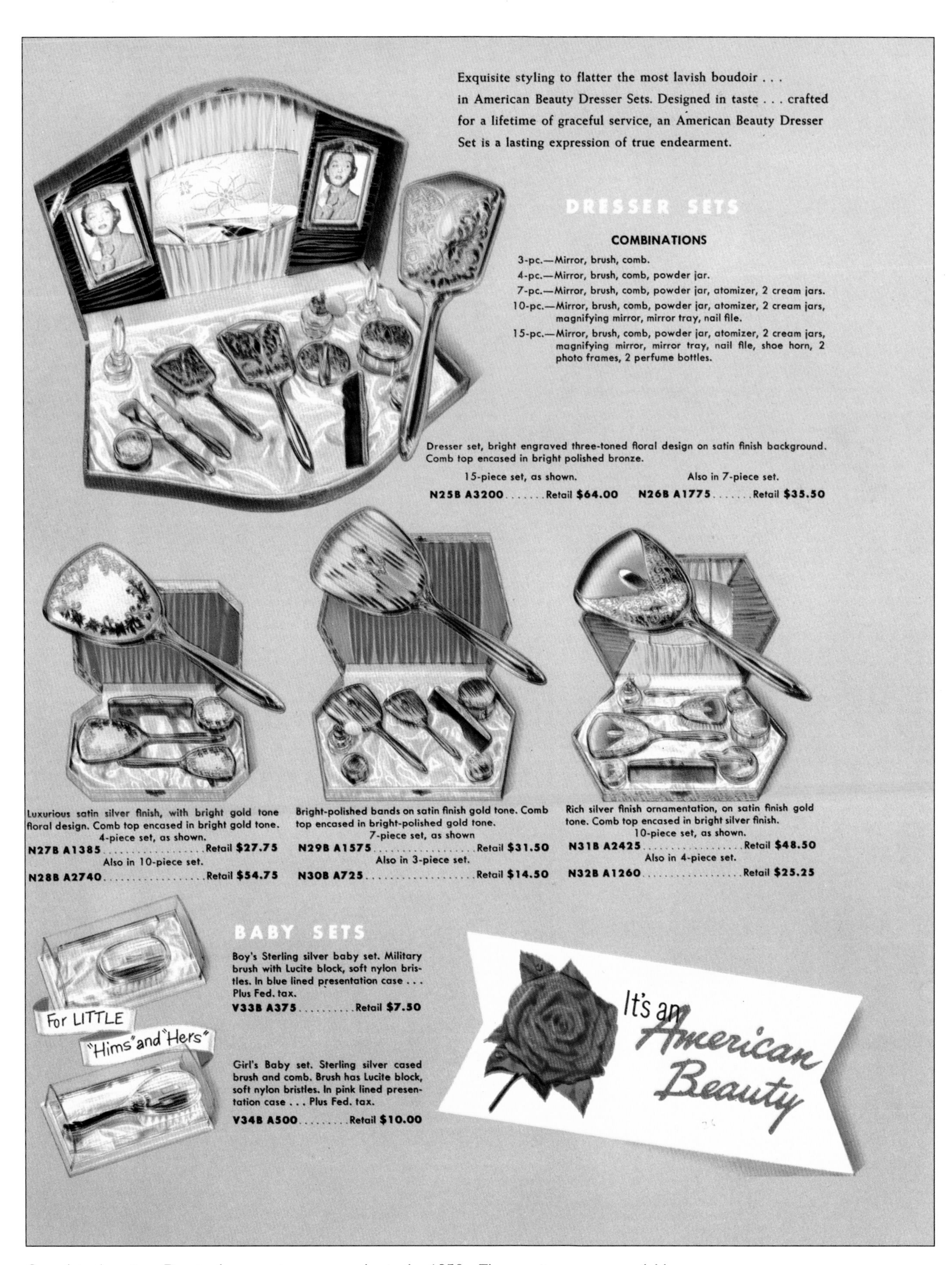

Exquisite styling to flatter the most lavish boudoir . . . in American Beauty Dresser Sets. Designed in taste . . . crafted for a lifetime of graceful service, an American Beauty Dresser Set is a lasting expression of true endearment.
DRESSER SETS
COMBINATIONS
3-pc.—Mirror, brush, comb.
4-pc.—Mirror, brush, comb, powder jar.
7-pc.—Mirror, brush, comb, powder jar, atomizer, 2 cream jars.
10-pc.—Mirror, brush, comb, powder jar, atomizer, 2 cream jars, magnifying mirror, mirror tray, nail file.
15-pc.—Mirror, brush, comb, powder jar, atomizer, 2 cream jars, magnifying mirror, mirror tray, nail file, shoe horn, 2 photo frames, 2 perfume bottles.
Dresser set, bright engraved three-toned floral design on satin finish background. Comb top encased in bright polished bronze.
15-piece set, as shown.
N25B A3200.......Retail $64.00
Also in 7-piece set.
N26B A1775.......Retail $35.50
Luxurious satin silver finish, with bright gold tone floral design. Comb top encased in bright gold tone.
4-piece set, as shown.
N27B A1385..................Retail $27.75
Also in 10-piece set.
N28B A2740..................Retail $54.75
Bright-polished bands on satin finish gold tone. Comb top encased in bright-polished gold tone.
7-piece set, as shown
N29B A1575..................Retail $31.50
Also in 3-piece set.
N30B A725..................Retail $14.50
Rich silver finish ornamentation, on satin finish gold tone. Comb top encased in bright silver finish.
10-piece set, as shown.
N31B A2425..................Retail $48.50
Also in 4-piece set.
N32B A1260..................Retail $25.25
BABY SETS
Boy's Sterling silver baby set. Military brush with Lucite block, soft nylon bristles. In blue lined presentation case . . . Plus Fed. tax.
V33B A375..........Retail $7.50
For LITTLE
"Hims" and "Hers"
Girl's Baby set. Sterling silver cased brush and comb. Brush has Lucite block, soft nylon bristles. In pink lined presentation case . . . Plus Fed. tax.
V34B A500........Retail $10.00
It's an American Beauty

Complete *American Beauty* dresser sets were popular in the 1950s. These patterns were available for sale in 1954. Combinations of bright gold and satin silver were desirable finishes.

Aluminum

For a short period of time, particularly around 1895, pure aluminum hollow ware was used to create handsome, lightweight boudoir items. Brush mirror and comb sets were satin engraved and extremely beautiful. Cologne bottles, puff boxes, atomizers and soap holders were common for women while shaving stick boxes, cloth brushes, shoe horns and tobacco boxes were some of the item manufactured for men. Just about any item that was manufactured in sterling silver or quadruple plate was mass-produced in aluminum. Aluminum was also less expensive. A sterling silver hairbrush with engraved decorations cost around $15.00 in 1895. An aluminum brush with similar decorations sold for $5.00. Aluminum offered a similar look at a fraction of the cost.

Rectangular-shaped footed box made of aluminum and designed to store gloves. The word "Gloves" is embossed on the hinged lid. A somewhat fancier aluminum box is also shown here with applied ornaments on each corner and fancy engraved designs. This box has the word "Jewels" embossed in the lid and the name "Hazel" is engraved. A smaller aluminum box, also footed with a hinged lid is embossed to read "Where's My Hairpins" and features an embossed hairpin on the lid. This box was personalized for "Margaret". Glove box $60-80; jewel box $75-100; hairpin box $45-65.

Fancy embossed aluminum brush and comb set from the 1890s still in the original presentation box. The comb was made of celluloid which resembles tortoiseshell. The embossed Victorian manicure set was made of sterling silver. Two salve jars, nail buffer and file make up this lovely set. Aluminum set $135-165; sterling silver set $145-195.

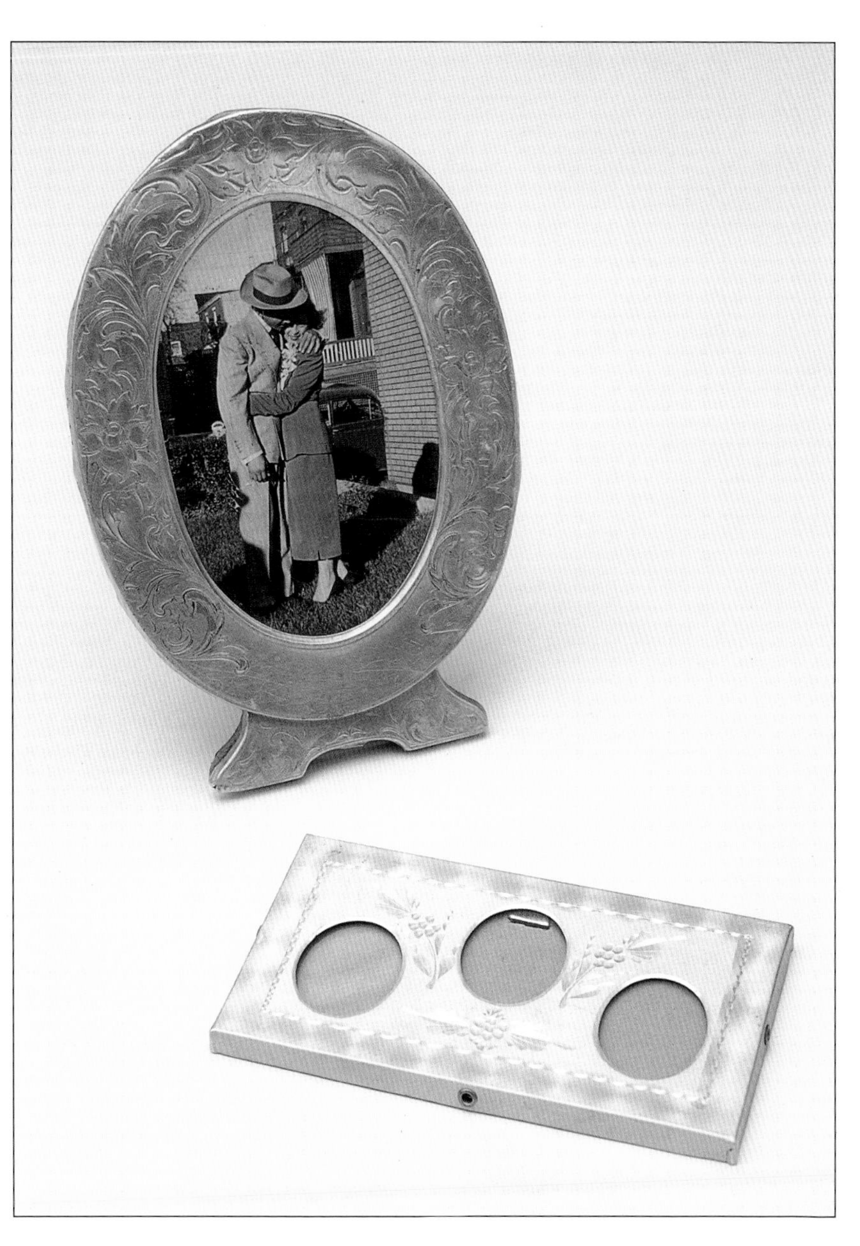

Oval mirror made of engraved aluminum dating back to the late 1890s. The 1950s photograph inside the frame is a picture of the author's parents, Marie and Vito Rodino. The rectangular frame, also made of engraved aluminum, contains three oval areas for pictures. $45-85 each.

Lovely embossed and engraved "Handkerchief" box also personalized for "Elizabeth". $70-90.

Inside view of aluminum hankie box.

PURE ALUMINUM HOLLOW WARE.

PRICES EACH.

BUSIEST HOUSE BHA IN AMERICA

Cut One-half Size.

No. 6472. TOOTHPICK or Match Holder, Satin Engraved [SNARL] $0 82

Cut One-half Size.

No. 6473. INKSTAND, Satin Finish....$1 30 [SNARLING.]

Cut One-third Size.

No. 6474. COMB, Engraved......[SNARLED] $2 50
No. 6474½. Comb, Satin........[SNARLEYOW] 2 25

Cut One-half Size.

No. 6475. TOOTHPICK or Match Holder, Engraved.....[SNEER] $0 45

Cut One-half Size.

No. 6476. TRAY, Satin Finish....$0 92 [SNOOZE.]

Cut One-third Size.

No. 6477. CLOTH BRUSH, Engraved......$5 00 [SNORE.]
No. 6477½. Cloth Brush, Satin......[SNORING] 4 00
Entire Handle of Aluminum. The bristles of best quality and fastened through Aluminum back.

Cut Three-fourths Size.

No. 6478. COLLAPSION CUP...$0 92 [SNUB.]

Cut One-half Size.

No. 6478½. TRAY [SNUBBER] $0 63

Cut One-third Size.

No. 6479. MIRROR, Engraved...............[SNUBBING] $6 50
No. 6479½. Mirror, Satin.......................[SNUBFUL] 5 00

Cut One-half Size.

No. 6480. SHAVING STICK BOX $1 34 [SINGLY.]

Cut One-half Size.

No. 6480½. SHAVING BRUSH......$2 16 [SNUGGLE.]

Cut One-third Size.

No. 6481. HAIR BRUSH, Engraved....[SNYDER] $4 50
No. 6481½. Hair Brush, Satin..........[SOAKING] 3 75
Entire back Aluminum, including part through which the bristles go.

195

Pure aluminum hollow ware was used to manufacture dresser accessories and other boudoir items in the 1890s. These engraved pieces shown here were available from the BHA Illustrated Catalog of 1895.

PURE ALUMINUM HOLLOW WARE.

PRICES EACH.

BUSIEST HOUSE BHA IN AMERICA

Cut One-third Size.

No. 6462. COLOGNE..[SMIRK] $4 10
Satin Finish, Hand Painted.

Cut One-half Size.

No. 6463. PAPER KNIFE ....[SMITE] $1 10

Cut One-half Size.

No. 6464. SHOE HORN ....[SMOULDER] $1 62

Cut One-third Size.

No. 6465. COLOGNE, [SMUDGE] $5 34
Satin Finish, Hand Painted.

Cut One-third Size.

No. 6466. MIRROR ......[SMUGGLE] $1 50
Applied Ornaments.

Cut One-half Size.

No. 6467. ATOMIZER .....[SMUT] $4 34
Satin, Knurled Borders.

Cut One-third Size.

No. 6468. WHISK BROOM ..[SNAG] $3 66
Engraved.

Cut One-half Size.

NO. 6469. PUFF BOX ....[SNAP] $1 25
Satin Finish, Hand Painted.

Cut One-half Size.

No. 6470. SOAP BOX..[SNAPPING] $1 62
Satin Finish.

Cut One-half Size.

No. 6471. PUFF BOX......[SNARE] $1 62
Engraved.

194

Satin finished and hand painted cologne bottles and puff boxes were made of pure aluminum hollow ware in 1895. Other dresser accessories, including atomizers, soap boxes, shoe horns and even handles on whisk brooms were made of aluminum.

Chapter Two

Plastic Dresser Sets

Celluloid

In the second half of the 19th century, a leading billiards manufacturer, Phelan and Collander, offered $10,000 in prize money to anyone who could successfully create a substitute for ivory in the manufacture of billiard balls. True ivory was becoming scarce, and something had to be done to eliminate the use of the genuine article.

New Yorker John Wesley Hyatt rose to the challenge to create a suitable substitute for ivory that could be molded and hardened for commercial use. He formed the Hyatt Billiard Ball Company in Albany, New York, in 1866. Hyatt experimented and gradually made improvements on his formula. One day, in his laboratory, Hyatt noticed a bottle of collodion that had been accidentally knocked over. Upon close examination, he realized that the substance had hardened when it dried and resembled the material he had been trying to create. By 1869, after more experimentation and improvement, Hyatt developed and patented a pyroxylin plastic called Celluloid which was a combination of cellulose and collodion. Hyatt then formed The Celluloid Manufacturing Company, in Albany, in 1871, but moved to Newark, New Jersey, in 1872, to a much larger building. The celluloid product was highly flammable and a few more moves became necessary over the years, due to a series of unfortunate fires. In 1873, celluloid became a registered trade name.

Even before John Hyatt began his experimentation, Alexander Parkes, in England, developed a substance he called Parkesine, which was actually the first man-made plastic made of organic cellulose. When heated, this material could be molded and its shape was retained when it was cooled.

In the early period of celluloid production, it was used for dentures, knife handles, eyeglass frames, collars, shirt cuffs, and piano keys besides billiard balls. By the mid-1870s, hairbrushes and other grooming aids were mass-produced in celluloid. As production methods and machinery improved, celluloid became the perfect substitute material for more expensive materials such as coral, jet, amber, tortoiseshell, and, of course, ivory. With the discovery of celluloid and other synthetic plastics, the prices of goods made from these materials dropped and the general public became able to afford items that previously had been purchased only by wealthy customers.

A process for graining the celluloid was invented in 1883, so that finished products could closely resembled real grained ivory. Five years later, celluloid was made to resemble tortoiseshell, amber, and jet; motion picture film was even made from it in the 1880s.

Large group of early celluloid dresser accessories with a variety of trade names. Some of the pieces are stamped *French Ivory*, while others are marked *Ivory Fiberloid, Ivory Pyralin and Ivoroid.* $25-100 each.

Advertisement for Ivory Fiberloid toilet articles as seen in popular magazines of the 1920s.

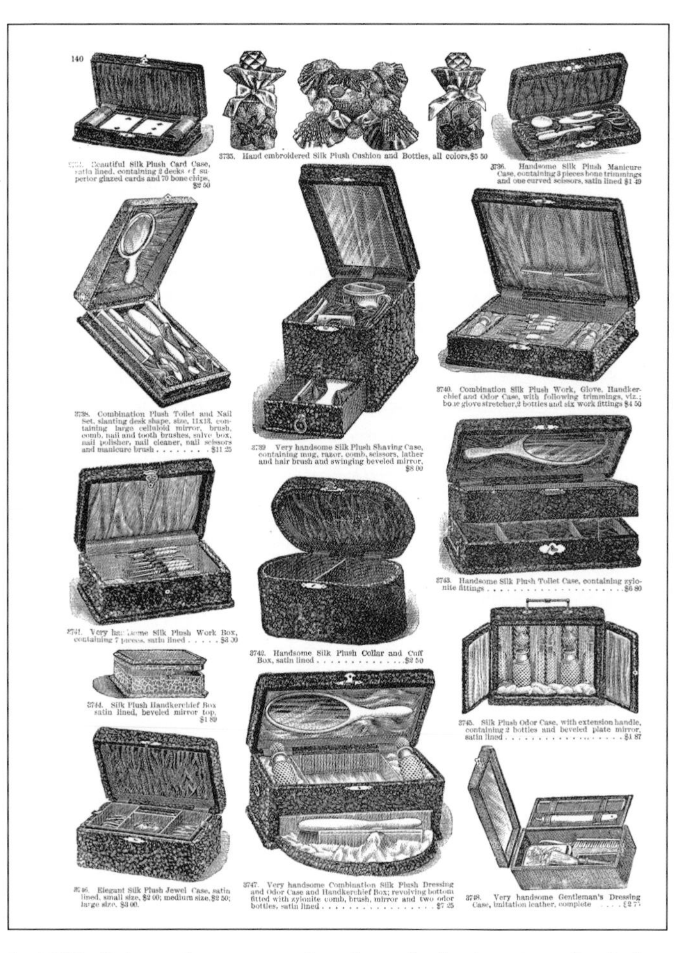

In 1886, *Zylonite* fittings were handsomely displayed in silk plush dressing cases. *Zylonite* was an early tradename for celluloid manufactured by the *American Zylonite Company* of North Adams, Massachusetts.

Celluloid became a popular material for ladies' and men's toilet articles and dresser sets in the late Victorian era. Dressing tables were filled with celluloid hairbrushes, hand mirrors, combs, powder jars, hair receivers, salve jars, cologne bottles, and manicure implements. A toilet set could consist of in excess of twenty pieces, depending on how fancy the consumer wanted to get or how much money he or she was willing to spend. Celluloid was expensive when it first came out; it was not intended to be a cheap imitation. A three-piece set could cost $75 in the 1880s!

Two lovely clear glass perfume bottles with unusual stoppers from the 1920s which fit snugly into their celluloid bases. One of the stoppers is clear faceted cut glass while the other is molded and frosted. No manufacturers' marks were found on either piece. $100-150 each.

As the demand for personal grooming articles made of celluloid increased, other companies began to market celluloid and similar pyroxylin thermoplastics under their own trade names. Different trade names were used by different companies for basically the same material; Zylonite, Pyralin and Fiberloid were three of the most common.

French Ivory

French Ivory, Ivory Fiberloid, Ivorine and Ivoris were common trade names for celluloid, ivory-colored plastics. Celluloid items were not always plain. Sometimes they were hand-painted, engraved or molded with fancy designs. Border designs with beveled edges were also popular. Many of the pieces that are found today were intricately engraved with a fancy monogram. This was a common practice in the 1920s. Upon closer examination of a piece of early celluloid, the trade name and manufacturers' name can sometimes be found on the piece.

In 1915, the Geo. T. Brodnax Company of gold and silversmiths in Memphis, Tennessee, advertised French Ivory toilet ware in their wholesale catalog. They mentioned in their ad that over the last two years since French Ivory was introduced that their assortment has increased dramatically due to the success of the line. French Ivory was imported into the United States from Canada in 1913. Because French Ivory closely resembled real ivory, it was a desirable look at a fraction of the cost of the real thing. Monogramming in script or old English lettering was offered free of charge. The engraving was usually filled in with blue ink as a nice contrast.

Dresser sets became extremely elaborate and contained many pieces. A hairbrush, hand mirror and comb were the basic components of a three-piece set. From that, however, the sets grew to include powder jars, hair receivers, cologne bottles, salve jars and atomizers. In addition to that, a complete set could also include jewel boxes, picture frames, flower vases and dresser clocks. Manicure implements were also designed as part of a complete set or as a separate set with up to 26 pieces. Celluloid boudoir items became so popular that by the 1920s, there were so many different patterns, that each pattern was assigned a different name. What was happening with the increased popularity of early plastics was the same thing that happened with the increased popularity of sterling silver and silver-plated dresser accessories. Each company that manufactured celluloid items had their own lines. Many companies had top-of-the-line sets with extravagant price tags as well as economy sets with cheaper price tags.

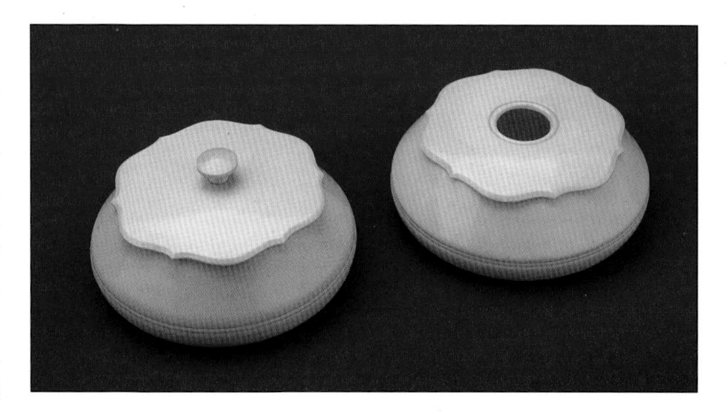

Puff box with lid and matching hair receiver made of Ivory Pyralin in the DuBarry pattern. $50-75 set.

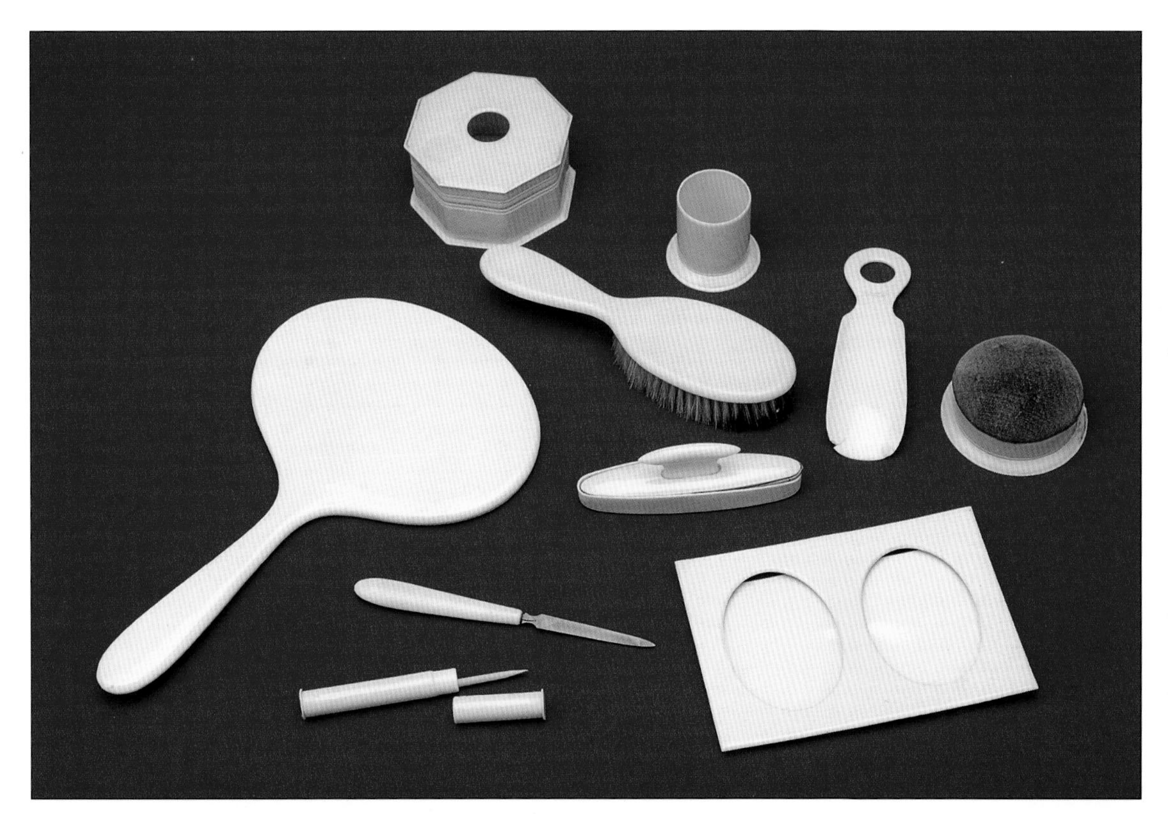

Diverse grouping of French Ivory celluloid accessories from the 1920s. Included is a large hand held mirror with an oval shape, hair brush, nail buffer, pincushion, jewel box, shoe horn, nail file, hair receiver, picture frame, cuticle tool holder and a celluloid holder for a glass perfume or scent bottle. The hairbrush is stamped *Ivory Fiberloid*; the shoe horn is stamped *Palmer's French Ivory*. $35-85 each.

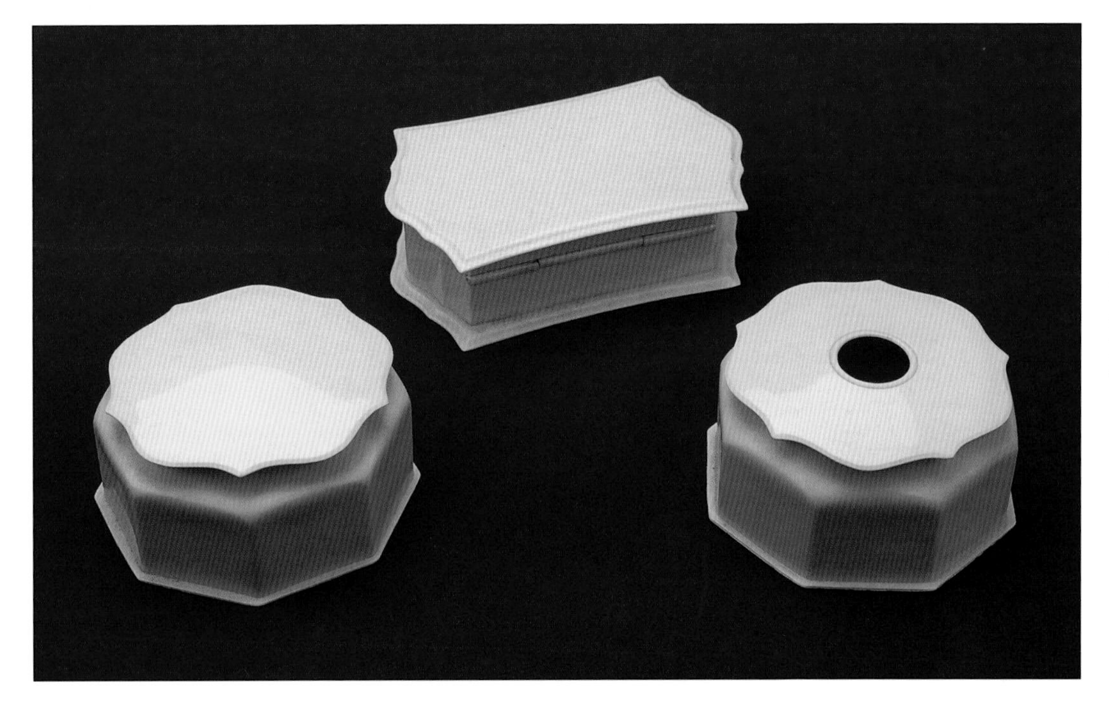

Three-piece Ivory Pyralin set in the DuBarry pattern consisting of a jewel casket, powder or puff box and hair receiver. $95-140 set.

A slight twist was added to this French Ivory set with a blue and gold pinstripe decorating the rim of the mirror, brush, shoe buttoner and nail file. The salve jar is not a part of this set. This set might have been designed for a child or young adult since it was made on a smaller scale than most sets. $125-145 set; salve jar $15-25.

Two French Ivory military brushes for men with monogrammed tops in fancy script lettering. Notice the stains in the material, which have been there for decades. Once stained, it is usually permanent and cannot be removed. $20-30 pair.

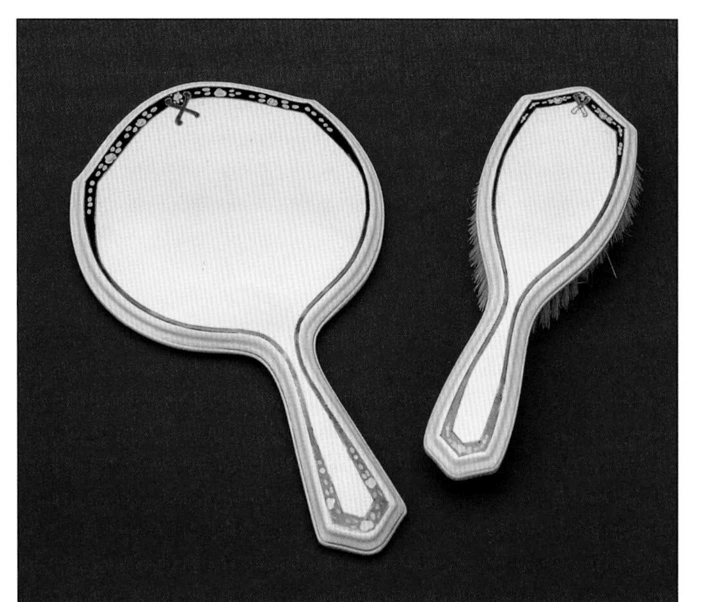

Hand-painted Ivoroid dresser set from the 1920s. A garland of leaves and flowers was executed by hand and applied to the black border, which runs around the rim of the mirror and brush adding more appeal to the set. $115-135 set.

102 GEO. T. BRODNAX, INC., GOLD AND SILVERSMITHS, MEMPHIS, TENN.

FRENCH IVORY TOILET WARE. (Illustrations one-half size.)

The goods of this character shown in our book for the last two years have met with such decided success that we have greatly increased the assortment. French Ivory closely resembles the real ivory in appearance, but excels it in the beauty of its finish and in durability. It makes an elegant set for a young lady's dresser.

We engrave without charge a three letter monogram on the mirrors and hair brushes, and a single letter, either Old English or Script, on the other pieces. Additional engraving will be charged extra. The engraving is filled in in colors, and the most pleasing contrast is made by using dark blue. We recommend this in preference to any other. In ordering be sure to state whether you prefer a Script or Old English letter on the small pieces.

26069	Handled Mirror, 14¼ inches long, 8 inches wide, attractive shape	$5.00	26250	Handled Mirror, 9½x5½	$2.00
26070	Ladies' Hair Brush, 9¼x2½, fine bristles	3.50	26251	Hair Brush, 8¼x2½	2.50
26071	Ladies' Comb, all coarse, 8¼ inches long	1.00	26252	Ladies' Comb, half coarse	.50
26072	Military Brushes, fine bristles, pair	4.00	26266	Handled Mirror, 11½x6¾	4.00
26073	Man's Comb	.50	26267	Vanity Mirror, 6¼x3½	1.00
26078	Shoe Hook, medium size	.50	26275	Small Salve Jar, glass lined	.50
26079	Flexible Nail File, medium size	.50	26276	Large flexible Nail File	.75
26080	Cuticle Knife, medium size	.50	26277	Large Button Hook	.75
26081	Nail Polisher and Tray	1.00	26278	Large Cuticle Knife	.75

All forwarding charges prepaid and safe delivery guaranteed.

French Ivory toiletware became increasingly popular in the early 20th century. It resembled real ivory but was more durable and less expensive. Many pieces were engraved with fancy monograms. Not only were the basic components of a dresser set available, but many other accessories for the boudoir were individually offered for sale in 1915.

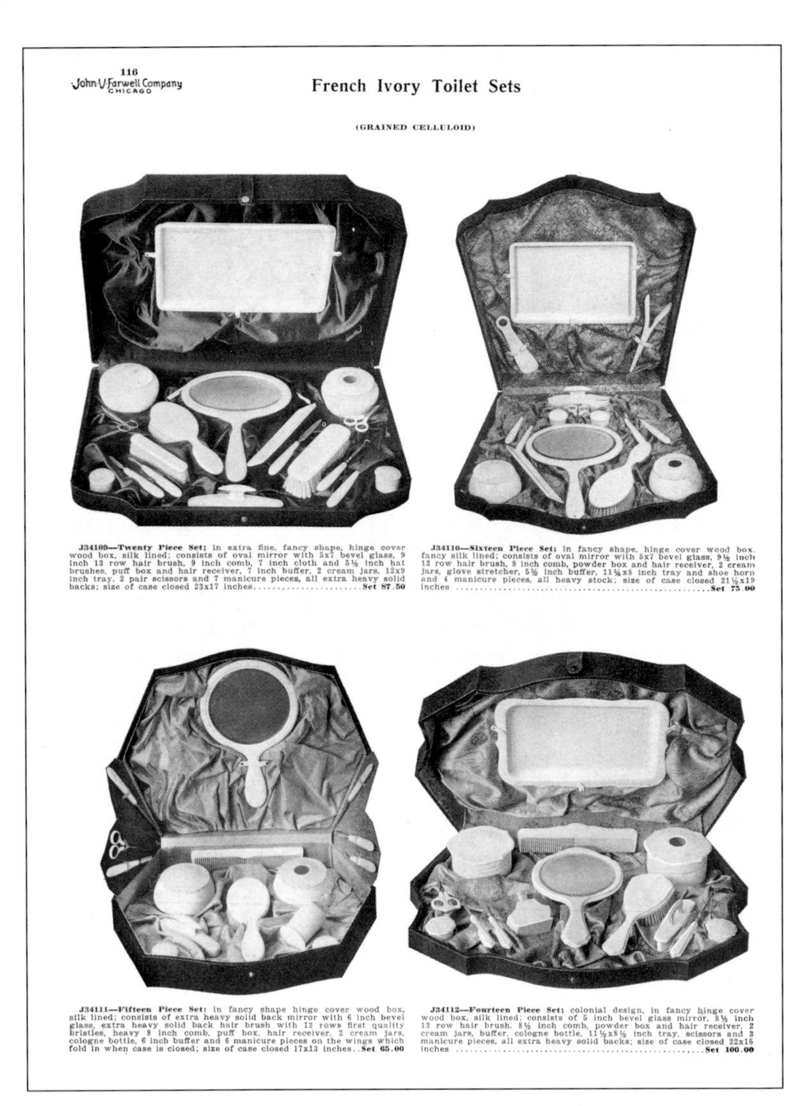

116
John V. Farwell Company
CHICAGO

French Ivory Toilet Sets

(GRAINED CELLULOID)

J34109—Twenty Piece Set; in extra fine, fancy shape, hinge cover wood box, silk lined; consists of oval mirror with 5x7 bevel glass, 9 inch 13 row hair brush, 9 inch comb, 7 inch cloth and 5½ inch hat brushes, puff box and hair receiver, 7 inch buffer, 2 cream jars, 12x9 inch tray, 2 pair scissors and 7 manicure pieces, all extra heavy solid backs; size of case closed 23x17 inches........................**Set 87.50**

J34110—Sixteen Piece Set; in fancy shape, hinge cover wood box, fancy silk lined; consists of oval mirror with 5x7 bevel glass, 9½ inch 13 row hair brush, 8 inch comb, powder box and hair receiver, 2 cream jars, glove stretcher, 5½ inch buffer, 11¼x8 inch tray and shoe horn and 4 manicure pieces, all heavy stock; size of case closed 21½x19 inches ....**Set 75.00**

J34111—Fifteen Piece Set; in fancy shape hinge cover wood box, silk lined; consists of extra heavy solid back mirror with 6 inch bevel glass, extra heavy solid back hair brush with 13 rows first quality bristles, heavy 8 inch comb, puff box, hair receiver, 2 cream jars, cologne bottle, 6 inch buffer and 6 manicure pieces on the wings which fold in when case is closed; size of case closed 17x13 inches..**Set 65.00**

J34112—Fourteen Piece Set; colonial design, in fancy hinge cover wood box, silk lined; consists of 5 inch bevel glass mirror, 8½ inch 13 row hair brush, 8½ inch comb, powder box and hair receiver, 2 cream jars, buffer, cologne bottle, 11½x8½ inch tray, scissors and 3 manicure pieces, all extra heavy solid backs; size of case closed 22x15 inches ....**Set 100.00**

Fourteen to twenty-piece French Ivory toilet sets, which was nothing more than grained Celluloid, were packaged in silk-lined hinged wooden boxes. A fourteen-piece set with a *Colonial* pattern, listed for $100.00 in 1920 from the John V. Farwell Company of Chicago.

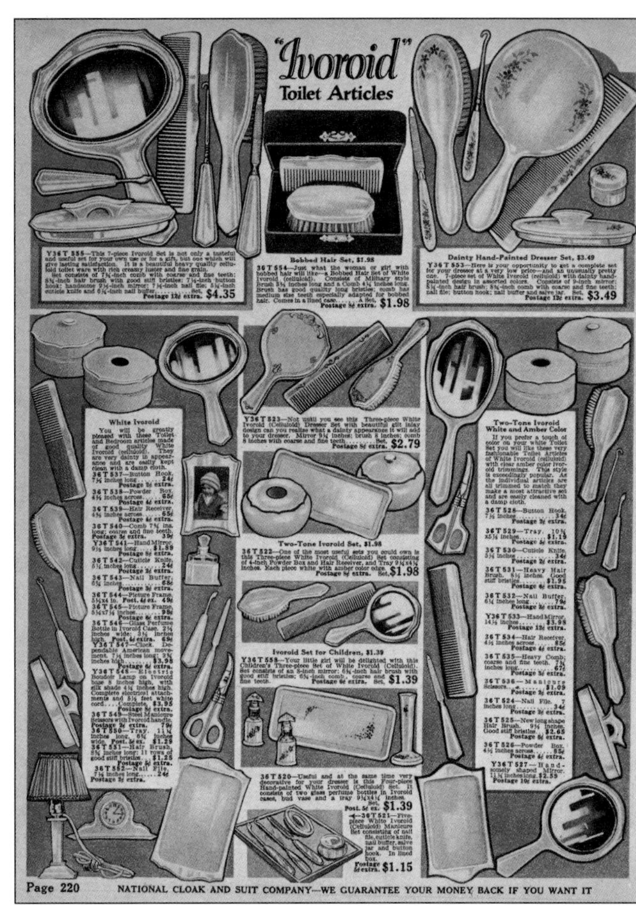

Ivoroid toilet ware, similar to French Ivory or Ivory Pyralin was a less expensive imitation. This assortment was available for sale from the National Cloak and Suit Company catalog in 1925. The ad referred to the items as follows: *You will be greatly pleased with these Toilet and Bedroom articles made of good quality Ivoroid (celluloid). They are very dainty in appearance and are easily kept clean with a damp cloth.* It became obvious while looking at this ad that many qualities of celluloid existed at that time, of varying qualities and under many different trade names. Prices varied accordingly. A hand mirror made of Ivoroid with beveled glass sold for $1.89 in 1925 while a similar one made of French Ivory sold for $19.50.

In 1920, the A. C. Becken Company, of Chicago, Illinois, advertised French Ivory toilet sets in their wholesale catalog. Popular priced French Ivory sets, consisting of thirteen pieces, wholesaled for $23.50. A better quality fine French Ivory in the *Fairfax* pattern sold individual pieces starting at $8.40 just for the hairbrush. The hand mirror in the same pattern was $11.00. The highest grade extra luster French Ivory was in a pattern called *Lavictoire*, which was a thin model with beveled edges. The hairbrush in this pattern cost $12.50. In another Chicago catalog in the same year, French Ivory toilet sets with a subtitle of grained celluloid were even more expensive. A thirteen-piece set in a silk-lined wooden box sold for $60.00; a twenty-piece set was $87.50. The *Colonial* pattern fourteen-piece set in a fancy hinged box was $100.00. These were wholesale prices in 1920!

Fine quality French Ivory easel back picture frames were highly fashionable boudoir accessories. Used for displaying pictures of loved ones, the frames were rectangular and oval with ball feet. Jewel boxes, glove boxes and pincushions were also a part of a lady's dressing table. Glass cologne bottles with hand-cut crystal stoppers were handsomely fitted into heavy French Ivory holders. Dresser clocks of all shapes and sizes were also offered for sale. Ivory polishing cream was available in tubes for preserving and cleaning the ivory. Unfortunately, if cologne or perfume was spilled on a celluloid dresser tray, it would leave a permanent mark on the French Ivory or celluloid if it was not immediately cleaned up.

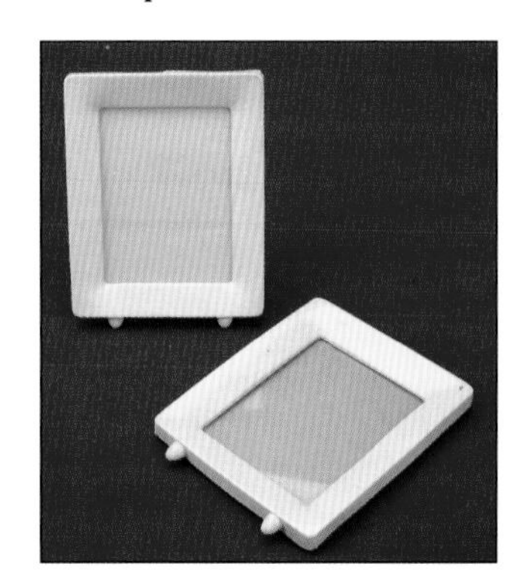

Pair of footed French Ivory picture frames from the 1920s. $35-45 each.

Individual brushes and combs were offered for sale made of French Ivory in 1920 from mail order catalogs.

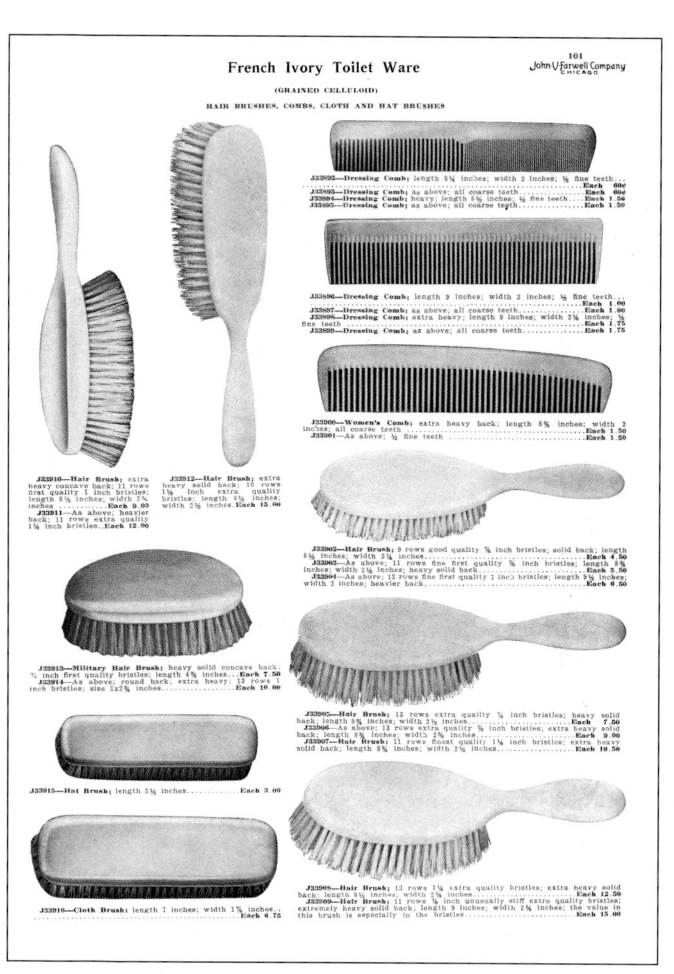

101

French Ivory Toilet Ware

John V. Farwell Company CHICAGO

(GRAINED CELLULOID)

HAIR BRUSHES, COMBS, CLOTH AND HAT BRUSHES

J33892—Dressing Comb; length 8¼ inches; width 2 inches; ½ fine teeth.......Each 60¢
J33893—Dressing Comb; as above; all coarse teeth.......Each 60¢
J33894—Dressing Comb; heavy; length 8½ inches; ½ fine teeth....Each 1.50
J33895—Dressing Comb; as above; all coarse teeth.......Each 1.50

J33896—Dressing Comb; length 9 inches; width 2 inches; ½ fine teeth.......Each 1.00
J33897—Dressing Comb; as above; all coarse teeth.......Each 1.00
J33898—Dressing Comb; extra heavy; length 9 inches; width 2¼ inches; ½ fine teeth.......Each 1.75
J33899—Dressing Comb; as above; all coarse teeth.......Each 1.75

J33900—Women's Comb; extra heavy back; length 8¾ inches; width 2 inches; all coarse teeth.......Each 1.50
J33901—As above; ½ fine teeth.......Each 1.50

J33910—Hair Brush; extra heavy concave back; 11 rows first quality 1 inch bristles; length 8½ inches; width 2¾ inches.......Each 9.00
J33911—As above; heavier back; 11 rows extra quality 1⅛ inch bristles..Each 12.00

J33912—Hair Brush; extra heavy solid back; 10 rows 1⅛ inch extra quality bristles; length 8½ inches; width 2½ inches.Each 15.00

J33902—Hair Brush; 9 rows good quality ⅞ inch bristles; solid back; length 8½ inches; width 2¼ inches.......Each 4.50
J33903—As above; 11 rows fine first quality ⅞ inch bristles; length 8¾ inches; width 2½ inches; heavy solid back.......Each 5.50
J33904—As above; 13 rows fine first quality 1 inch bristles; length 9½ inches; width 3 inches; heavier back.......Each 6.50

J33913—Military Hair Brush; heavy solid concave back; ¾ inch first quality bristles; length 4⅜ inches...Each 7.50
J33914—As above; round back; extra heavy; 13 rows 1 inch bristles; size 5x2¾ inches.......Each 10.00

J33905—Hair Brush; 13 rows extra quality ⅞ inch bristles; heavy solid back; length 8¾ inches; width 2½ inches.......Each 7.50
J33906—As above; 13 rows extra quality ⅞ inch bristles; extra heavy solid back; length 8¾ inches; width 2⅝ inches.......Each 9.00
J33907—Hair Brush; 11 rows finest quality 1⅛ inch bristles; extra heavy solid back; length 8¾ inches; width 2½ inches.......Each 10.50

J33915—Hat Brush; length 5½ inches.......Each 3.00

J33916—Cloth Brush; length 7 inches; width 1⅞ inches.......Each 6.75

J33908—Hair Brush; 13 rows 1⅛ extra quality bristles; extra heavy solid back; length 8½ inches; width 2⅝ inches.......Each 12.50
J33909—Hair Brush; 11 rows ⅞ inch unusually stiff extra quality bristles; extremely heavy solid back; length 9 inches; width 2⅝ inches; the value in this brush is especially in the bristles.......Each 15.00

105

Fine French Ivory

John V. Farwell Company CHICAGO

(GRAINED CELLULOID)

PUFF BOXES, HAIR RECEIVERS, BUD VASES, ETC.

For Ivory Engraving Chart see page 109.

J33966—Puff Box; diameter 4 inches; depth 2 inches.......Each 2.25
J33967—As above; **Hair Receiver**.......Each 2.25

J33968—Hair Pin Holder; heavy stock; 5 compartments; length 5½ inches; height 3⅛ inches.......Each 2.00

J33969—Puff Box; heavy stock; diameter 3½ inches; depth 1½ inches.......Each 1.50
J33970—As above; **Hair Receiver** to match.......Each 1.50

J33971—Puff Box; heavy stock; diameter 4¼ inches; height 2¾ inches.......Each 2.00
J33972—Hair Receiver to match; as above.......Each 2.00

J33974—Bud Vase or Hat Pin Holder; height 5½ inches; heavy stock.......Each 2.50

J33973—Bud Vase or Hat Pin Holder; heavy stock; height 7¼ inches.......Each 3.00

J33975—Bud Vase; extra heavy; 8 inches high.......Each 2.25

J33976—Puff Box; extra heavy; diameter 4 inches; height 2 inches.......Each 2.50
J33977—As above; **Hair Receiver** to match.......Each 2.50

Use Ivory Polishing Cream for preserving and cleaning ivory; put up in tubes.......Dozen 8.00

J33979—Puff Box; extra heavy stock; diameter 5 inches; height 2¼ inches; matches J33978.......Each 4.50

J33978—Hair Receiver; extra heavy stock; diameter 5 inches; height 2¼ inches; matches J33979.......Each 4.50

J33980—Puff Box and Puff; just the right depth to hold the puff and powder; diameter 5 inches; depth 1½ inches; complete with lambs wool puff.......Each 3.00

J33981 — Elephant Hair Receiver; heavy stock; has small cover for opening with French ivory statuette elephant handle; diameter 5 inches; height 2¼ inches (not including elephant).......Each 2.75
J33982—As above; **Puff Box** to match.......Each 2.75

J33983 — Bonnet Brush; solid back and handle; 2¾ inch bristles; length 6½ inches; width 3½ inches.......Each 2.50

J33984—Soap Box; heavy; 3¾x2¾ inches....Each 1.50

J33985 — Flower Vase; French ivory holder; fluted thin glass vase; height 6 inches..Each 2.00

J33986—Bud Vase; French ivory holder; thin glass vase; height 8½ inches.......Each 1.75

Puff boxes, hair receivers and bud vases were also made of fine French Ivory. Ivory polishing cream was made available to preserve and clean the French Ivory.

French Ivory Clocks

(GRAINED CELLULOID).

EXCEPT WHERE SPECIFIED ALL HAVE ONE DAY AMERICAN MOVEMENTS.

Use Ivory Polishing Cream for preserving and cleaning ivory..............**Dozen 8.00**

J33953—Extra heavy Solid French Ivory Clock; better quality; one day movement; the case is made from solid stock; height 3½ inches; width 5 inches..........**Each 13.50**

J33954—Desk Clock and Calendar; extra heavy French ivory case with place for stamps under calendar which drops down; it lays flat on desk with clock facing up; length 5 inches; width 4½ inches........**Each 9.50**

J33955—French Ivory Clock; extra heavy case; height 3½ inches; width 4¼ inches.....................................**Each 8.50**

J33956—French Ivory Clock; height 4¼ inches; width 7 inches...........**Each 7.50**

J33957—One Day Alarm Clock; in heavy French ivory case; Jr. tatoo intermittent alarm; width 7¾ inches; height 3¾ inches.....................................**Each 13.50**

J33958—French Ivory Clock; extra heavy case; height 5 inches; width of base 6 inches.....................................**Each 9.00**

J33959—French Ivory Clock; height 4 inches; width 4 inches................**Each 5.75**

J33960—French Ivory Clock; heavy case; height 3¾ inches; width 7½ inches.....................................**Each 7.50**

J33961—As above; with white radiolite luminous dial ....**Each 9.00**

J33962—French Ivory Clock; ball feet; extra heavy case; height 4¼ inches; width 4¼ inches ....**Each 8.50**

J33963—French Ivory Clock; height 4¼ inches; width 4 inches...........**Each 5.75**

J33964—French Ivory Clock; extra heavy case; 8 day movement; width 6½ inches; height 4¾ inches; depth 3½ inches.....................................**Each 12.50**

J33965—French Ivory Clock; with 8 day New Haven movement; height 5 inches; width of base 4½ inches...............**Each 12.50**

French Ivory dresser clocks made in a variety of shapes and sizes were offered for sale in 1920. Prices ranged from $5.75 to $13.50 for a better quality made of extra heavy solid stock.

Fine French Ivory

(GRAINED CELLULOID)

COLOGNES, MANICURE ARTICLES AND CANDLESTICKS

For Ivory Engraving Chart see page 109.

Use Ivory Polishing Cream for Preserving and Cleaning Ivory.

Put up in tubes. Dozen 8.00

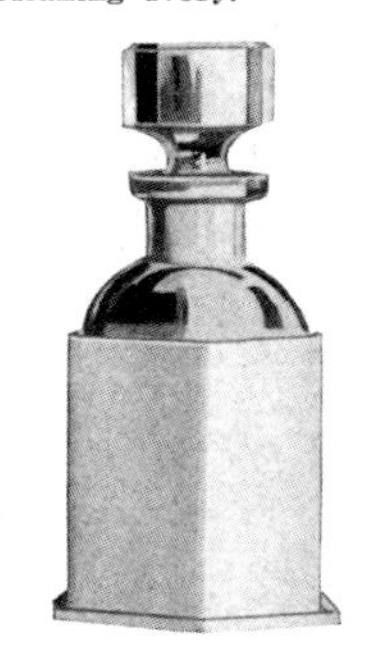

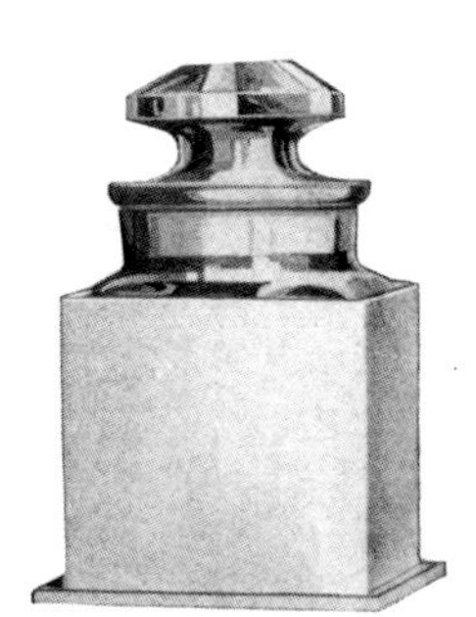

J33987—1 oz. Cologne Bottle; heavy French ivory holder; height 4½ inches.. ....Each 1.00

J33988 — Cologne; heavy French ivory holder; cut stopper; height 5 inches... ....Each 2.50

J33989—Double Cologne; heavy French ivory holder; ground glass stoppers; height 4 inches; length 3¼ inches ....Each 2.25

J33990 — Cologne; heavy French ivory six sided holder; height 4 inches.... ....Each 2.50

J33991 — Salts Bottle; French ivory holder; cut stopper; height 4 inches; 2½ inches square......... ....Each 2.00

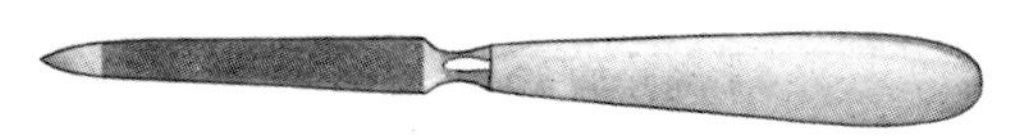

Manicure articles; extra quality; steel fittings.

J33995—File ....Each 50¢ **J33997—Corn Knife**..Each 50¢
J33996—Cuticle ......Each 50¢ **J33998—Shoe Hook**..Each 50¢
Length of file 7¼ inches.

Manicure articles; extra quality; steel fittings.

J33999—File ....Each 1.00
J34000—Corn Knife ....Each 1.00
J34001—Cuticle ....Each 1.00
J34002—Shoe Hook ....Each 1.00
Length of handles 3½ inches.

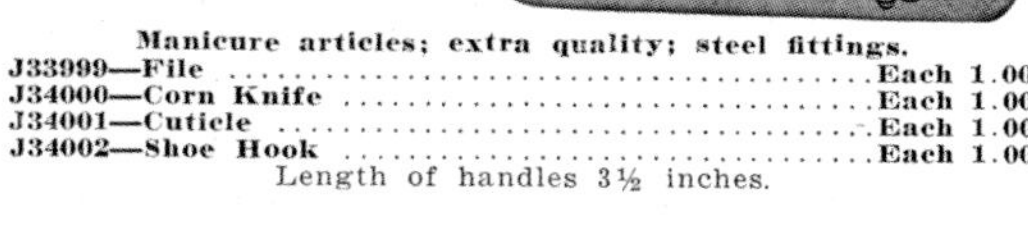

Manicure articles; extra quality; steel fittings.

J34003—File ....Each 75¢
J34004—Cuticle ....Each 75¢
J34005—Shoe Hook ....Each 75¢
Length of file 6¾ inches.

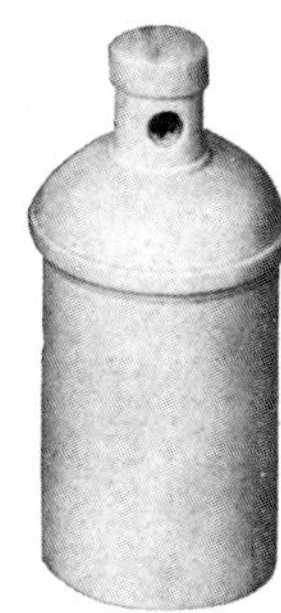

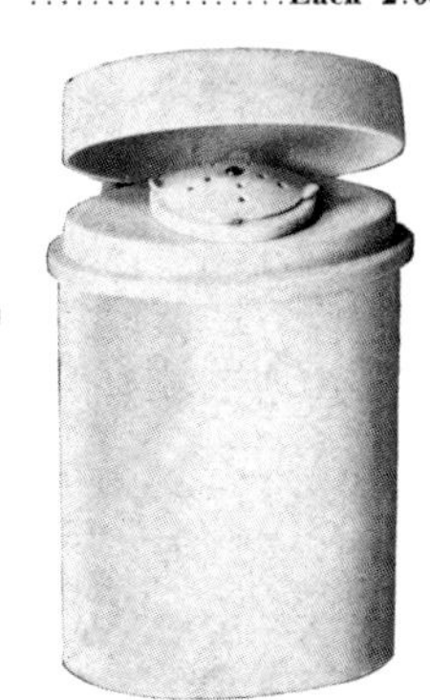

J33992 — Cologne; heavy French ivory holder; ground glass stopper; height 4 inches..Each 1.25

J33993 — Tooth Powder Bottle; all heavy French ivory; height 4 inches..Each 1.50

J33994 — Talcum Powder Box; all French ivory; extra heavy; height 4 inches; width 2¾ inches.Each 2.50

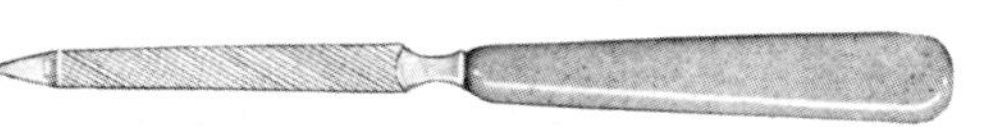

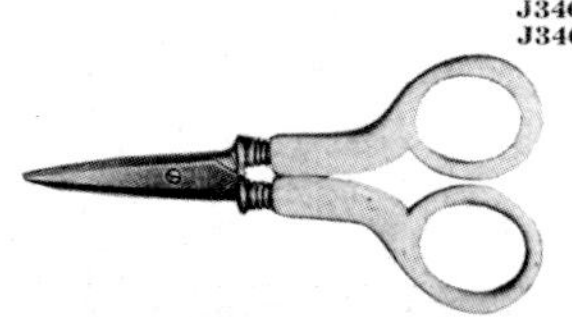

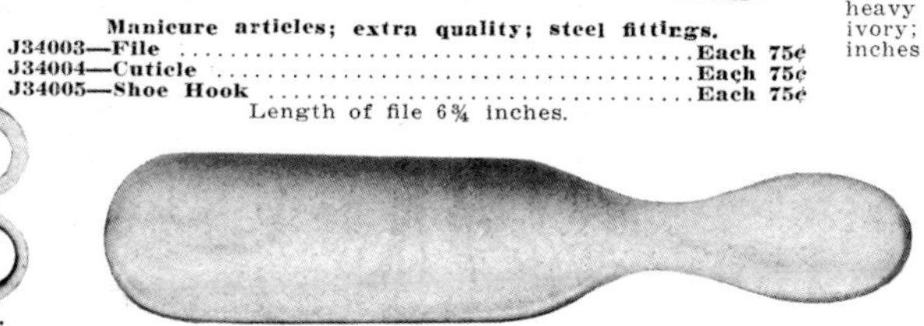

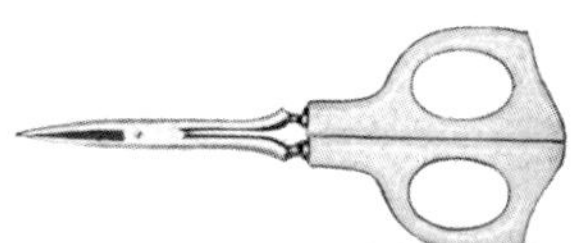

J34006—Manicure Scissors; French ivory handle; length 8⅜ inches ....Each 1.50

J34008—Shoe Horn; all French ivory; length 7½ inches..... ....Each 1.00

J34007—Manicure Scissors; French ivory handles; length 5 inches; matches the Le Pompadour pattern. ....Each 2.50

J34009—Buffer and Tray; extra heavy French ivory; interchangeable chamois; length 5½ inches..Each 2.50

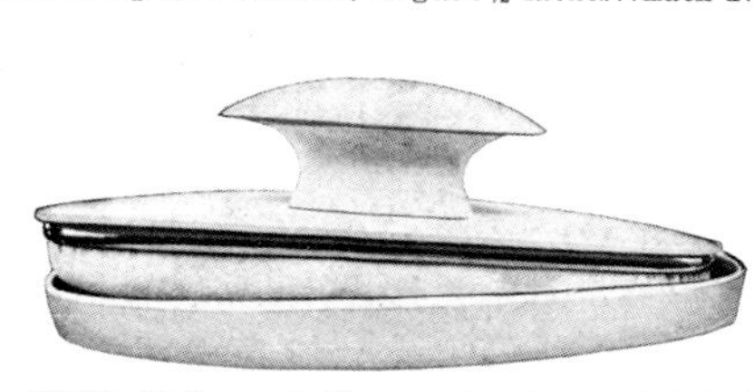

J34013—Electric Boudoir Lamp; French ivory base; 6½ inch old rose silk shade; height 14 inches; complete with silk cord and plug.. ....Each 13.50

J34014—Electric Boudoir Lamp; French ivory base; 6½ inch old rose silk shade; height 14 inches; complete with silk cord and plug.. ....Each 16.00

J34010—Buffer and Tray; extra heavy detachable chamois; length 7½ inches...............Each 2.00
J34011—As above; length 5¾ inches......Each 1.50
J34012—Buffer and Tray; as above; length 4⅞ inches ....Each 1.25

J34015 — Candlestick; French ivory stand; complete with candle, shade and holder; old rose shade..... ....Each 2.50
J34015½—As above; heavier holder; colonial pattern, matches toilet ware on page 100........Each 5.00

Cologne bottles, manicure implements, candlesticks and electric boudoir lamps were also made of French Ivory and popular in 1920.

107
John V. Farwell Company
CHICAGO

Fine French Ivory

(GRAINED CELLULOID)

JEWEL BOXES AND CUSHIONS

For Ivory Engraving Chart see page 109.

J34016—Oval Jewel Box: heavy French ivory; silk velvet lined; length 5 inches; height 6⅞ inches.................. **Each 5.25**

J34017—Jewel Box: dome top; velvet lined; has tray with two compartments and large compartment below; size 6x4x3 inches........................ **Each 7.50**

J34018—Heart Shape Jewel Box: heavy French ivory; silk velvet lined; one large compartment and place at back for rings; diameter 4½ inches; height 1⅞ inches........................ **Each 6.75**

J34019—Rosary Case or Jewel Box: velvet lined; diameter 2¾ inches.......... **Each 2.50**

J34020—Jewel or Trinket Box: dome top heavy French ivory; silk velvet lined; size 4¼x2¼x1¾ inches .... **Each 3.00**
J34021—As above; size 5½x2¾x2¼........................ **Each 5.25**

J34022—Silk velvet cushion with compartment underneath cushion for jewelry or trinkets; heavy French ivory; diameter 4½ inches .... **Each 3.75**

J34023—Jewel Box: dome top; heavy French ivory; silk velvet lined; size 4x3x1¼ inches. .... **Each 4.50**
J34024—As above; larger; size 4½x3¾x1¾ inches .... **Each 5.25**

J34025—Trinket or Jewel Box: heavy French ivory throughout; size 2⅝x2⅝ inches........................ **Each 1.25**

J34026—Jewel Box: extra heavy French ivory; dark blue silk velvet lined; size 5x3x2 inches .... **Each 5.25**

J34027—Footed Jewel or Trinket Box: extra heavy French ivory; silk velvet lined; with compartment for rings; size 5x5x2½ inches .... **Each 7.50**

J34028—Jewel Box and Cushion, with drawer; cushion is of fine silk velvet and drawer is lined with satin; size 4½x4⅛ inches........................ **Each 7.50**

J34029—Jewel Box: extra heavy stock; silk velvet lining; division for rings at back; hinge cover. size 5¾x4x2 inches.. **Each 6.00**

J34030—Glove Box: dome top; extra heavy French ivory throughout; silk velvet lined; size 9¾x4x2½ inches; matches Handkerchief Box J34031.......... **Each 12.50**

J34031—Handkerchief Box: dome top; extra heavy French ivory throughout; silk velvet lined; size 6 inches square; depth 2¾ inches; matches Glove Box J34030 .... **Each 12.50**

Assortment of jewel and trinket boxes and pincushions made of fine French Ivory and offered for sale in 1920.

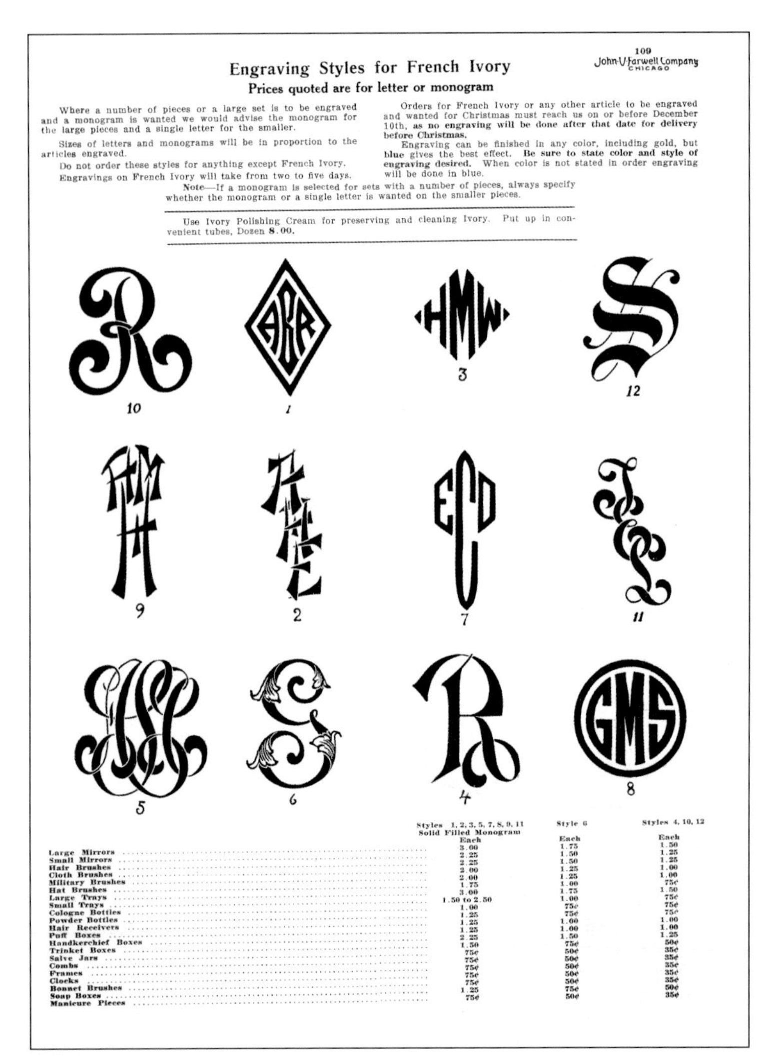

109
John V. Farwell Company
CHICAGO

Engraving Styles for French Ivory

Prices quoted are for letter or monogram

Where a number of pieces or a large set is to be engraved and a monogram is wanted we would advise the monogram for the large pieces and a single letter for the smaller.

Sizes of letters and monograms will be in proportion to the articles engraved.

Do not order these styles for anything except French Ivory.

Engravings on French Ivory will take from two to five days.

Orders for French Ivory or any other article to be engraved and wanted for Christmas must reach us on or before December 10th, **as no engraving will be done after that date for delivery before Christmas.**

Engraving can be finished in any color, including gold, but **blue** gives the best effect. **Be sure to state color and style of engraving desired.** When color is not stated in order engraving will be done in blue.

Note—If a monogram is selected for sets with a number of pieces, always specify whether the monogram or a single letter is wanted on the smaller pieces.

Use Ivory Polishing Cream for preserving and cleaning Ivory. Put up in convenient tubes, Dozen **8.00**.

	Styles 1, 2, 3, 5, 7, 8, 9, 11 Solid Filled Monogram Each	Style 6 Each	Styles 4, 10, 12 Each
Large Mirrors	3.00	1.75	1.50
Small Mirrors	2.25	1.50	1.25
Hair Brushes	2.25	1.50	1.25
Cloth Brushes	2.00	1.25	1.00
Military Brushes	2.00	1.25	1.00
Hat Brushes	1.75	1.00	75¢
Large Trays	3.00	1.75	1.50
Small Trays	1.50 to 2.50	1.00	75¢
Cologne Bottles	1.00	75¢	75¢
Powder Bottles	1.25	75¢	75¢
Hair Receivers	1.25	1.00	1.00
Puff Boxes	1.25	1.00	1.00
Handkerchief Boxes	2.25	1.50	1.25
Trinket Boxes	1.50	75¢	50¢
Salve Jars	75¢	50¢	35¢
Combs	75¢	50¢	35¢
Frames	75¢	50¢	35¢
Clocks	75¢	50¢	35¢
Bonnet Brushes	75¢	50¢	35¢
Soap Boxes	1.25	75¢	50¢
Manicure Pieces	75¢	50¢	35¢

This catalog advertisement displays wonderful engraving styles for French Ivory that were popular in 1920.

As the 1920s progressed, new techniques were applied to French Ivory and the ivory grained celluloid. One short-lived technique was a laminate of gold or silver glitter over the ivory celluloid. This pattern was called *Goldaleur* or *Silvaleur*. Problems arose with this technique and sometimes a chemical reaction occurred causing a green corrosion called glitter verdigris. Once this happens, the item will eventually crumble. Also, if an item begins to corrode, the corrosion spreads like a disease to the other pieces in the set. If you notice any corrosion, isolate the infected piece immediately before further damage occurs.

Seven-piece ivory grained celluloid set laminated with a thin sheet of gold glitter. This pattern, referred to as *Goldaleur,* was manufactured in the 1920s by the Celluloid Corporation. This set includes a round ring mirror, nail file, nail buffer, toothbrush holder, glass-lined salve jar, jewel box and shoe horn. $195-240 set.

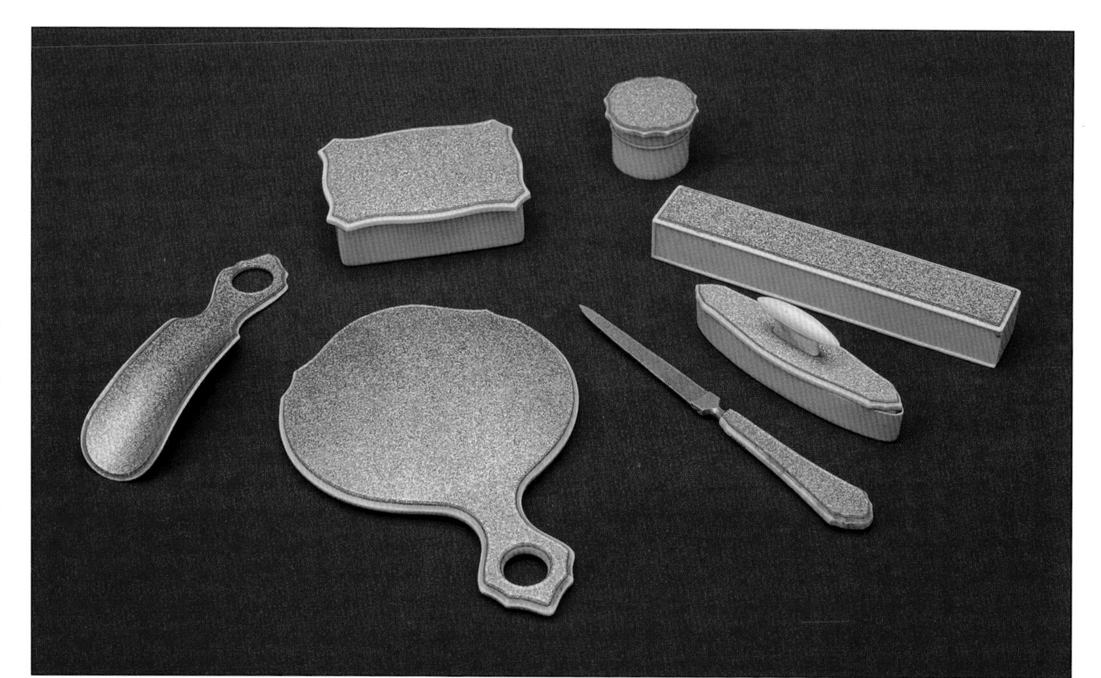

Another ivory grained and gold glitter celluloid set manufactured by the Celluloid Corporation. This set was called *Silvaleur* and contains ten pieces. I cannot really see much of a difference between the Goldaleur and the Silvaleur patterns. They look identical. $245-295 set.

Unfortunately, this type of application to the celluloid did not always last. Two the original pieces of the Goldaleur dresser set have begun to deteriorate. This green corrosion is called *glitter verdigris*. If these two pieces were not isolated from the rest of the set, this decomposition would have affected the remaining pieces of the set. It would have actually spread like a disease.

Rectangular mirror and matching brush made of DuPont Pyralin made to look like tortoiseshell.$85-110.

Hair brush and matching hand mirror set made of a thin sheet of celluloid made to resemble tortoiseshell. The rims are trimmed in a plated base metal. $95-125.

Six-piece manicure set made of demi-shell Pyrolin which was similar to celluloid and made to resemble tortoiseshell. This was popular around 1920. Some companies spelled Pyrolin with an *o* while others spelled it with an *a*. $80-100.

In 1920, Demi-Shell Pyrolin toilet ware was offered for sale. As a handsome substitute for expensive tortoiseshell, demi-shell pyrolin was an aforadable alternative. Pyrolin was similar to celluloid, but with a tortoiseshell appearance instead of an ivory appearance. The *DuBarry* pattern was popular in this material and a three-piece set wholesaled for $24.75.

In 1925, Ivoroid toilet articles were offered for sale by the National Cloak & Suit Company. Either the quality of this grained ivory was poor or the prices were really beginning to drop because a seven-piece set was listed for only $4.35. A dainty hand-painted set, also with seven pieces, listed at $3.49. These were very low prices compared to some of the other companies that offered articles made of similar materials but with different trade names. Ads for Ivory Fiberloid were very popular in leading magazines in the 1920s. One advertisement described Fiberloid as, "Sterling on silver and guarantees solid material, style and beautiful workmanship."

99

Demi Shell Pyrolin Toilet Ware

John V. Farwell Company
CHICAGO

THE DU BARRY PATTERN

Pyrolin is a composition similar to celluloid; has the exact appearance of tortoise shell and is unusually handsome.
Every piece is extra heavy.

J33850—Powder Box; diameter 5 inches........Each 4.75

J33852—Cold Cream Jar; diameter 3 inches........Each 3.50

J33851—Hair Receiver; diameter 5 inches........Each 4.75

J33853—File; length 7¼ inches; extra quality flexible file...Each 2.00

J33854—Cuticle Knife; length 5 inches........Each 2.50

J33855—Shoe Hook; length 7½ inches........Each 2.00

J33856—Buffer and Tray; length 6 inches........Each 5.75

J33858—Cloth Brush; length 7 inches........Each 10.50

J33859—Comb; length 7½ inches........Each 2.25
J33859½—As above; all coarse teeth........Each 2.25

J33857—Mirror; length 10¾ inches; 5½ inch bevel glass....Each 12.00

J33860—Hair Brush; length 8½ inches; 12 rows extra quality bristles........Each 10.50

Celluloid toilet ware made to look like real tortoiseshell was also popular in the early 20th century. *Demi Shell Pyrolin* toilet ware, made in the *DuBarry* pattern was offered for sale from John V. Farwell in 1920.

Ivory Pyralin

The Arlington Company of Arlington, New Jersey, marketed its ivory-white toilet ware under the trade name of Ivory Py-ra-lin. The Arlington Company had been manufacturing Ivory Py-ra-lin since 1883 and was made of a cellulose product called "py-r(ox)a-lin." In the early 1920s, it was advertised as being made of "solid stock and patiently aged for months and months. Every article has been given a graining so delicate and true that you would think it could only have come from the gleaming tusks of some fine old elephant." The Arlington Company was responsible for the *DuBarry* pattern, which seemed to be the most popular pattern in Ivory Pyralin toilet ware. Two-and three-piece sets were offered for sale in addition to fourteen-and fifteen-piece sets neatly arranged in fancy cloth-lined presentation boxes. A fourteen-piece toilet and manicure set consisted of a beveled edge mirror, fine quality bristle hairbrush, powder puff box, cloth brush, comb, two salve jars, nail polisher, shoe horn, button hook, glove stretcher, nail file, cuticle knife and cuticle scissors. This set sold for $54.00. Infants' sets were also offered for sale. They contained a hairbrush, comb, powder box, soapbox and a combination rattler and teething ring all made of Ivory Pyralin. The consumer was given the option of being able to pick and choose the pieces. You did not have to buy complete sets if you didn't want too. All types of brushes were available. Anything from a regular hairbrush to a whiskbroom was offered for sale. Clocks, perfume bottles, hand mirrors, nail buffers, soap boxes, jewel boxes and talcum powder boxes were made of Ivory Pyralin and available from the Arlington Company.

A green and gold vertical stripe is laminated over a layer of ivory pyralin. The set is monogrammed in Old English lettering. The set is marked *Azurtone DuBarry Py-ra-lin*. $136-160.

Wonderful black Art Deco designs accent this three-piece maize-colored pyralin dresser set. $145-175.

This amber-colored celluloid dresser set was enhanced with blue and clear paste stones. Even incomplete, this set has thirteen pieces. This type of celluloid was sometimes referred to as *Fiberloid.* It was popular around 1925. A patent for setting rhinestones into celluloid was issued in 1902 to a man named Martin Brown. $300-350.

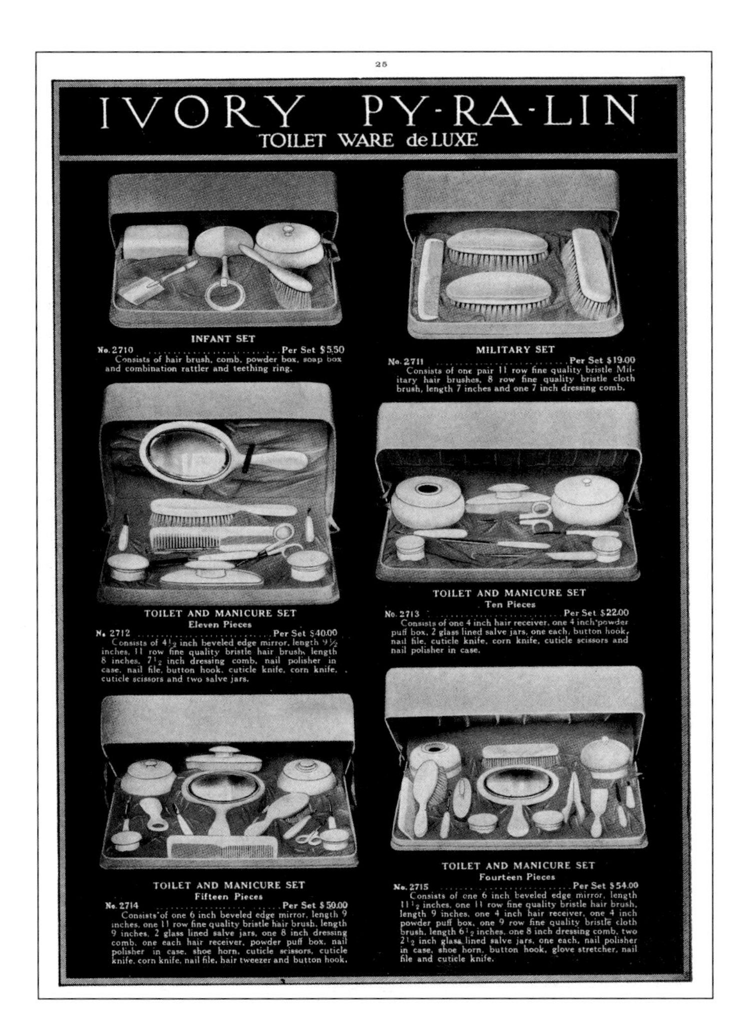

28

IVORY PY-RA-LIN
TOILET WARE deLUXE

INFANT SET

No. 2710 ....Per Set $5.50
Consists of hair brush, comb, powder box, soap box and combination rattler and teething ring.

MILITARY SET

No. 2711 ....Per Set $19.00
Consists of one pair 11 row fine quality bristle Military hair brushes, 8 row fine quality bristle cloth brush, length 7 inches and one 7 inch dressing comb.

TOILET AND MANICURE SET
Eleven Pieces

No. 2712 ....Per Set $40.00
Consists of 4½ inch beveled edge mirror, length 9½ inches, 11 row fine quality bristle hair brush, length 8 inches, 7½ inch dressing comb, nail polisher in case, nail file, button hook, cuticle knife, corn knife, cuticle scissors and two salve jars.

TOILET AND MANICURE SET
Ten Pieces

No. 2713 ....Per Set $22.00
Consists of one 4 inch hair receiver, one 4 inch powder puff box, 2 glass lined salve jars, one each, button hook, nail file, cuticle knife, corn knife, cuticle scissors and nail polisher in case.

TOILET AND MANICURE SET
Fifteen Pieces

No. 2714 ....Per Set $50.00
Consists of one 6 inch beveled edge mirror, length 9 inches, one 11 row fine quality bristle hair brush, length 9 inches, 2 glass lined salve jars, one 8 inch dressing comb, one each hair receiver, powder puff box, nail polisher in case, shoe horn, cuticle scissors, cuticle knife, corn knife, nail file, hair tweezer and button hook.

TOILET AND MANICURE SET
Fourteen Pieces

No. 2715 ....Per Set $54.00
Consists of one 6 inch beveled edge mirror, length 11½ inches, one 11 row fine quality bristle hair brush, length 9 inches, one 4 inch hair receiver, one 4 inch powder puff box, one 9 row fine quality bristle cloth brush, length 6½ inches, one 8 inch dressing comb, two 2½ inch glass lined salve jars, one each, nail polisher in case, shoe horn, button hook, glove stretcher, nail file and cuticle knife.

Toilet and manicure sets, military sets and even infant sets were made of Ivory Pyralin and presented in fabric-lined boxes. A fourteen-piece toilet and manicure set sold for $54.00 in 1923 and included everything from a hairbrush to a glove stretcher!

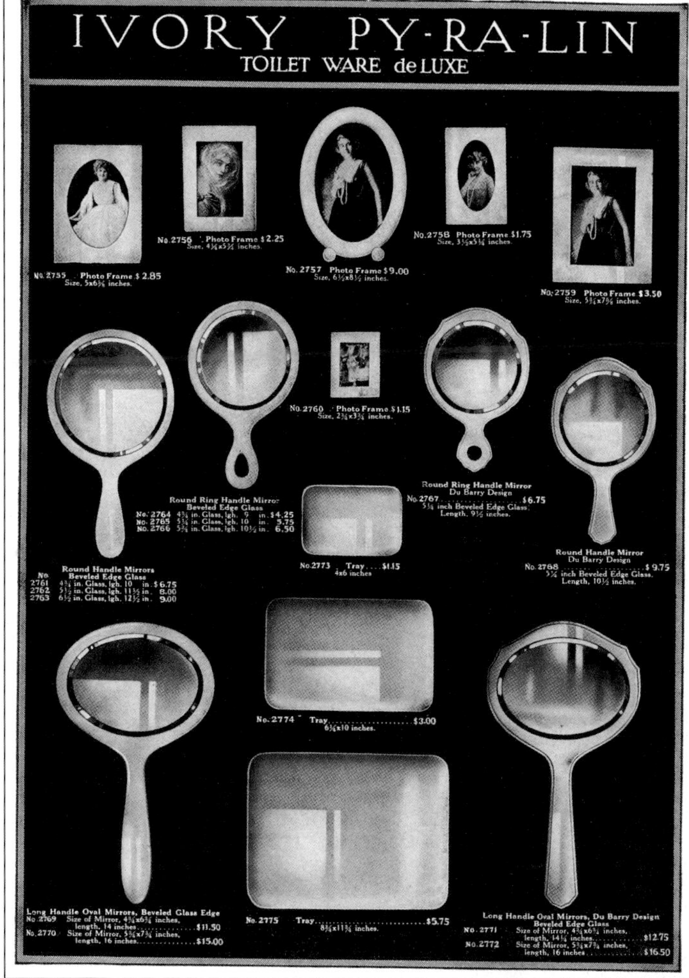

29

IVORY PY-RA-LIN
TOILET WARE deLUXE

No. 2755 Photo Frame $2.85
Size, 5x6⅝ inches.

No. 2756 Photo Frame $2.25
Size, 4¼x5¾ inches.

No. 2757 Photo Frame $9.00
Size, 6½x8½ inches.

No. 2758 Photo Frame $1.75
Size, 3½x5¼ inches.

No. 2759 Photo Frame $3.50
Size, 5¾x7⅝ inches.

No. 2760 Photo Frame $1.15
Size, 2¾x3¾ inches.

Round Ring Handle Mirror
Beveled Edge Glass

No. 2764 4¾ in. Glass, lgh. 9 in. $4.25
No. 2765 5¼ in. Glass, lgh. 10 in. 5.75
No. 2766 5¾ in. Glass, lgh. 10½ in. 6.50

Round Ring Handle Mirror
Du Barry Design

No. 2767 ....$6.75
5¼ inch Beveled Edge Glass.
Length, 9½ inches.

Round Handle Mirrors
Beveled Edge Glass

No.		
2761	4¾ in. Glass, lgh. 10 in.	$6.75
2762	5½ in. Glass, lgh. 11½ in.	8.00
2763	6½ in. Glass, lgh. 12½ in.	9.00

No. 2773 Tray....$1.15
4x6 inches

Round Handle Mirror
Du Barry Design

No. 2768 ....$9.75
5¼ inch Beveled Edge Glass.
Length, 10½ inches.

No. 2774 Tray.............................$3.00
6½x10 inches.

Long Handle Oval Mirrors, Beveled Glass Edge

No. 2769 Size of Mirror, 4¾x6¾ inches, length, 14 inches.............$11.50
No. 2770 Size of Mirror, 5¾x7¾ inches, length, 16 inches.............$15.00

No. 2775 Tray.............................$5.75
8¾x11¾ inches.

Long Handle Oval Mirrors, Du Barry Design
Beveled Edge Glass

No. 2771 Size of Mirror, 4¾x6¾ inches, length, 14¼ inches.............$12.75
No. 2772 Size of Mirror, 5¾x7¾ inches, length, 16 inches.............$16.50

Assorted hand mirrors, photo frames and trays made of ivory Pyralin.

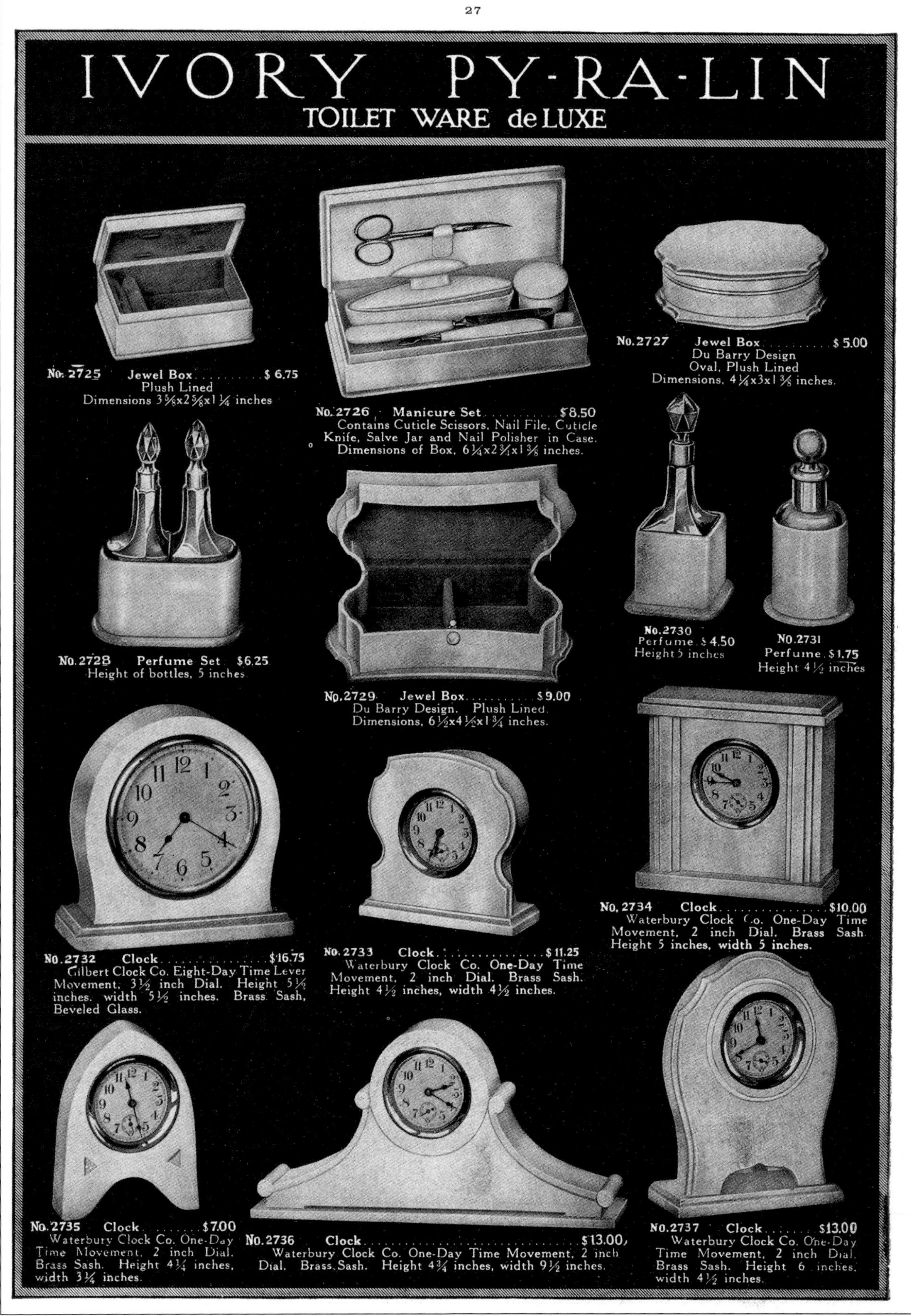

The Arlington Manufacturing Company of Arlington, New Jersey, manufactured its line of celluloid under the trade name of *Ivory Pyralin*. This line of deluxe toilet ware was sold in 1923.

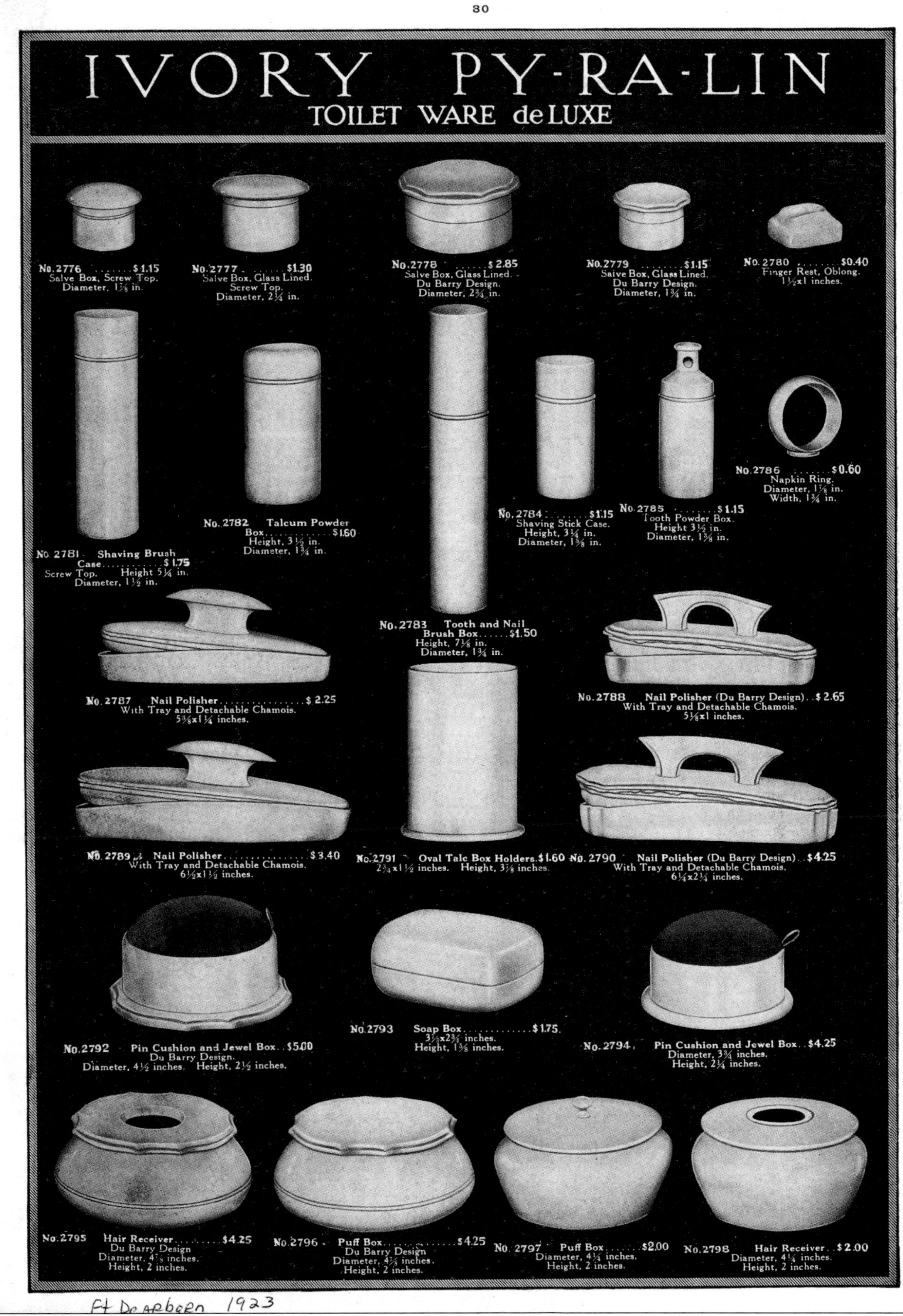

Nail polishers, hair receivers, puff boxes, talcum powder boxes, salve boxes and tooth and nail brush boxes made of Ivory Pyralin in 1923.

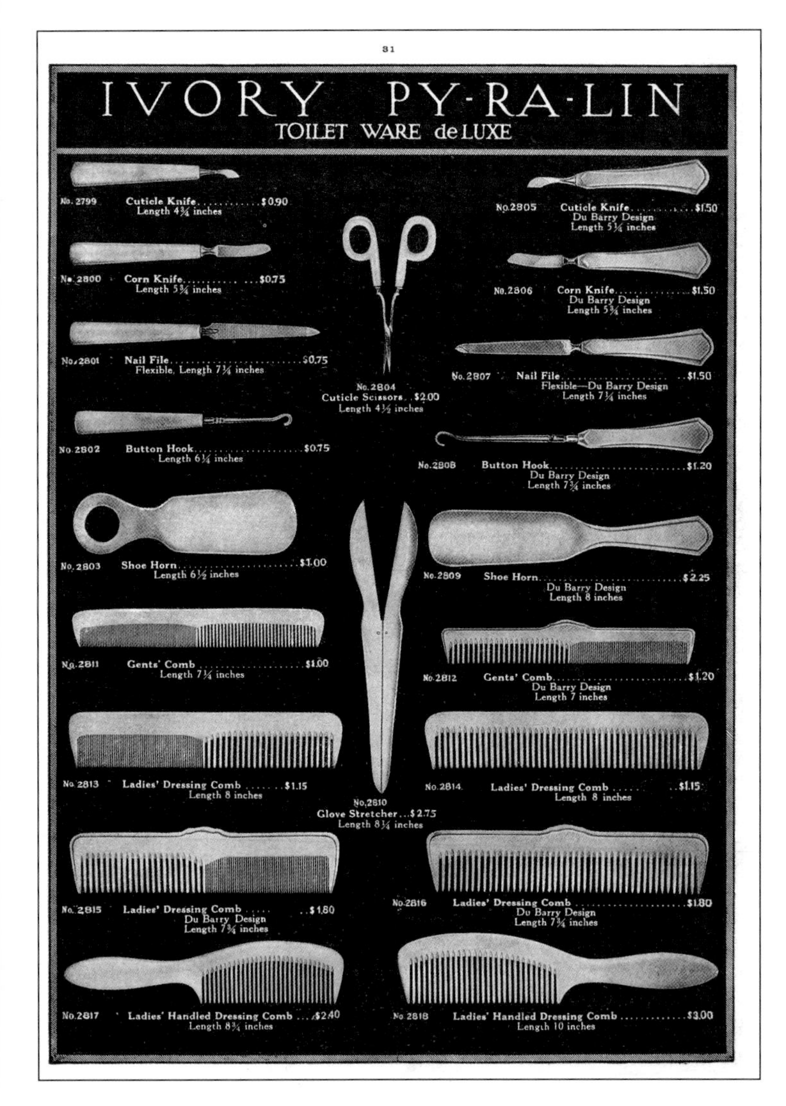

Ladies' and gents' dressing combs and assorted manicure implements made of Ivory Pyralin and offered for sale in 1923.

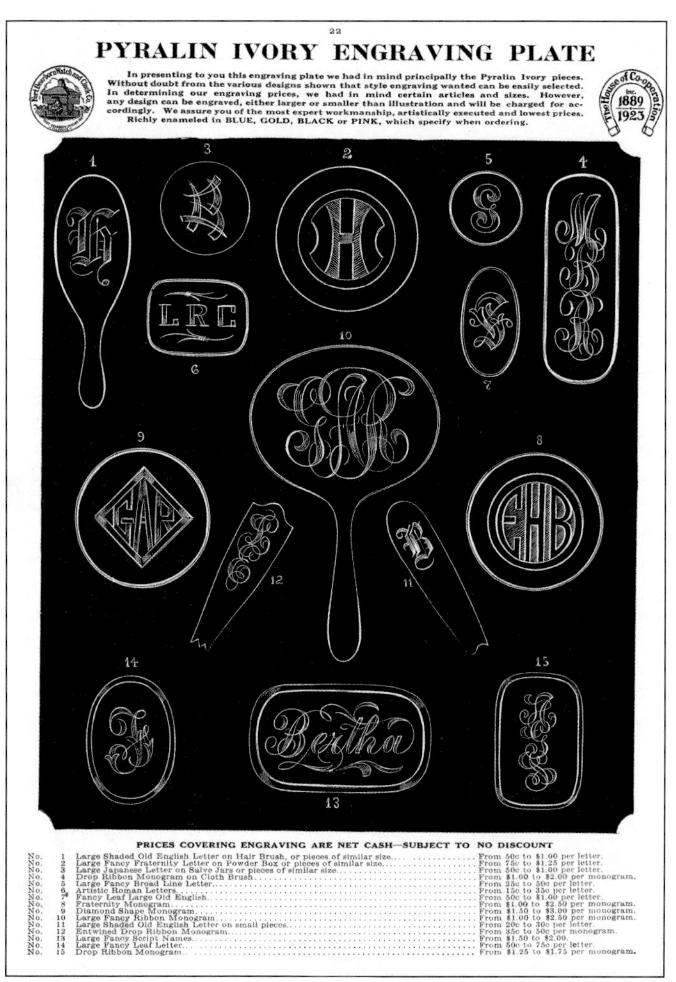

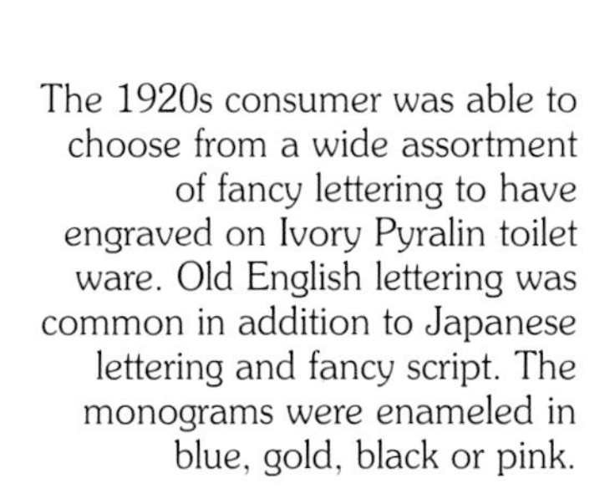
The 1920s consumer was able to choose from a wide assortment of fancy lettering to have engraved on Ivory Pyralin toilet ware. Old English lettering was common in addition to Japanese lettering and fancy script. The monograms were enameled in blue, gold, black or pink.

As early as 1884, a patent was issued to George M. Mowbray, of Massachusetts, for the use of flaked or ground mica and pyroxlin to create an effect that resembled a luminous pearl. This pearl-like appearance was achieved prior to this by the use of fish scales. Forty years later, a synthetic pearlescence resulted from the direct application of warm mercury to celluloid. This particular look became popular in the 1920s and 1930s. Demand increased dramatically to the point where many other companies began marketing toilet ware with the pearl look.

In 1925, the Celluloid Company introduced its new creation which was called Arch-Amerith. Bright jewel-tone plastics were skillfully created into "dresser sets for the most refined and delicate taste." The *Morene* pattern had a soft wavy look that it reminded one of the lustrous sheen of moiré fabric. Rose, green and orchid were the colors offered in this pattern. The *Charmond* pattern was described as having a "deep translucence which gives visibility to the under surface grain." The handles were made to resemble Brazilian onyx. The quartz-like pattern was transparent flecked with snow-white flakes and ornamented with a silver design. The *Dorchester* pattern had a surface, which looked like mother of pearl and accented with a colonial decoration in black. The *Plymouth* pattern was also mother-of-pearl in peach, green, and maize with black and gold designs. These were the patterns offered for sale in 1932. Six years later, Amerith dresser sets had more of a modern look. The *Sicily* pattern was made in Pierretone, a "specially designed material of fine tonal qualities. It is ornamented with genuine cloisonné decorations and with embossed gold-plated handles". Another modern pattern was the *Sherlain* pattern with colored crystal handles.

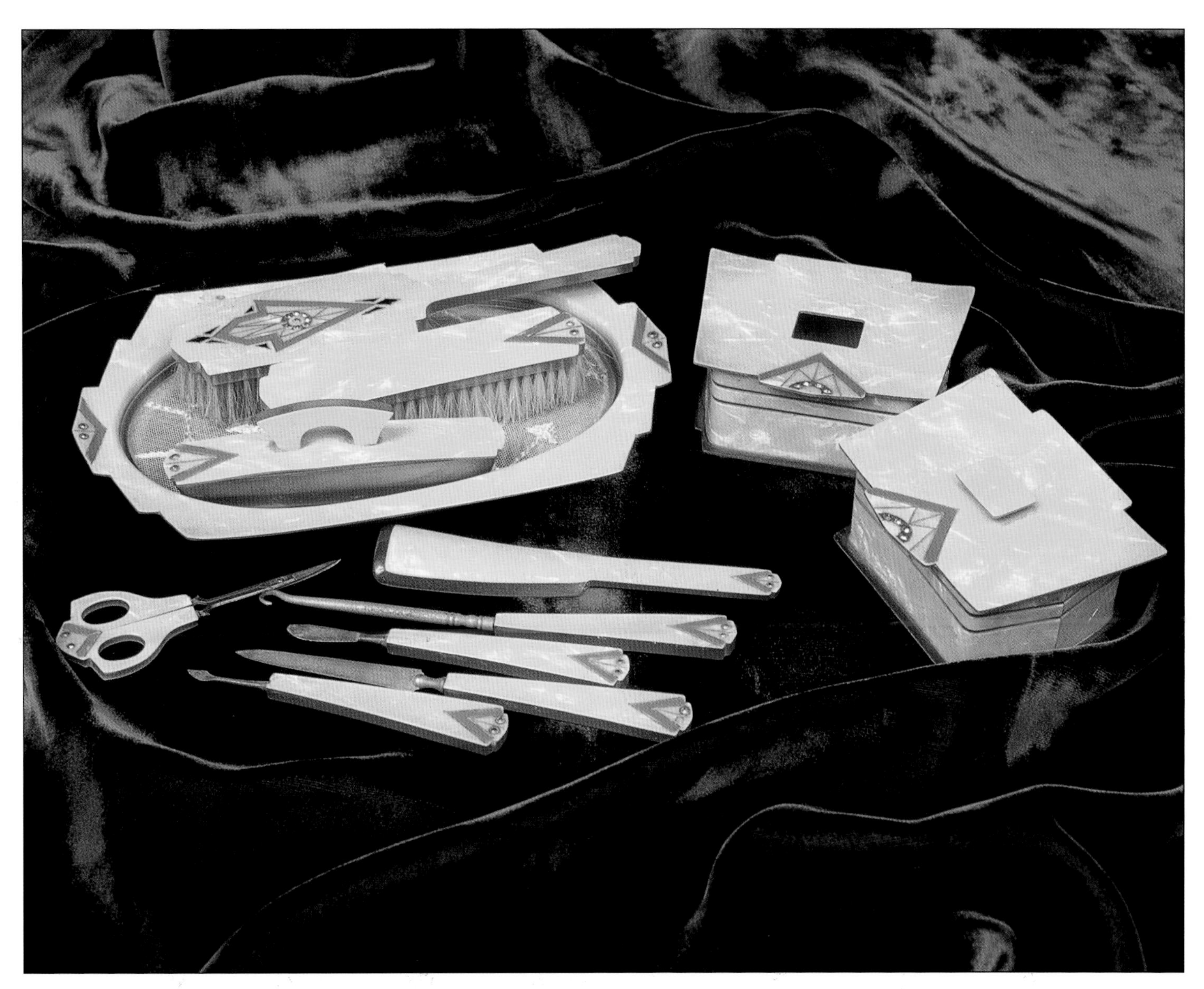

Maize pearl-on-amber Pyralin dresser set accented with detailed Art Deco designs and glass stones. These double composite sets were the rage in fashionable boudoir accessories in the early 1930s made by the DuPont Viscoloid Company. This set includes twelve pieces. $245-295 set.

This unusual Pyralin set is pearl-on-amber with black. The set includes a tray, powder box and divided box both with separate lids accented with metal knobs and applied decorations in black and gold. The set is marked *Pyralin* and it is the *Sheraton* pattern. The tray, which originally had lace under glass, was replaced with some colored fabric. $95-125 set.

This has to be the most complete dresser set that I have ever found. Made of jade pearl-on-jet Pyralin, the set includes twenty-six pieces. Some of the pieces are made of a thin layer of the jade pearl Pyralin, while other pieces are double and some even triple composities of jade pearl-on-jet-on-pearl. $395-450 set.

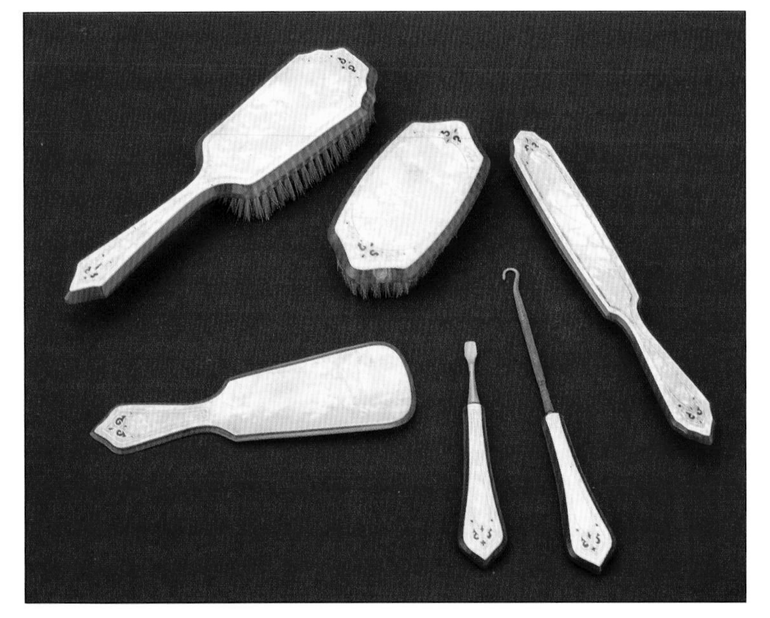

Pearl-on-amber Pyralin set with applied gold and black decoration. This partial set includes a hair brush, military brush, nail buffer, shoe horn, shoe buttoner and cuticle tool. $115-145.

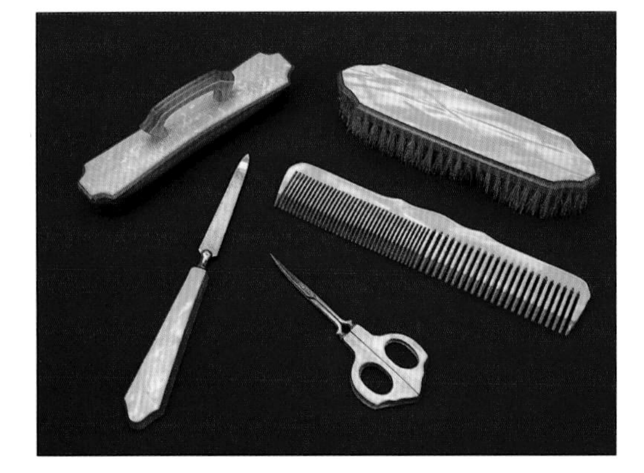

Cloth brush, nail buffer, comb, file and scissors make up this maize pearl-on-amber Pyralin set. $90-120.

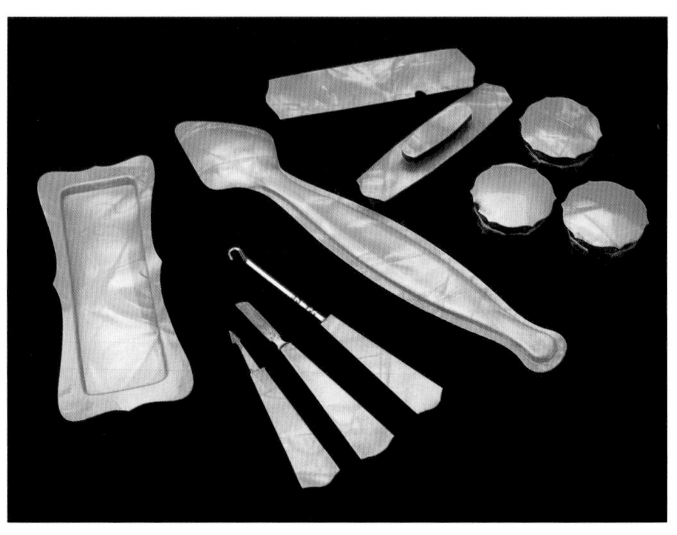

A partial ten-piece set of jade pearl-on-jet Pyralin. The jars have amber glass bottoms. $135-155.

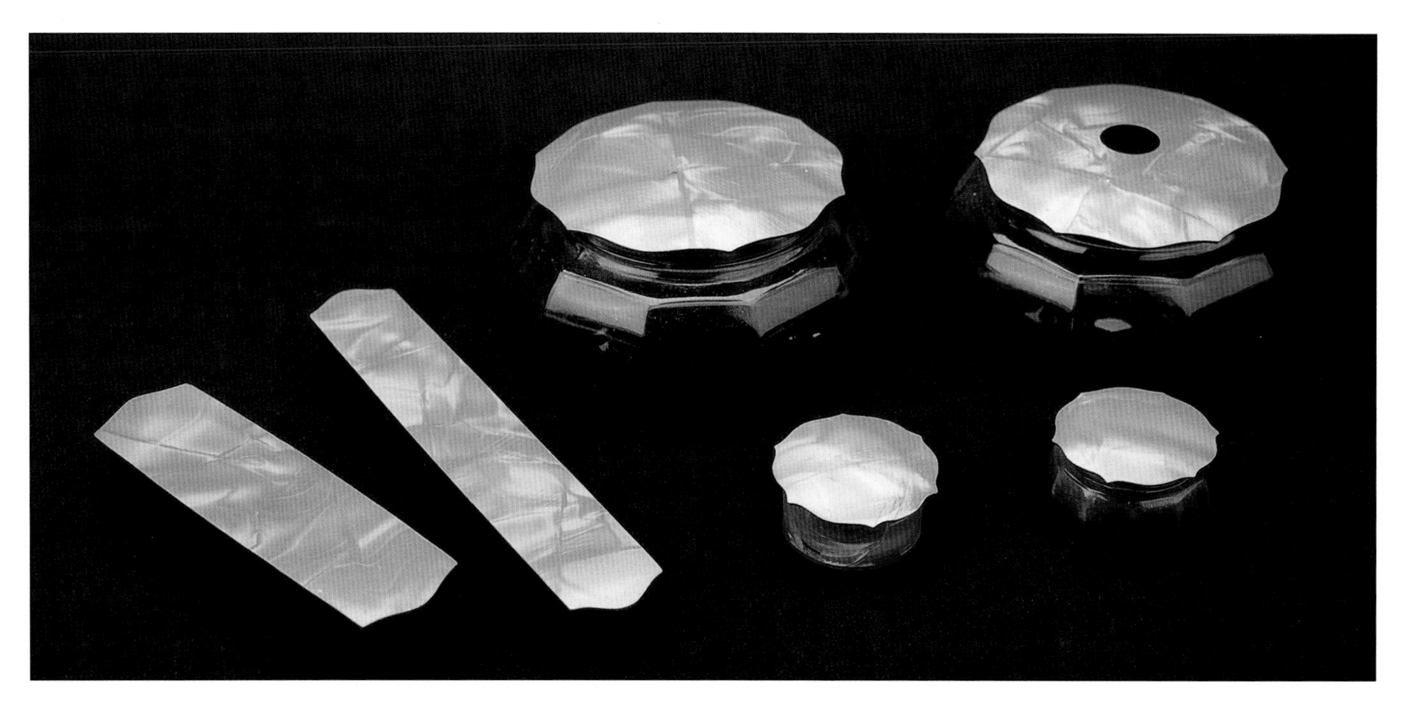

Six more pieces of jade pearl-on-jet Pyralin. The jars have amber glass bottoms. Jade pearl Pyralin was extremely popular in the 1930s. $145-165 set.

Here are odd pieces of three different dresser sets. The clock is made of pearl-on-amber Pyralin; the nail buffer and lids are pearl-on-jet Pyralin while the scissors are a triple composite of pearl-on-jet-on-pearl. Clock $75-100; all other pieces $20-35 each.

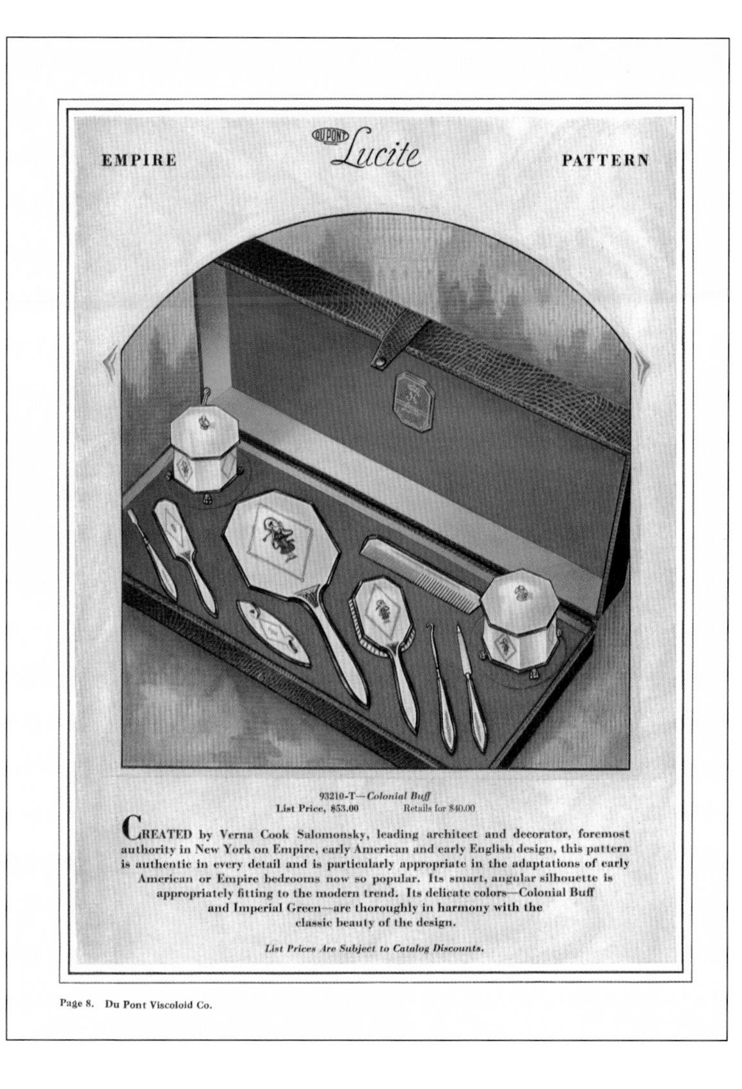

EMPIRE DU PONT *Lucite* PATTERN

93210-T—*Colonial Buff*
List Price, $53.00 Retails for $40.00

CREATED by Verna Cook Salomonsky, leading architect and decorator, foremost authority in New York on Empire, early American and early English design, this pattern is authentic in every detail and is particularly appropriate in the adaptations of early American or Empire bedrooms now so popular. Its smart, angular silhouette is appropriately fitting to the modern trend. Its delicate colors—Colonial Buff and Imperial Green—are thoroughly in harmony with the classic beauty of the design.

List Prices Are Subject to Catalog Discounts.

Page 8. Du Pont Viscoloid Co.

The DuPont Viscoloid Company created Lucite boudoir accessories and toilet sets in 1928. An array of different patterns was sold through trade and mail-order catalogs. The *Empire* pattern was created by Verna Cook Salomonsky in 1930, who was a leading architect and decorator.

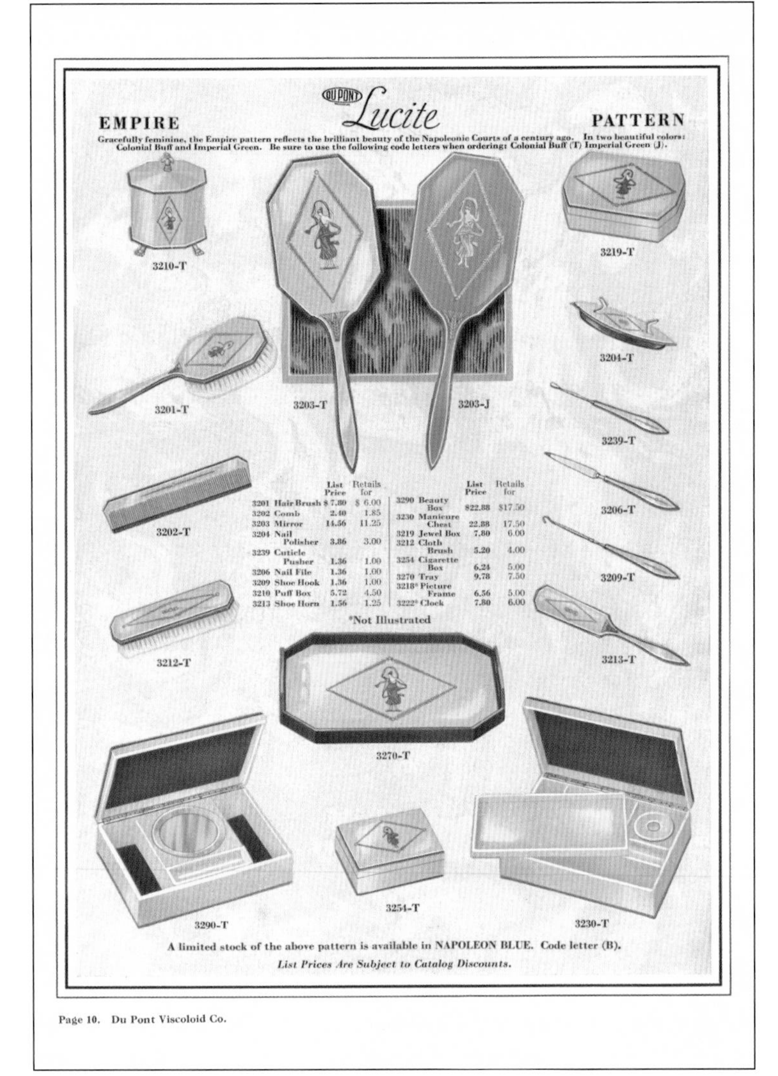

EMPIRE *Lucite* PATTERN

Gracefully feminine, the Empire pattern reflects the brilliant beauty of the Napoleonic Courts of a century ago. In two beautiful colors: Colonial Buff and Imperial Green. Be sure to use the following code letters when ordering: Colonial Buff (T) Imperial Green (J).

	List Price	Retails for
3201 Hair Brush	$7.80	$ 6.00
3202 Comb	2.40	1.85
3203 Mirror	14.56	11.25
3204 Nail Polisher	3.86	3.00
3239 Cuticle Pusher	1.36	1.00
3206 Nail File	1.36	1.00
3209 Shoe Hook	1.36	1.00
3210 Puff Box	5.72	4.50
3213 Shoe Horn	1.56	1.25

	List Price	Retails for
3290 Beauty Box	$22.88	$17.50
3230 Manicure Chest	22.88	17.50
3219 Jewel Box	7.80	6.00
3212 Cloth Brush	5.20	4.00
3254 Cigarette Box	6.24	5.00
3270 Tray	9.78	7.50
3218* Picture Frame	6.56	5.00
3222* Clock	7.80	6.00

*Not Illustrated

A limited stock of the above pattern is available in NAPOLEON BLUE. Code letter (B).

List Prices Are Subject to Catalog Discounts.

Page 10. Du Pont Viscoloid Co.

The *Empire* pattern was available in Colonial Buff, Imperial Green and Napoleon Blue.

Pyralin and Early Lucite

The DuPont Viscoloid Company, of New York City, launched a wonderful campaign in the late 1920s and continued it throughout the 1930s marketing their new creations in boudoir accessories made of Lucite and Pyralin. Lucite was introduced into the market in 1928. The campaign proved extremely successful. In the fall of 1929, DuPont stated that women of the day wanted a completely new type of toilet ware, which "harmonized with the modern boudoir." This new look prompted DuPont and other companies to expand on an already thriving industry and create what the consumer desired. Set after set was offered for sale in both wholesale and retail mail order catalogs. Each line of toilet ware was given a specific pattern name, available in assorted colors. The choices were endless. It is hard to imagine today being offered what the women of the 1920s and 1930s were offered. It had to have been mind-boggling. These early plastic sets were indeed, modern, and a new vogue in toiletries.

The demand for boudoir accessories reached its peak in the 1930s. An advertisement for DuPont boudoir accessories of Pyralin and Lucite by the DuPont Viscoloid Company stated, "For years a new vogue in boudoir accessories was urgently needed – something so radically different, so smart, so modern – that people who owned toilet sets of the old conventional type would recognize that they were out of date and buy these new up-to-date creations." The ad went on to say that that was the reason that Lucite was created in 1928 and that Pyralin was restyled employing the materials and effects most popular in this type of merchandise during the past year.

An unusual Lucite set, marketed in 1930 by the DuPont Viscoloid Company, was the *Wedgwood* pattern. "Skillfully adapted from the work of Josiah Wedgwood, the famous creator of Staffordshire pottery in the 18th century – the Wedgwood Blue in this new Lucite set is an exact reproduction of his favorite color and the decoration is typical of his well-known style." The *Watteau* pattern, also made of Lucite, was decorated with garden scenes in the French style of the eminent painter Watteau. The *Orchis* pattern, designed by Nash, a well-known stylist of the day, was decorated with exotic blooms. The *Empire* pattern represented the graceful beauty of the Napoleonic courts of the early 1800s. This pattern was created by Verna Cook Salomonsky, a leading architect and decorator knowledgeable on Empire, early American, and early English design.

Another beautiful Lucite pattern made by DuPont in 1930 was the *Ming* pattern. The inspiration for the lovely oriental motif was a hawthorn blossom design found on a rare vase from the Ming Dynasty. The colors available

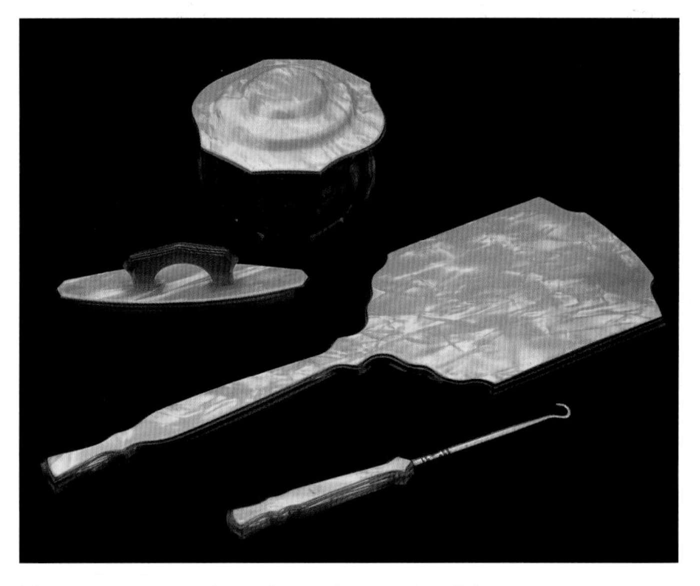

The powder jar in this jade pearl-on-amber Arlton set was made of green glass. The *Arlton* line by DuPont was available in jade-pearl-on-amber, rose pearl-on-amber, orchid pearl-on-amber and maize pearl-on-amber. The amber layer underneath the top pearl layer beveled out on all sides creating a slightly different look and adding more depth to the pieces. The product was popular throughout the 1930s. The mirror on this set is stamped *DuPont Arlton U.S.A.* $140-170.

in the pattern were Sea Jade, Mandarin Red, and Jet Black. This was an especially rich-looking pattern.

Simplicity was the impetus behind the *Crystal* pattern of Lucite from DuPont. Jewel-like colors called Sunset Rose, Mist Blue and Golden Dawn were simple yet elegant and created a modern look of the times. The *Diane* and *Venetia* patterns were distinctive with the pieces angled instead of rounded, like most other sets. The squared appearance created a geometric look. The new *Venetia* pattern was described as "definitely modern but not Modernistic. Its simplicity of line and treatment reflect the true fundamentals of modern art and make it distinctly smart and in good taste." The colors available for this pattern were Sea Jade and Mandarin Red, with harmonizing gold and silver decorations.

This double composite eight-piece dresser set is made with a layer of jade pearl over a layer of jade Lucite. Applied gold decorations create an elegant look. The set consists of a tray, brush, mirror, powder box, jewel box, hair receiver, shoehorn and nail buffer. I honestly do not know where the rubber band came from! $150-195.

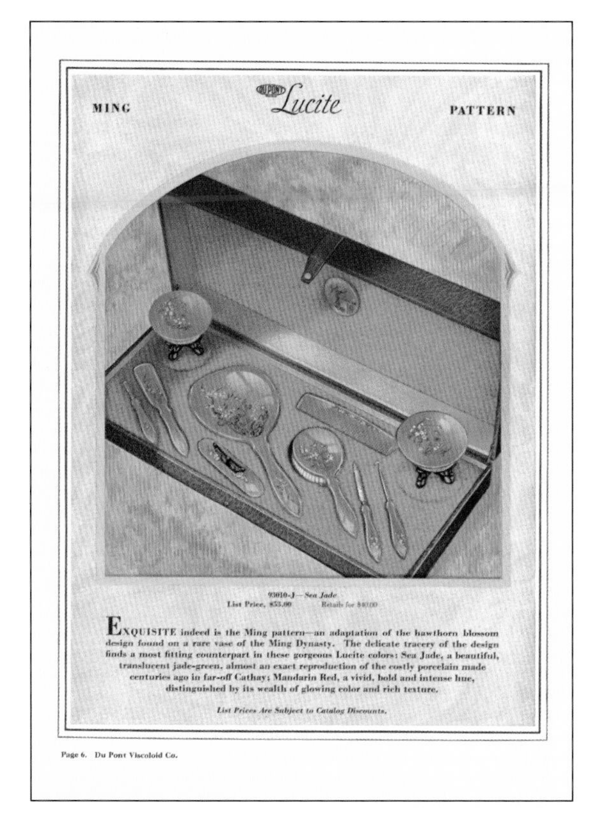

MING — DuPont Lucite — PATTERN

93010-J—Sea Jade
List Price, $53.00 Retails for $40.00

EXQUISITE indeed is the Ming pattern—an adaptation of the hawthorn blossom design found on a rare vase of the Ming Dynasty. The delicate tracery of the design finds a most fitting counterpart in these gorgeous Lucite colors: Sea Jade, a beautiful, translucent jade-green, almost an exact reproduction of the costly porcelain made centuries ago in far-off Cathay; Mandarin Red, a vivid, bold and intense hue, distinguished by its wealth of glowing color and rich texture.

List Prices Are Subject to Catalog Discounts.

Page 6. Du Pont Viscoloid Co.

The DuPont *Ming* pattern, made of Lucite, was another 1930s model. It was adapted from a Ming Dynasty vase, which was decorated with a hawthorn blossom.

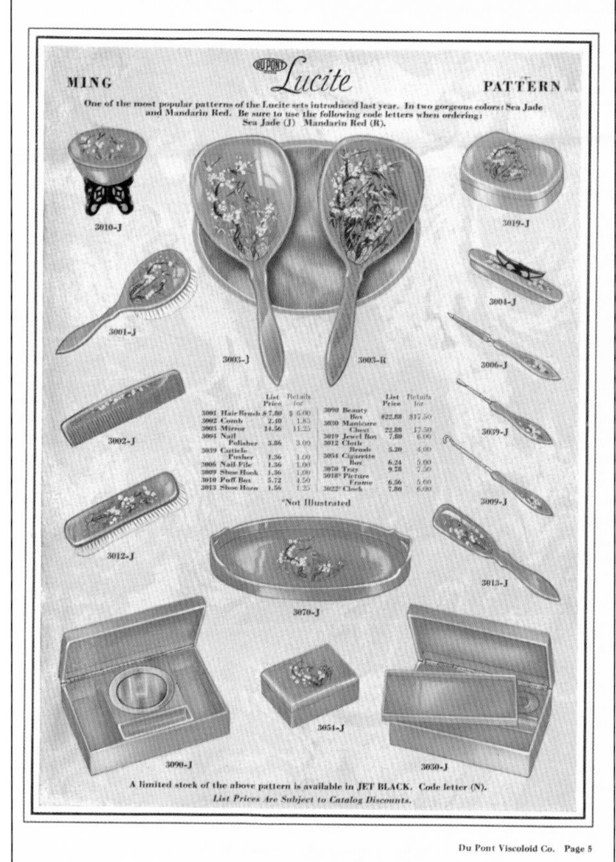

MING — DuPont Lucite — PATTERN

One of the most popular patterns of the Lucite sets introduced last year. In two gorgeous colors: Sea Jade and Mandarin Red. Be sure to use the following code letters when ordering: Sea Jade (J) Mandarin Red (R).

3010-J 3019-J 3004-J 3001-J 3003-J 3003-R 3006-J 3002-J 3039-J 3009-J 3012-J 3013-J 3070-J 3054-J 3090-J 3030-J

		List Price	Retails for
3001	Hair Brush	$7.80	$6.00
3002	Comb	2.40	1.85
3003	Mirror	14.56	11.25
3004	Nail Polisher	3.86	3.00
3039	Cuticle Pusher	1.36	1.00
3006	Nail File	1.36	1.00
3009	Shoe Hook	1.36	1.00
3010	Puff Box	5.72	4.50
3013	Shoe Horn	1.56	1.25
3090	Beauty Box	$22.88	$17.50
3030	Manicure Chest	22.88	17.50
3019	Jewel Box	7.80	6.00
3012	Cloth Brush	5.20	4.00
3054	Cigarette Box	6.24	5.00
3070	Tray	9.78	7.50
3018*	Picture Frame	6.56	5.00
3022*	Clock	7.80	6.00

*Not Illustrated

A limited stock of the above pattern is available in JET BLACK. Code letter (N).

List Prices Are Subject to Catalog Discounts.

Du Pont Viscoloid Co. Page 5

The *Ming* pattern was available in Sea Jade and Mandarin Red and a limited supply of Jet Black.

The *Crystal* pattern by DuPont had a jewel-like appearance and was available in Sunset Rose, Mist Blue and Golden Dawn.

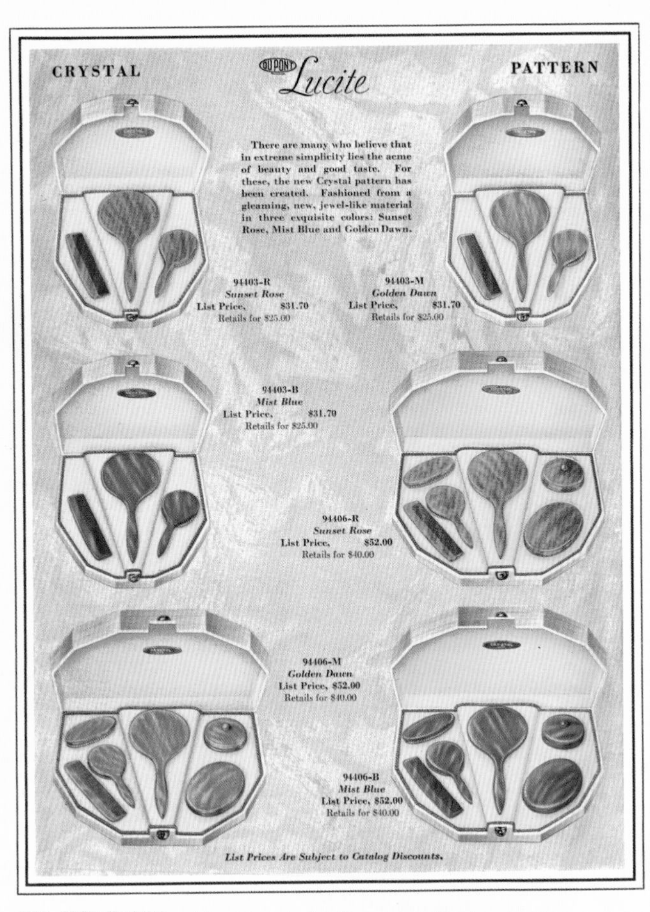

CRYSTAL — DU PONT Lucite — PATTERN

There are many who believe that in extreme simplicity lies the acme of beauty and good taste. For these, the new Crystal pattern has been created. Fashioned from a gleaming, new, jewel-like material in three exquisite colors: Sunset Rose, Mist Blue and Golden Dawn.

94403-R
Sunset Rose
List Price, $31.70
Retails for $25.00

94403-M
Golden Dawn
List Price, $31.70
Retails for $25.00

94403-B
Mist Blue
List Price, $31.70
Retails for $25.00

94406-R
Sunset Rose
List Price, $52.00
Retails for $40.00

94406-M
Golden Dawn
List Price, $52.00
Retails for $40.00

94406-B
Mist Blue
List Price, $52.00
Retails for $40.00

List Prices Are Subject to Catalog Discounts.

Page 4. Du Pont Viscoloid Co.

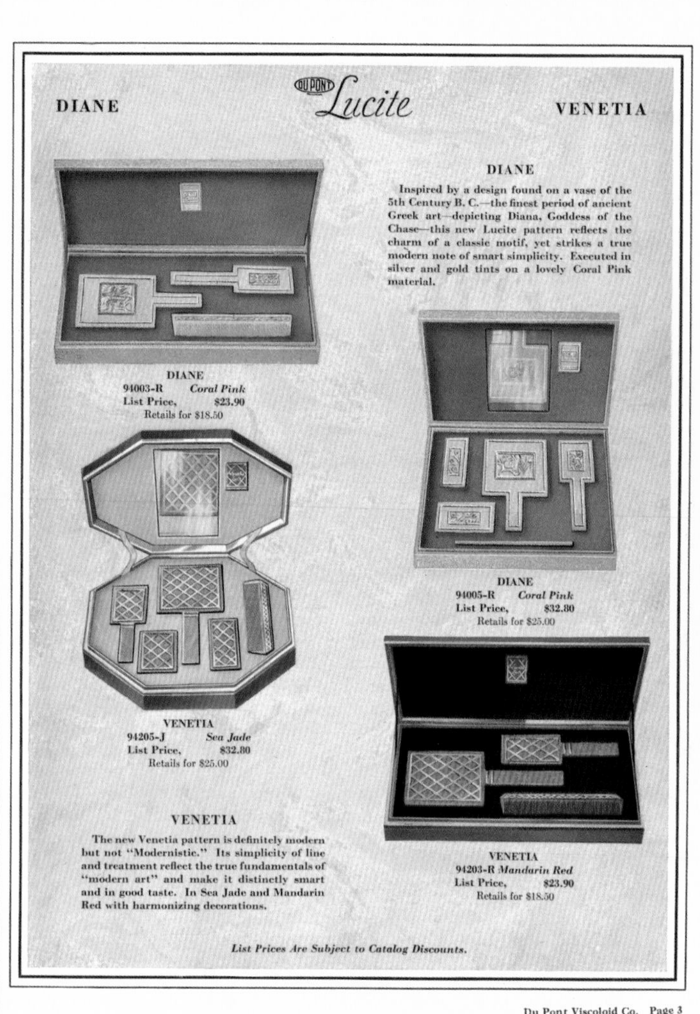

DIANE — DU PONT Lucite — VENETIA

DIANE

Inspired by a design found on a vase of the 5th Century B. C.—the finest period of ancient Greek art—depicting Diana, Goddess of the Chase—this new Lucite pattern reflects the charm of a classic motif, yet strikes a true modern note of smart simplicity. Executed in silver and gold tints on a lovely Coral Pink material.

DIANE
94003-R *Coral Pink*
List Price, $23.90
Retails for $18.50

DIANE
94005-R *Coral Pink*
List Price, $32.80
Retails for $25.00

VENETIA
94205-J *Sea Jade*
List Price, $32.80
Retails for $25.00

VENETIA

The new Venetia pattern is definitely modern but not "Modernistic." Its simplicity of line and treatment reflect the true fundamentals of "modern art" and make it distinctly smart and in good taste. In Sea Jade and Mandarin Red with harmonizing decorations.

VENETIA
94203-R *Mandarin Red*
List Price, $23.90
Retails for $18.50

List Prices Are Subject to Catalog Discounts.

Du Pont Viscoloid Co. Page 3

DuPont's *Venetia* and *Diane* patterns were advertised as *definitely modern but not modernistic.* These pieces were designed with squared edges and more in tune to the geometric look of the period.

FOR TODAY'S LOVELY BOUDOIRS

DU PONT Lucite

New du Pont Accessories that meet every preference for color, design and price.

Today women are more style-conscious than ever before. In their home furnishings, and particularly in accessories for the boudoir, they demand the newest and smartest colors and fashions.

The 1931 du Pont patterns, therefore, have been styled by experts to conform to the latest fashion trends and are presented in fancy gift boxes that provide the eye-appeal that speeds sales.

TRIANON PATTERN

94603-J
Springtime Green
List Price, $26.00

94603-M
Peach Antoinette
List Price, $26.00

94610-M—Peach Antoinette
List Price, $43.70

94610-J
Springtime Green
List Price, $43.70

Designed by Burton Keeler, Trianon is an authentic French pattern. It is decorated with the feathers and plumes so beloved by Marie Antoinette and the delicate scrolls so highly prized by the designers of the Louis XVI period. In two colors—Springtime Green and Peach Antoinette with harmonizing decorations.

The *Trianon* pattern of DuPont Lucite was popular in 1930 and available in Springtime Green and Peach Antoinette. This authentic French pattern, depicting feathers, plumes and scroll work was designed by Burton Keeler.

Each year, DuPont introduced new patterns for toilet ware; in 1932 Lucite patterns were flower-oriented. The *Marigold* pattern in rose, maize, and jade was designed with an inlaid marigold flower decoration. The *Sweet Pea* pattern had a reproduction of a live flower. Sets were available with three pieces, nine pieces, or even ten-piece sets. The Lucite was described as, "a permanent decoration that cannot be marred, made by a patented method and exclusively by DuPont". A slightly different look was the *Old Lace* pattern, described as exquisite Renaissance lace embellished with garlands of rose buds and enhanced by Lustrous Pearl on Jade Lucite base. The patterns were becoming complex, with a variety of techniques incorporated.

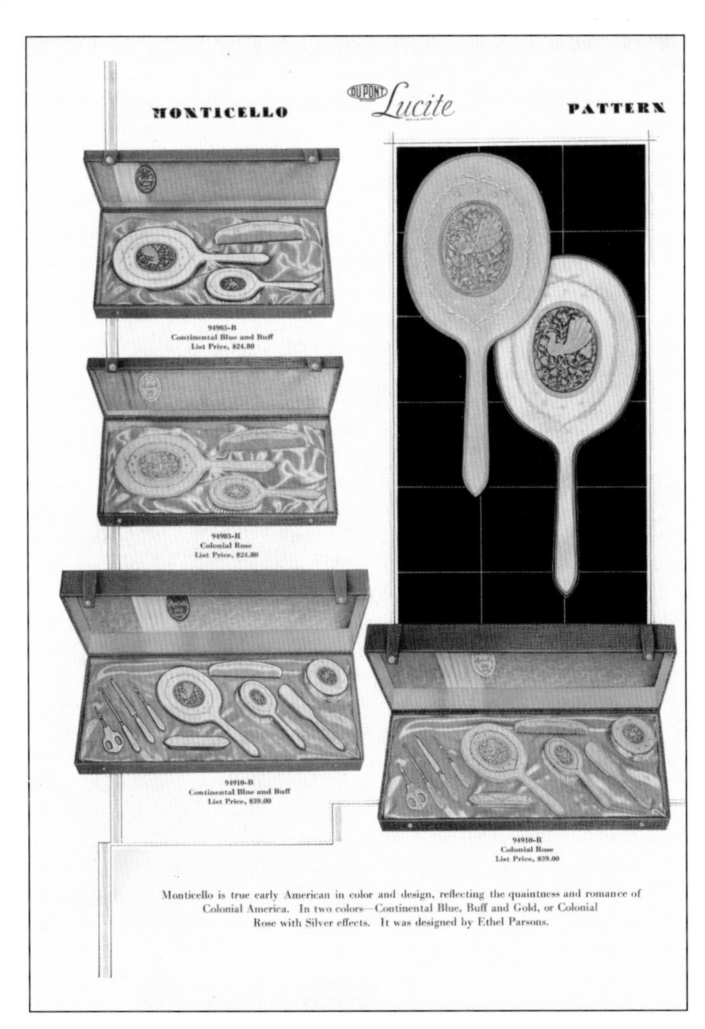

DuPont's *Monticello* pattern made of Lucite was available in Colonial Rose, with Silver effects or Continental Blue, Buff and Gold. It was designed as *true early American in color and design reflecting the quaintness and romance of Colonial America.* The designer was Ethel Parsons.

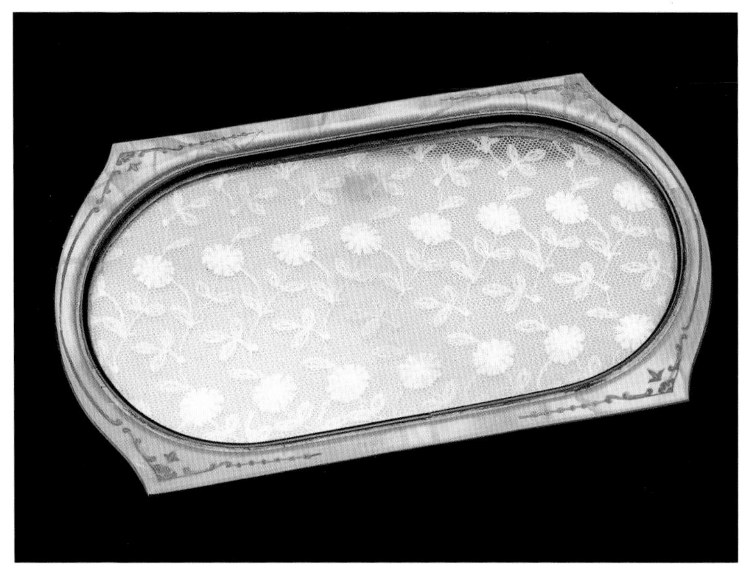

This lace under glass dresser tray was originally part of a much larger set. Made of a double composite of jade pearl-on-jet, it was manufactured by DuPont, given the tradename *Arlton*, and was extremely popular throughout the 1930s. $55-70.

Powder box and matching hair receiver made of jade pearl-on-amber Pyralin with jeweled and applied decorations. $75-95 set.

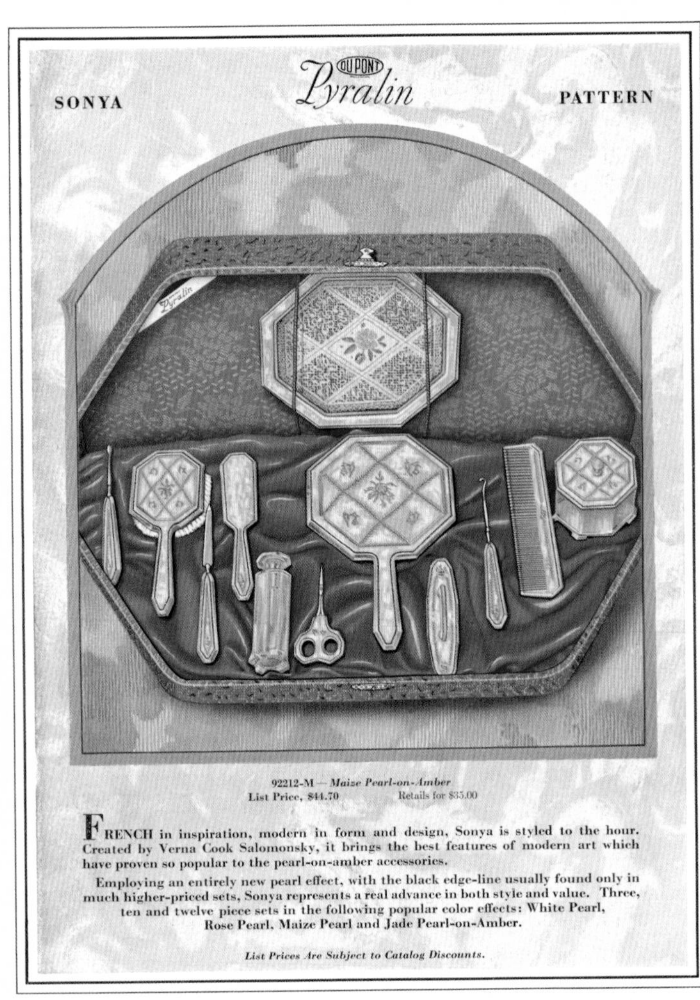

SONYA **DU PONT Pyralin** PATTERN

92212-M — *Maize Pearl-on-Amber*
List Price, $44.70 Retails for $35.00

FRENCH in inspiration, modern in form and design, Sonya is styled to the hour. Created by Verna Cook Salomonsky, it brings the best features of modern art which have proven so popular to the pearl-on-amber accessories.

Employing an entirely new pearl effect, with the black edge-line usually found only in much higher-priced sets, Sonya represents a real advance in both style and value. Three, ten and twelve piece sets in the following popular color effects: White Pearl, Rose Pearl, Maize Pearl and Jade Pearl-on-Amber.

List Prices Are Subject to Catalog Discounts.

Page 12. Du Pont Viscoloid Co.

DuPont's *Sonya* pattern from 1930 was made of rose, maize, white or jade pearl-on-amber Pyralin with a black-edge line trim usually only offered on more expensive sets.

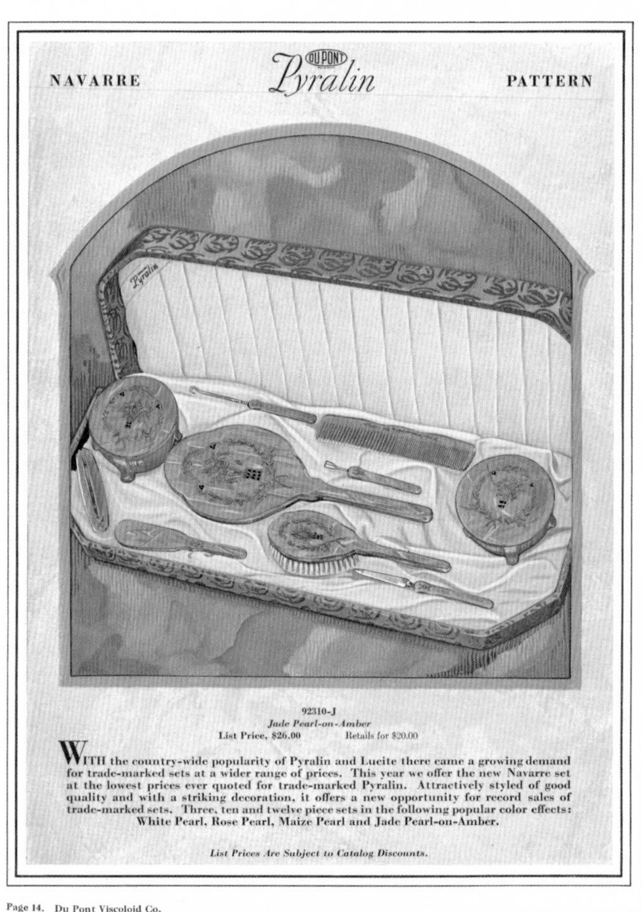

NAVARRE **DU PONT Pyralin** PATTERN

92310-J
Jade Pearl-on-Amber
List Price, $26.00 Retails for $20.00

WITH the country-wide popularity of Pyralin and Lucite there came a growing demand for trade-marked sets at a wider range of prices. This year we offer the new Navarre set at the lowest prices ever quoted for trade-marked Pyralin. Attractively styled of good quality and with a striking decoration, it offers a new opportunity for record sales of trade-marked sets. Three, ten and twelve piece sets in the following popular color effects: White Pearl, Rose Pearl, Maize Pearl and Jade Pearl-on-Amber.

List Prices Are Subject to Catalog Discounts.

Page 14. Du Pont Viscoloid Co.

The *Navarre* pattern of trade-marked Pyralin was fashionable in 1930.

The DuPont Viscoloid Company also offered dresser sets made of Pyralin. The Pyralin sets differed from the Lucite sets in that the Pyralin was layered or made in composite form. For example, most sets featured a top layer of a pearlized color and a bottom layer of an amber color. A green Pyralin set would be called *jade pearl-on-amber* or a pink Pyralin set could be referred to as *rose pearl-on-amber,* and a golden color set was called *maize pearl-on-amber.* There was even a white pearl-on-amber Pyralin pattern. Just like the many patterns of Lucite, even more patterns of Pyralin existed in the 1930s. Pattern names like *Madelon, Lustris, Navarre,* and *Sonya* were a few of the most popular lines in the early 1930s. The *Madelon* pattern was unique in that it was designed with a triple bevel edge and distinctive center decorations.

By 1935, DuPont Pyralin patterns became more elaborate and the sets were becoming larger. The *Lamant* pattern was available in jade, rose and maize pearl-on-amber with inlaid decorations of gold and black. A twenty-piece set, which included a mirrored tray and electric clock, in a spectacular presentation box listed for $45.90. In 1936, the *Noiray* pattern was absolutely exquisite. It was rich jet black Pyralin with chromium-finished metal shields for monogramming and silver colored inlaid decorations. The *Kongo* pattern, which was advertised a year earlier, was described as "one of the most striking sets in Jet Color and Silver Color combinations. The outline is streamline without being extreme." A slightly different look was achieved with the *Ebonette* pattern, which had pearlized Pyralin on both sides of a jet-colored Pyralin body and an applied center decoration of a hammered gold vase filled with floral sprays in gold and black.

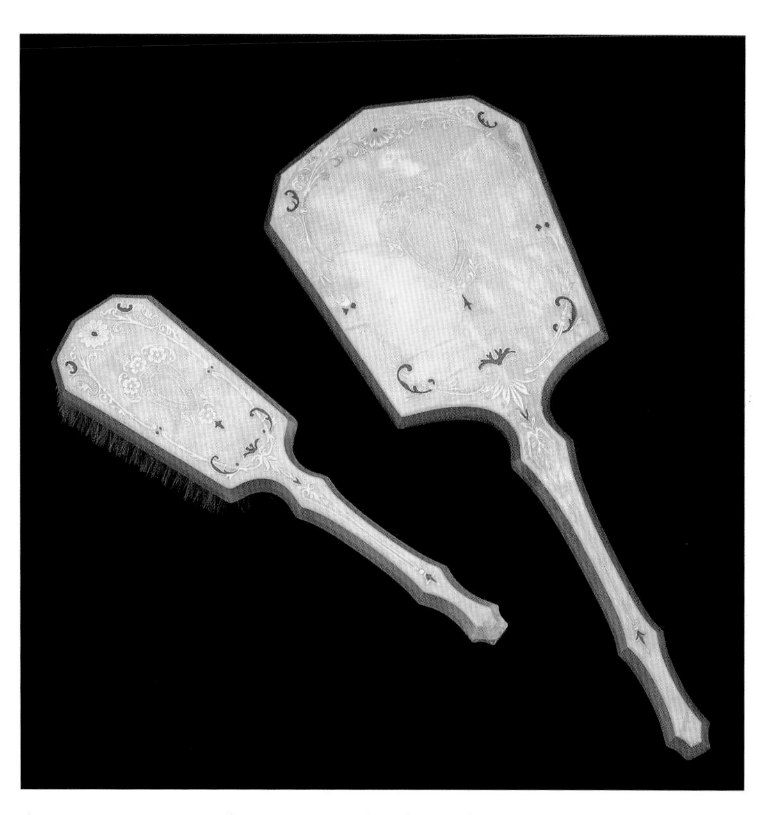

Attractive two-piece dresser set made of a triple composite of pearl-on-amber Pyralin with applied black and gold decorations. The set is stamped *Pyramid.* $95-135 set.

Rose pearl-on-amber Pyralin dresser set enhanced with green rhinestones and applied decorations. The clock movement was manufactured by The Lux Clock Company. $100-150.

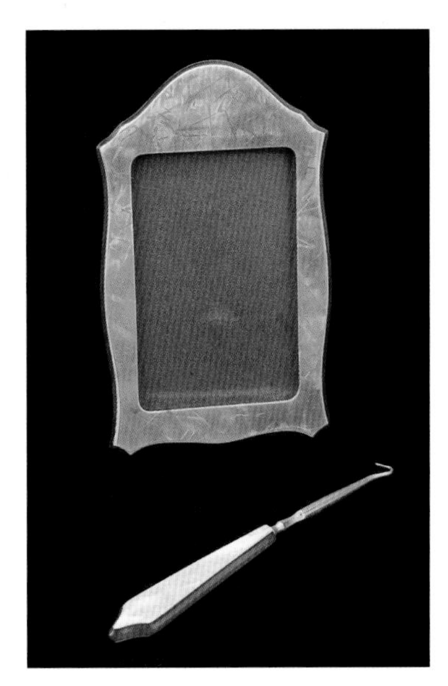

Although this rose pearl-on-amber picture frame and shoe buttoner appear to be DuPont Arlton, it is stamped *Laurelton Arch Amerith* $40-50 set.

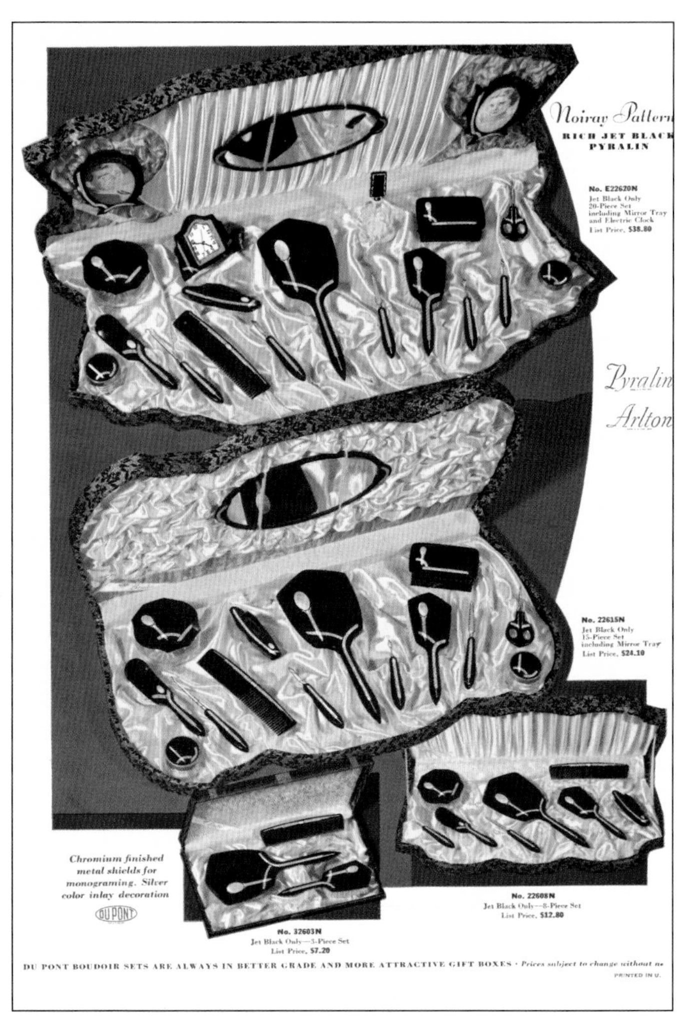

Another stunning jet black Pyralin set by DuPont in the mid 1930s was the *Noiray* pattern with chromium finished metal shields for monogramming and silver inlay decorations.

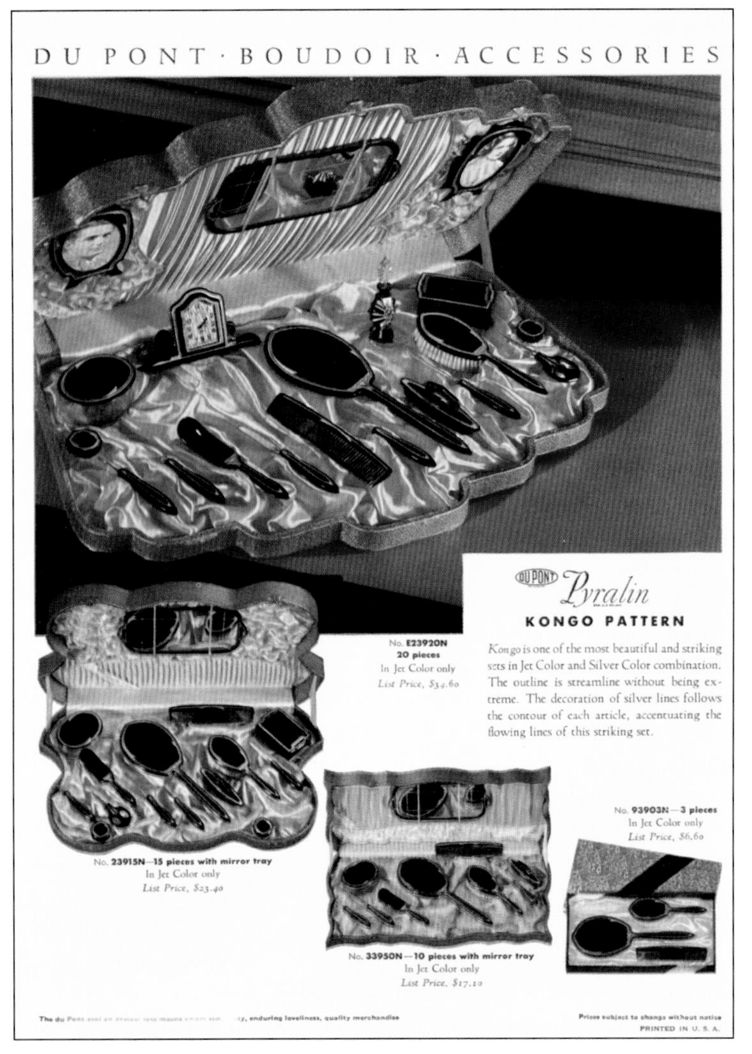

The *Kongo* pattern of DuPont Pyralin was absolutely stunning. Jet black pyralin trimmed in silver created a striking combination. This was a popular look in 1935.

DuPont was continually introducing new types of plastic to fill the demand for these wonderful sets. In 1936, a translucent Pyralin was introduced. Featured as the *Caress* pattern, rose, jade and maize were the colors employed and the inlaid border decorations were done in silver and black. Round, fluted clear crystal Pyralin handles created a very stunning look. In the same year, DuPont's Lucite pattern *Adoray* was a satin, pearl-on-clear crystal Pyralin. The centers were decorated with chromium-plated metal shields used for monogramming. As the 1930s progressed, more and more transparent Lucite and Pyralin was being used in the manufacture of dresser sets.

The *Marbray* pattern was another new look. It was offered in extra white marble Pyralin with black veins or black marble Pyralin with white and brown veins running through it. Bright chromium-plated sections separated the handle from the rest of the piece. The *Patrician* pattern of 1939 was a cloudlike Pyralin with veins of contrasting color and clear Lucite handles. This set was offered in colors called Hyacinth and Tea Rose. The *Deauville* pattern also displayed a cloudlike appearance with hand-carved colored crystal handles with a hexagon shape. This set was offered in colors like Chartreuse and Temple of Heaven Blue. An unusual twist was visible in composite pieces in 1939, with the introduction of the *Windsor* pattern. Rich solid colors, like green, blue, red and tan, were layered on an Old Ivory Pyralin base, reminiscent of the earlier French Ivory Celluloid sets. A carved decoration was situated directly in the center.

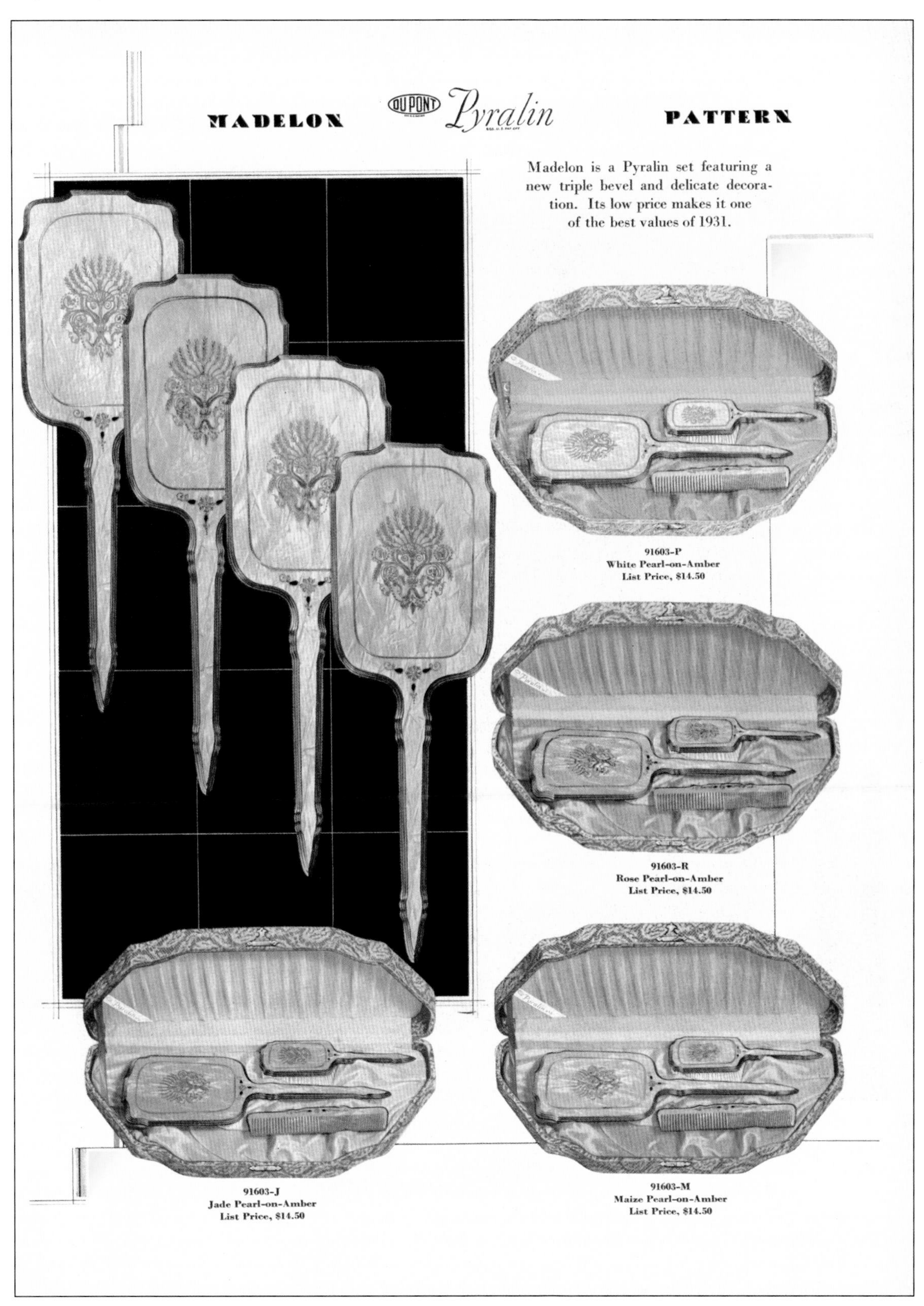

MADELON
DU PONT Pyralin
PATTERN
Madelon is a Pyralin set featuring a new triple bevel and delicate decoration. Its low price makes it one of the best values of 1931.
91603-P
White Pearl-on-Amber
List Price, $14.50
91603-R
Rose Pearl-on-Amber
List Price, $14.50
91603-J
Jade Pearl-on-Amber
List Price, $14.50
91603-M
Maize Pearl-on-Amber
List Price, $14.50

The *Madelon* pattern is a double composite Pyralin with a triple bevel along the edges and a striking center decoration accented with gold. It was available in maize, jade, rose or white pearl-on amber pyralin and fashionable in 1931.

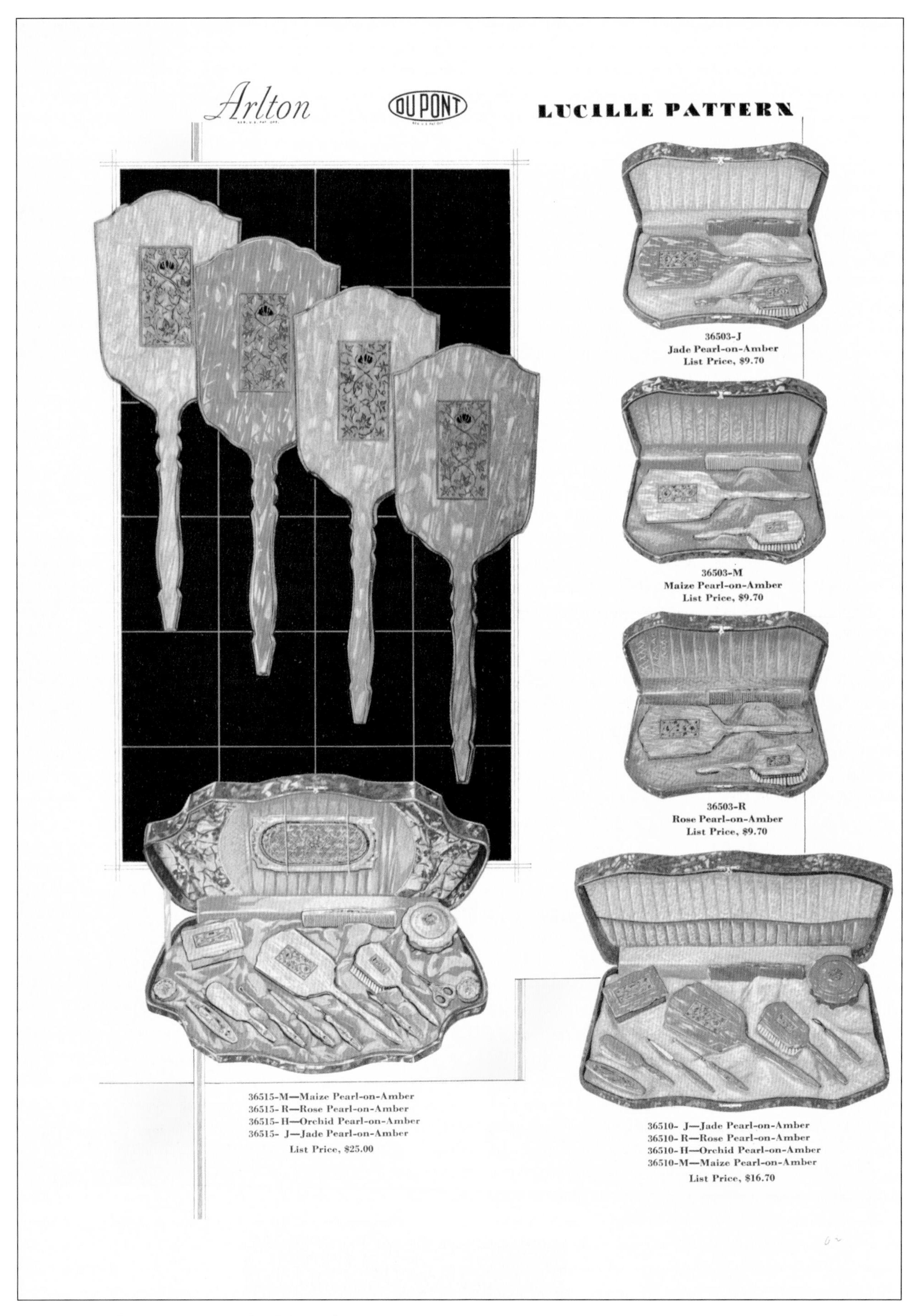

DuPont's *Lucille* pattern was made of the trade-name *Arlton*. It was also a double composite.

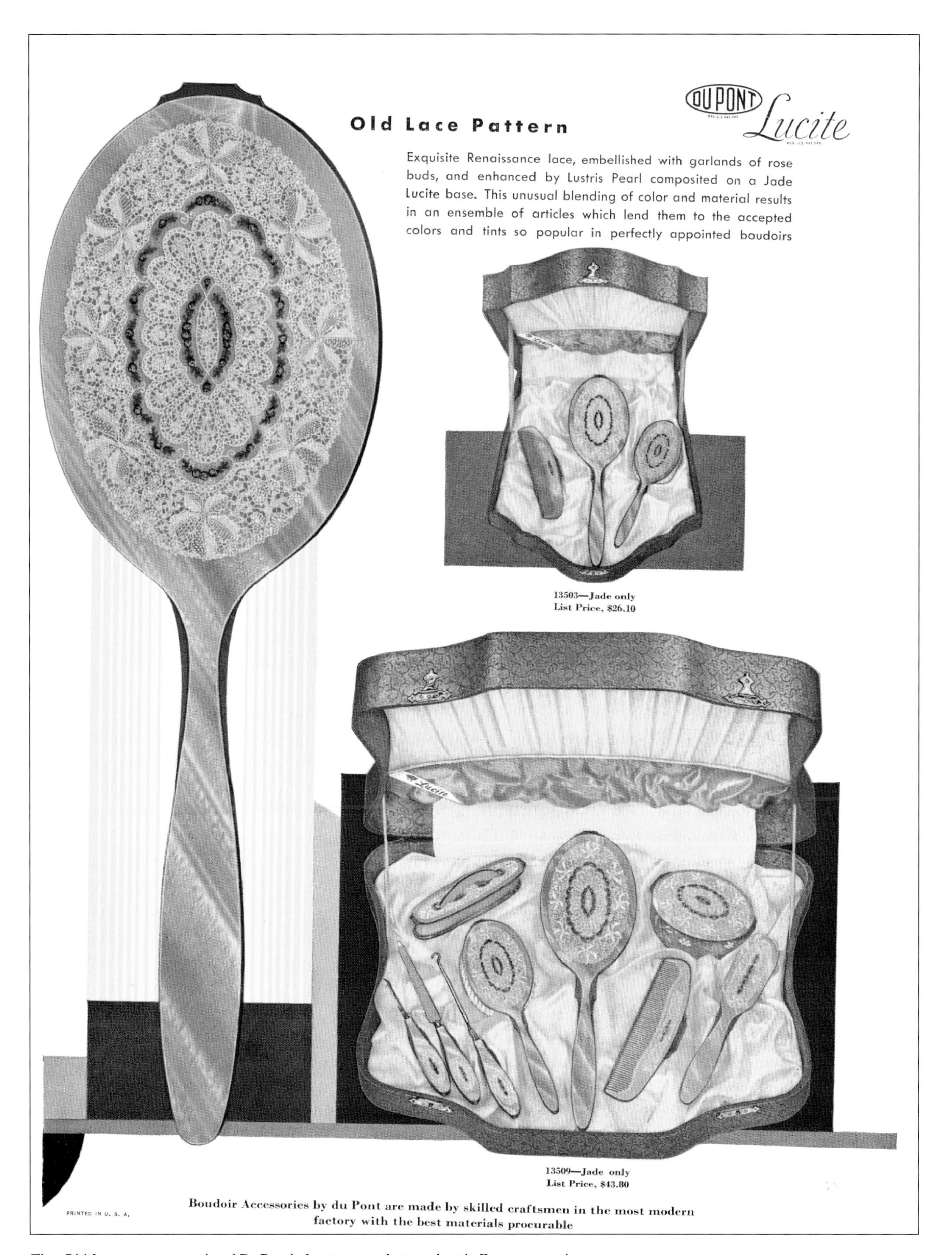

Old Lace Pattern
DUPONT Lucite
Exquisite Renaissance lace, embellished with garlands of rose buds, and enhanced by Lustris Pearl composited on a Jade Lucite base. This unusual blending of color and material results in an ensemble of articles which lend them to the accepted colors and tints so popular in perfectly appointed boudoirs
13503—Jade only
List Price, $26.10
13509—Jade only
List Price, $43.80
PRINTED IN U. S. A.
Boudoir Accessories by du Pont are made by skilled craftsmen in the most modern factory with the best materials procurable

The *Old Lace* pattern made of DuPont's Lucite was designed with *Renaissance lace embellished with garlands of rose buds and enhanced by Lustris pearl composited on a Jade Lucite base.* This was an extremely desirable set in the early 1930s.

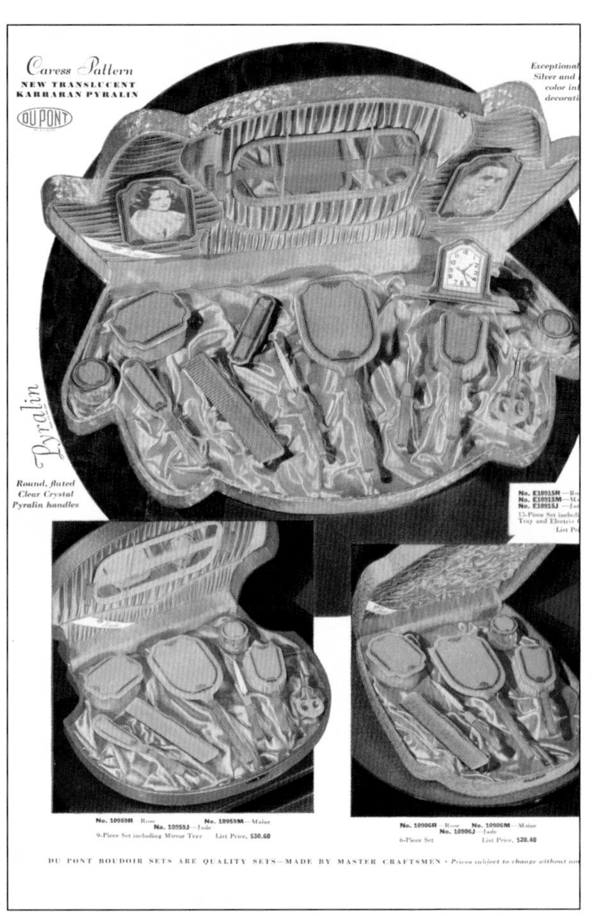

The *Caress* pattern of DuPont Pyralin was unique in that the pyralin was translucent maize, rose or jade and the handles were round, fluted and clear. Silver and black inlaid decorations added the finishing touches.

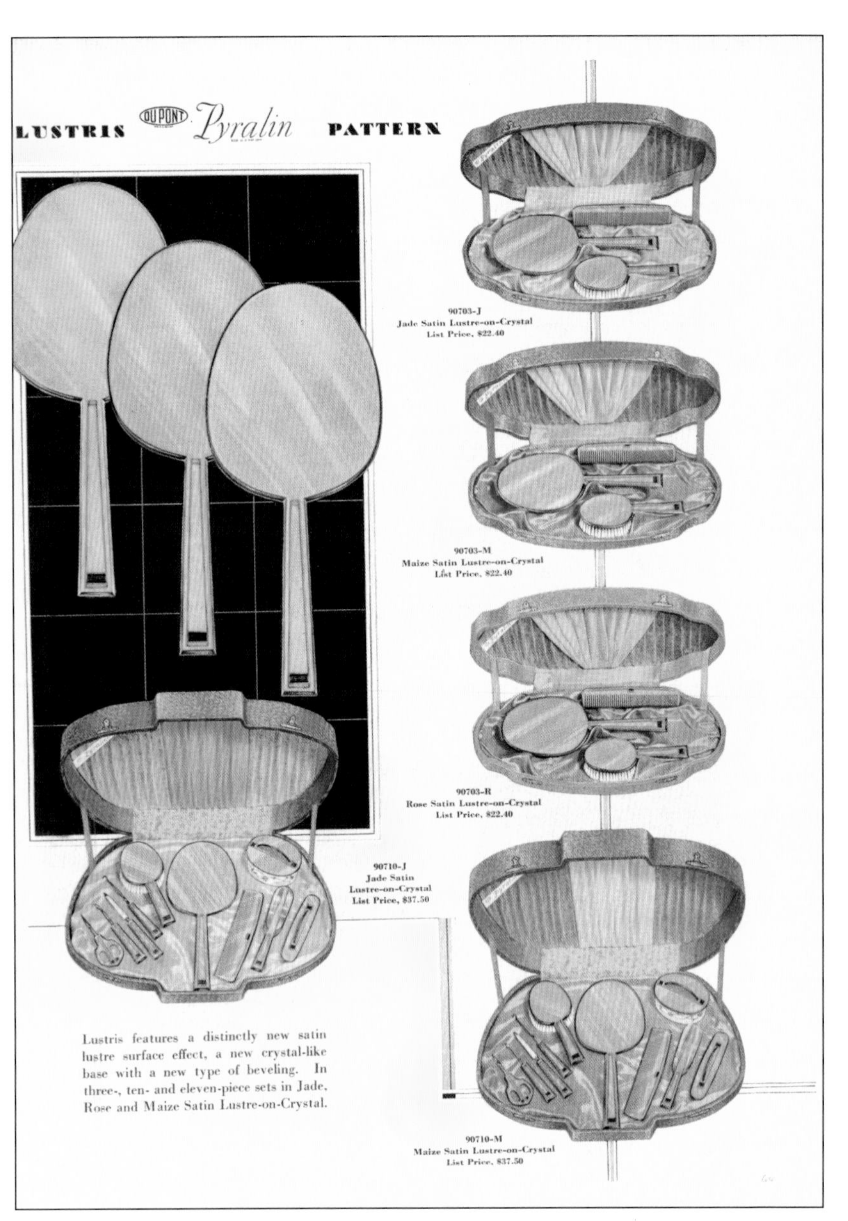

The *Lustris* pattern from DuPont was a double composite of a satin lustre Pyralin on a new crystal-like base with a new type of beveling. The *Nulustris* pattern was a also a satin luster surface with a translucent-like base. These were popular patterns in the early 1930s.

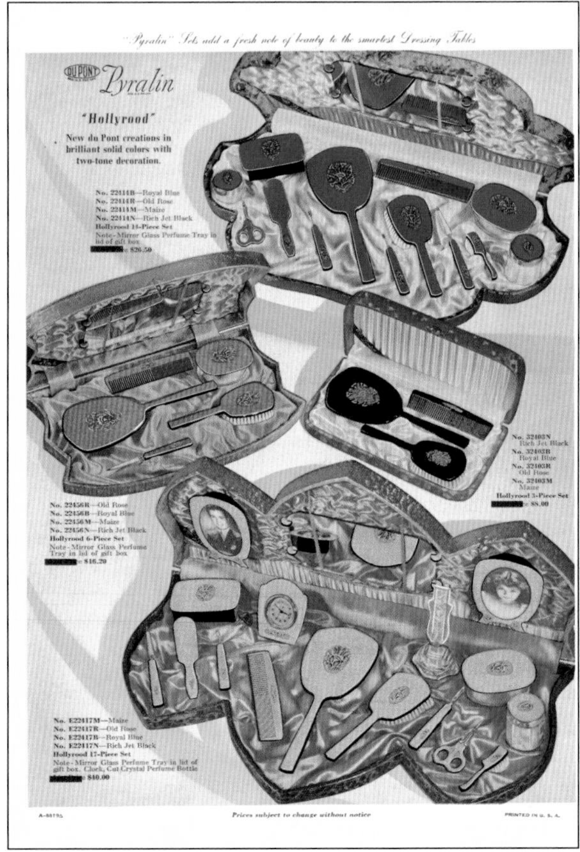

In 1939, the DuPont dresser set pattern called *Hollyrood* was made of a bright solid color Pyralin accented with two-tone decorations. Colors available were maize, old rose, royal blue and rich jet black. A seventeen-piece set, which included a glass perfume tray and a cut crystal perfume bottle listed for $40.00.

In the early 1930s, The Celluloid Corporation marketed its toilet ware under the trade name ART-Y-ZAN. Beautiful ten to twenty-piece sets were available in colors called Orchid, Canary, Rose, Jade and Beige. Inlaid decorations were found in the centers of certain sets and along the borders in other sets. A fifteen-piece set in the *Victoria* pattern listed for $49.20. The *Victoria* pattern was a reproduction of *Italian Onyx*. Other patterns were referred to as *Fairmont*, *Clairmont*, and *Sylvia*.

Jeweled pearl-on-amber dresser set consisting of a powder box, toothbrush holder, tooth powder box and salve jar. Intricate engraving is enhanced with enamel and colored rhinestones. $115-155 set.

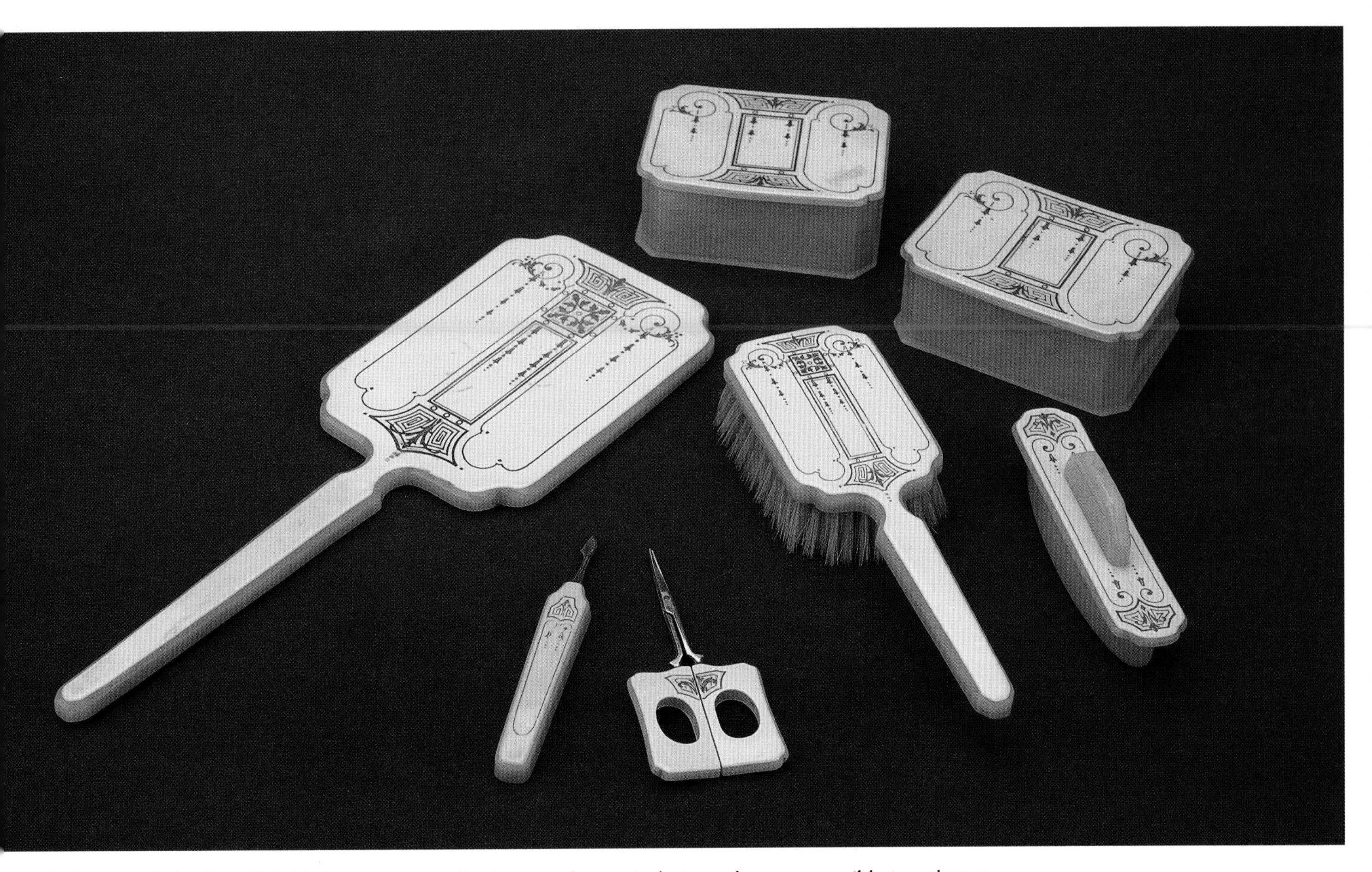

As the popularity for celluloid dresser accessories increased, new techniques became possible to enhance the plain French Ivory. This seven-piece set was elaborately engraved and then gilded. This set was made with a thin layer of cream-colored celluloid on top of butterscotch, pearlized plastic. The set includes a jewel box, powder box, brush, mirror, nail buffer, cuticle knife and manicure scissors. The mirror is stamped *ART-Y-ZAN*. This pattern is called *Rosemont* and it was described as *a modern design in Old Ivory Tone*. Popular in 1933, a ten-piece set in a satin-lined box listed for $36.20. $165-225 set.

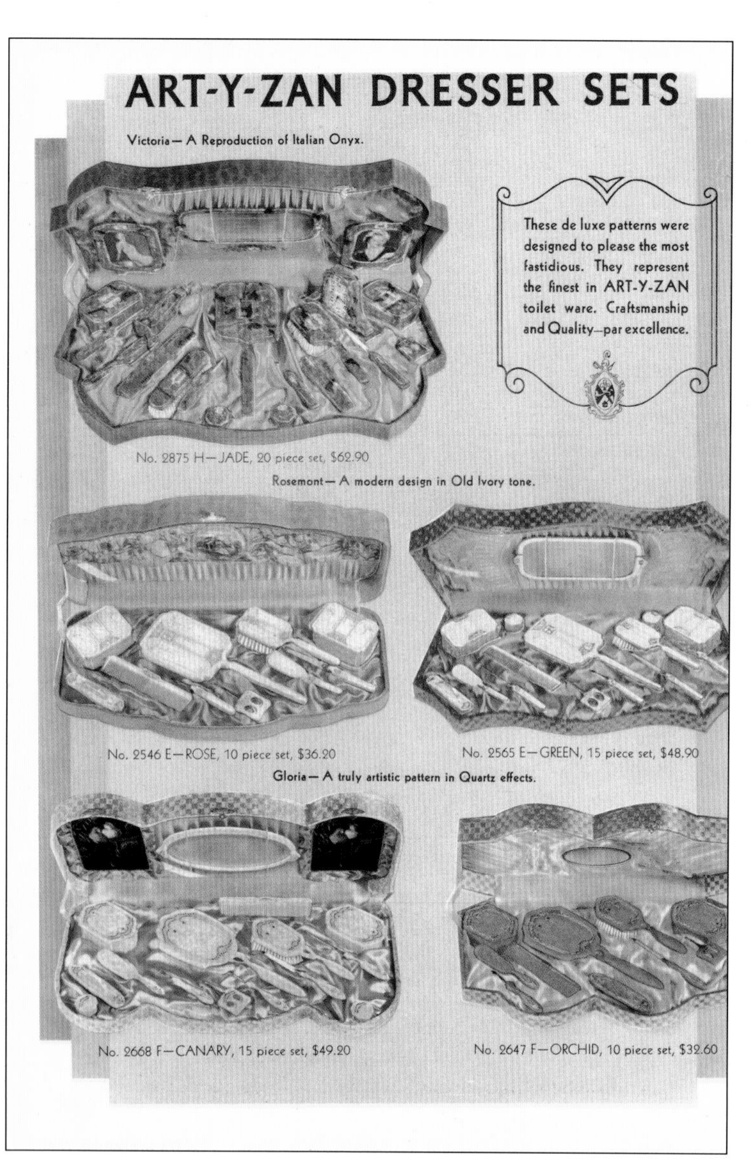

ART-Y-ZAN DRESSER SETS
Victoria— A Reproduction of Italian Onyx.
These de luxe patterns were designed to please the most fastidious. They represent the finest in ART-Y-ZAN toilet ware. Craftsmanship and Quality—par excellence.
No. 2875 H—JADE, 20 piece set, $62.90
Rosemont— A modern design in Old Ivory tone.
No. 2546 E—ROSE, 10 piece set, $36.20
No. 2565 E—GREEN, 15 piece set, $48.90
Gloria— A truly artistic pattern in Quartz effects.
No. 2668 F—CANARY, 15 piece set, $49.20
No. 2647 F—ORCHID, 10 piece set, $32.60

ART-Y-ZAN Dresser sets were marketed by The Celluloid Corporation. The *Victoria* pattern was a reproduction of Italian onyx. The *Rosemont* pattern was a modern rendition of French Ivory and the *Gloria* pattern resembled quartz. These sets were fashionable in 1933. Orchid, canary, green, rose and jade were popular colors.

Right:
AMERITH Dresser sets offered for sale in 1938 manufactured by The Celluloid Corporation. Patterns included the *Sicily* pattern made of Pierretone which was advertised as *"a specially designed material of fine tonal qualities ornamented with genuine cloisonné decorations and embossed gold plated handles"*. The *Dorchester* pattern was *"Mother of pearl laminated on both surfaces of the Amerith crystal material with etched medallions and gold embossed handle."* The *Sherlain* pattern was *"featured pastel shades with translucent effects with colored crystal handles and etched gold metal ornaments."* Finally, the *Fairmount* pattern was *"translucent colors with gold metal ornaments and gold metal handles"*.

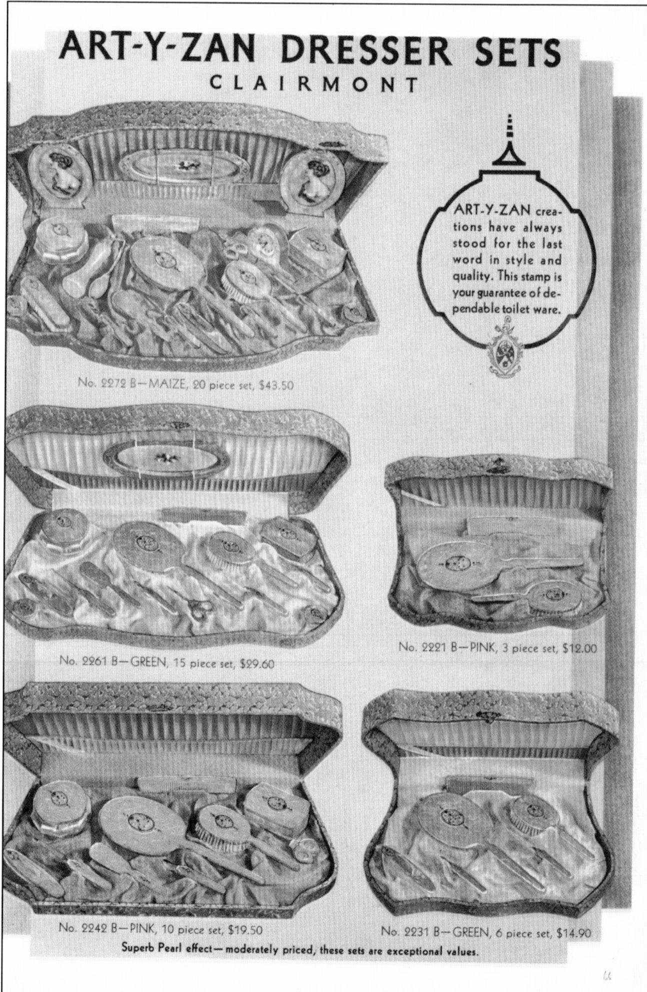

ART-Y-ZAN DRESSER SETS
CLAIRMONT
ART-Y-ZAN creations have always stood for the last word in style and quality. This stamp is your guarantee of dependable toilet ware.
No. 2272 B—MAIZE, 20 piece set, $43.50
No. 2261 B—GREEN, 15 piece set, $29.60
No. 2221 B—PINK, 3 piece set, $12.00
No. 2242 B—PINK, 10 piece set, $19.50
No. 2231 B—GREEN, 6 piece set, $14.90
Superb Pearl effect— moderately priced, these sets are exceptional values.

The *Clairmont* pattern of *ART-Y-ZAN* was described as having a *superb Pearl effect.* It was a popular style in 1933.

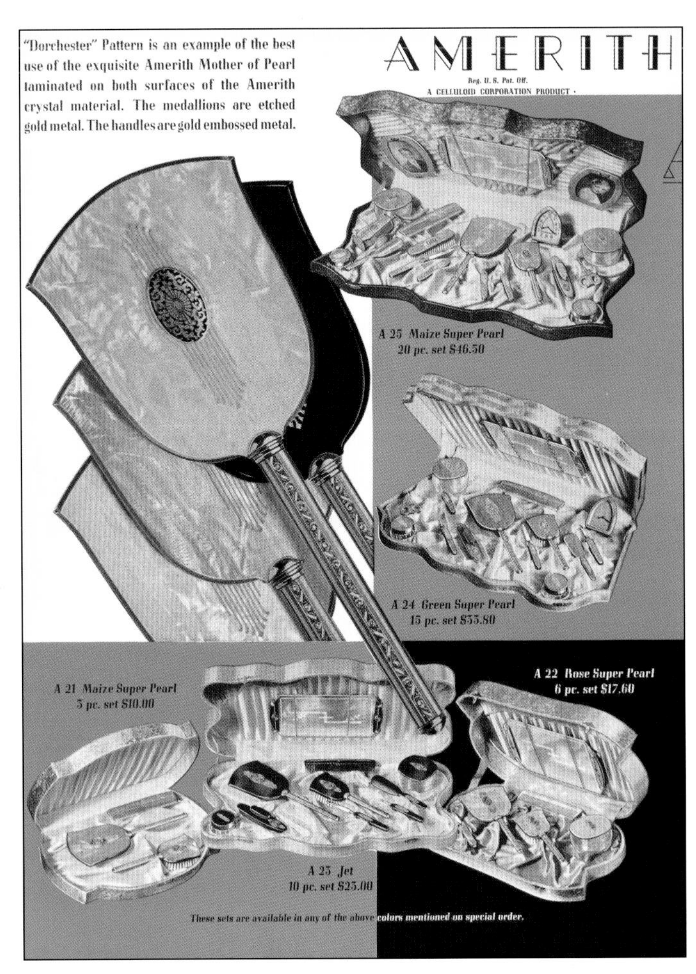
"Dorchester" Pattern is an example of the best use of the exquisite Amerith Mother of Pearl laminated on both surfaces of the Amerith crystal material. The medallions are etched gold metal. The handles are gold embossed metal.
AMERITH
Reg. U. S. Pat. Off.
A CELLULOID CORPORATION PRODUCT
A 25 Maize Super Pearl
20 pc. set $46.50
A 24 Green Super Pearl
15 pc. set $35.80
A 21 Maize Super Pearl
5 pc. set $10.00
A 22 Rose Super Pearl
6 pc. set $17.60
A 23 Jet
10 pc. set $25.00
These sets are available in any of the above colors mentioned on special order.

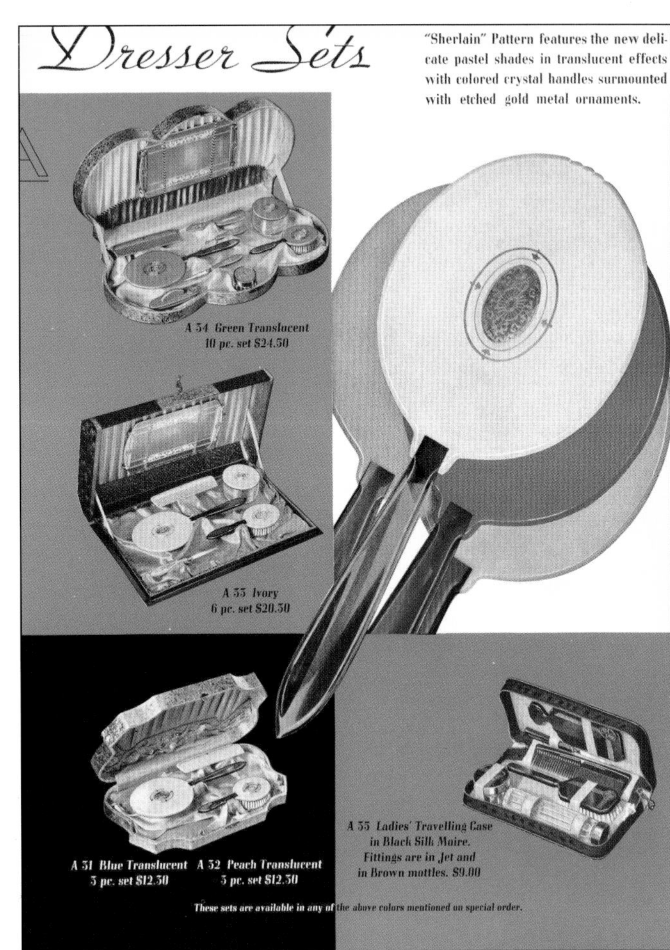
Dresser Sets
"Sherlain" Pattern features the new delicate pastel shades in translucent effects with colored crystal handles surmounted with etched gold metal ornaments.
A 34 Green Translucent
10 pc. set $24.50
A 33 Ivory
6 pc. set $20.30
A 31 Blue Translucent
5 pc. set $12.30
A 32 Peach Translucent
5 pc. set $12.30
A 35 Ladies' Travelling Case
in Black Silk Moire.
Fittings are in Jet and
in Brown mottles. $9.00
These sets are available in any of the above colors mentioned on special order.

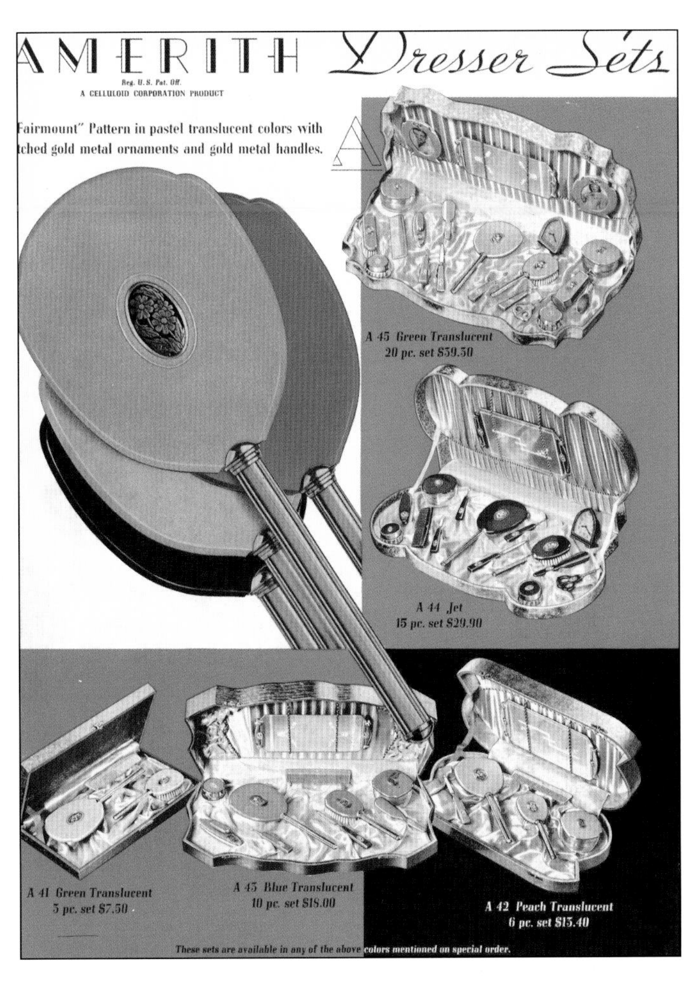
AMERITH Dresser Sets
Reg. U. S. Pat. Off.
A CELLULOID CORPORATION PRODUCT
Fairmount" Pattern in pastel translucent colors with tched gold metal ornaments and gold metal handles.
A 45 Green Translucent
20 pc. set $39.50
A 44 Jet
15 pc. set $29.90
A 41 Green Translucent
5 pc. set $7.50
A 43 Blue Translucent
10 pc. set $18.00
A 42 Peach Translucent
6 pc. set $13.40
These sets are available in any of the above colors mentioned on special order.

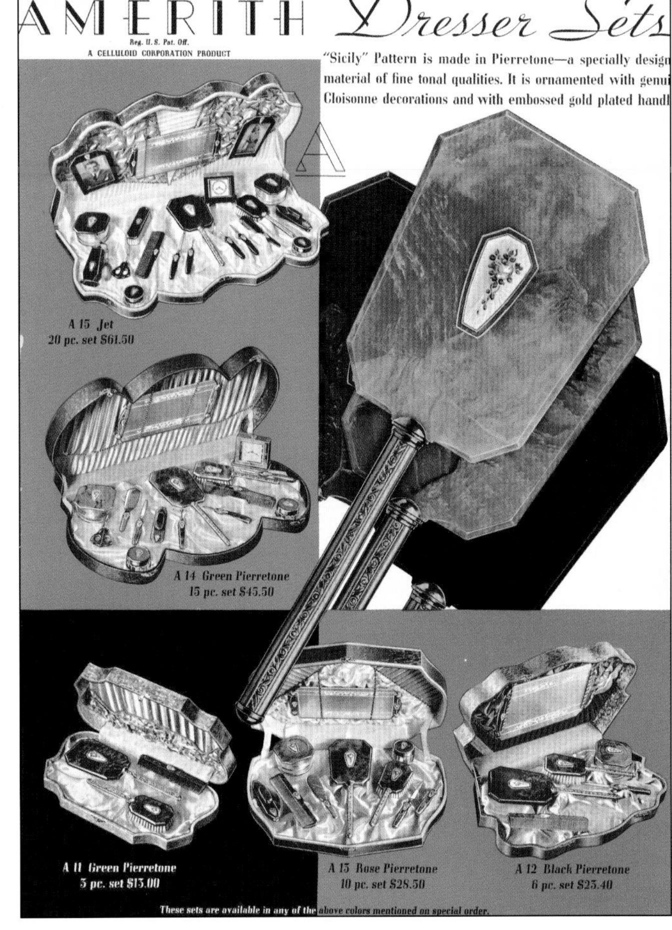
AMERITH Dresser Sets
Reg. U. S. Pat. Off.
A CELLULOID CORPORATION PRODUCT
"Sicily" Pattern is made in Pierretone—a specially desig material of fine tonal qualities. It is ornamented with genu Cloisonne decorations and with embossed gold plated handl
A 15 Jet
20 pc. set $61.50
A 14 Green Pierretone
15 pc. set $45.50
A 11 Green Pierretone
5 pc. set $13.00
A 13 Rose Pierretone
10 pc. set $28.50
A 12 Black Pierretone
6 pc. set $23.40
These sets are available in any of the above colors mentioned on special order.

Later Lucite

Most of us today think of Lucite as being only a clear acrylic plastic. When DuPont first introduced Lucite toilet ware in 1928, it was opaque and similar to Pyralin. But DuPont was constantly improving their products and developing new ones. Lucite, derived from petroleum, not nitro-cellulose like Celluloid and Pyralin, became a more important material in the automobile industry and for aircraft windshields, gunner turrets, and nose cones for bombers and fighters during World War II.

When the war ended, Lucite became popular for uses in the home, for decorative and utilitarian products as well as personal grooming aids. Lucite's crystal-clear appearance and strength made it a better material than cellulose.

In the 1940s, ladies' Lucite dresser sets, military sets for men, and other toilet items were available through mail-order and wholesale catalogs. Advertisements also appeared in leading magazines of the day. One popular style dresser set, made of clear Lucite, featured twisted borders. The mirror and brush backs were sometimes engraved and occasionally hand painted. In 1942, as seen in the Carson Pirie Scott catalog, a two-piece genuine clear Lucite brush and comb set retailed for $5.95. A man's club style brush and fine comb to match also made of clear Lucite listed for $7.95. A bath brush with nine rows of medium Exton bristles was $7.50. By 1946, a clear Lucite set consisting of three pieces retailed anywhere from $8.00 to $18.00 per set.

In 1948, Jewelite Gift Sets by Pro-phy-lac-tic, makers of the famous Pro-phy-lac-tic toothbrush, offered lovely transparent dresser sets in jewel colors. Sets for women and men came in Topaz, Ruby, Emerald, Sapphire and of course, Crystal. The brushes were made with Prolon bristles which were advertised as being very resilient. By the early 1950s, Jewelite sets were known as "America's Most Famous Brush & Comb Set".

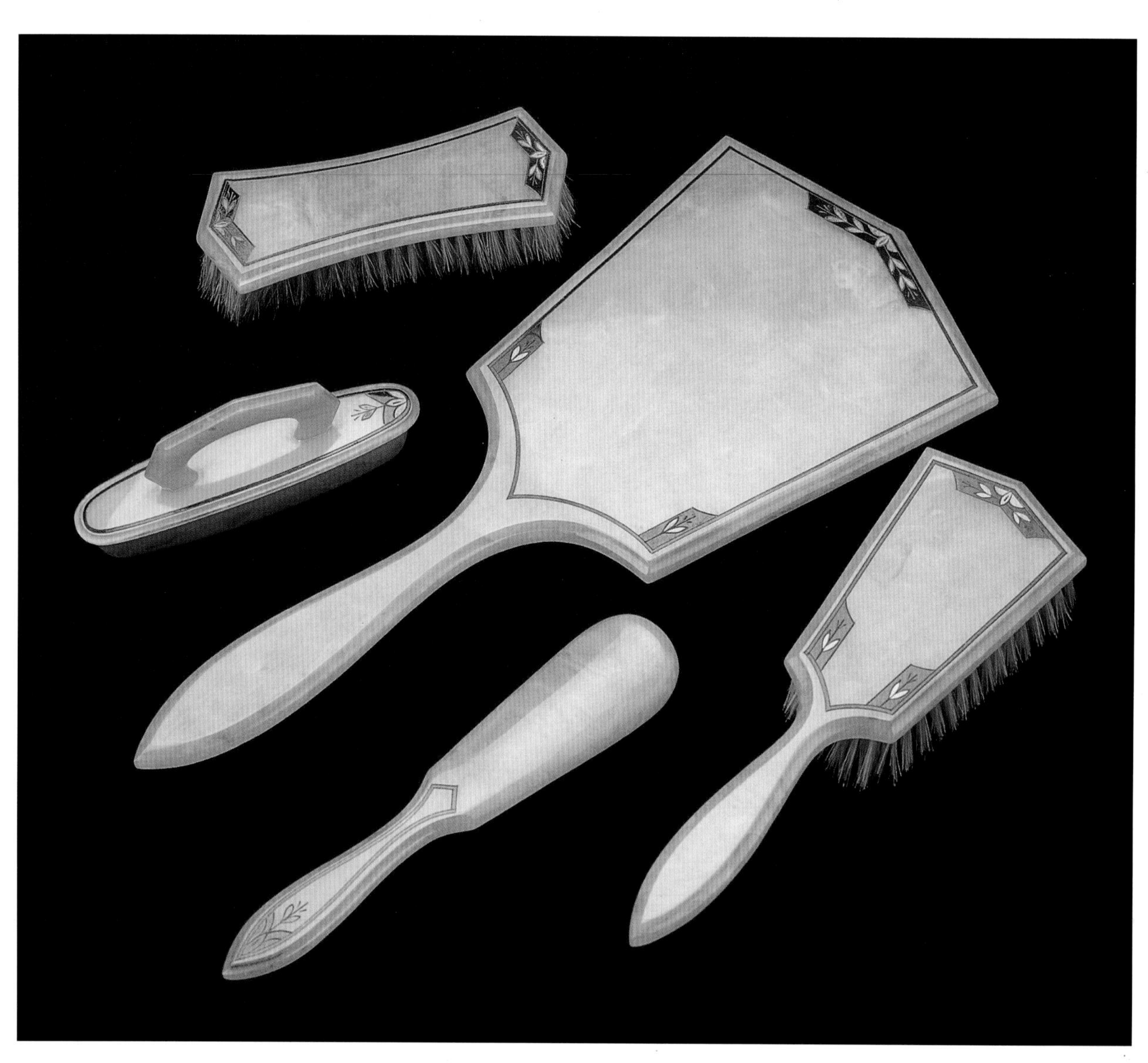

Lovely five-piece set consisting of large hand mirror, hair brush, shoe horn, hat brush and nail buffer made of a pearlized green plastic trimmed in black. The pattern is called *Lady Betty* and the set is stamped *PYROCEL*. $200-250 set.

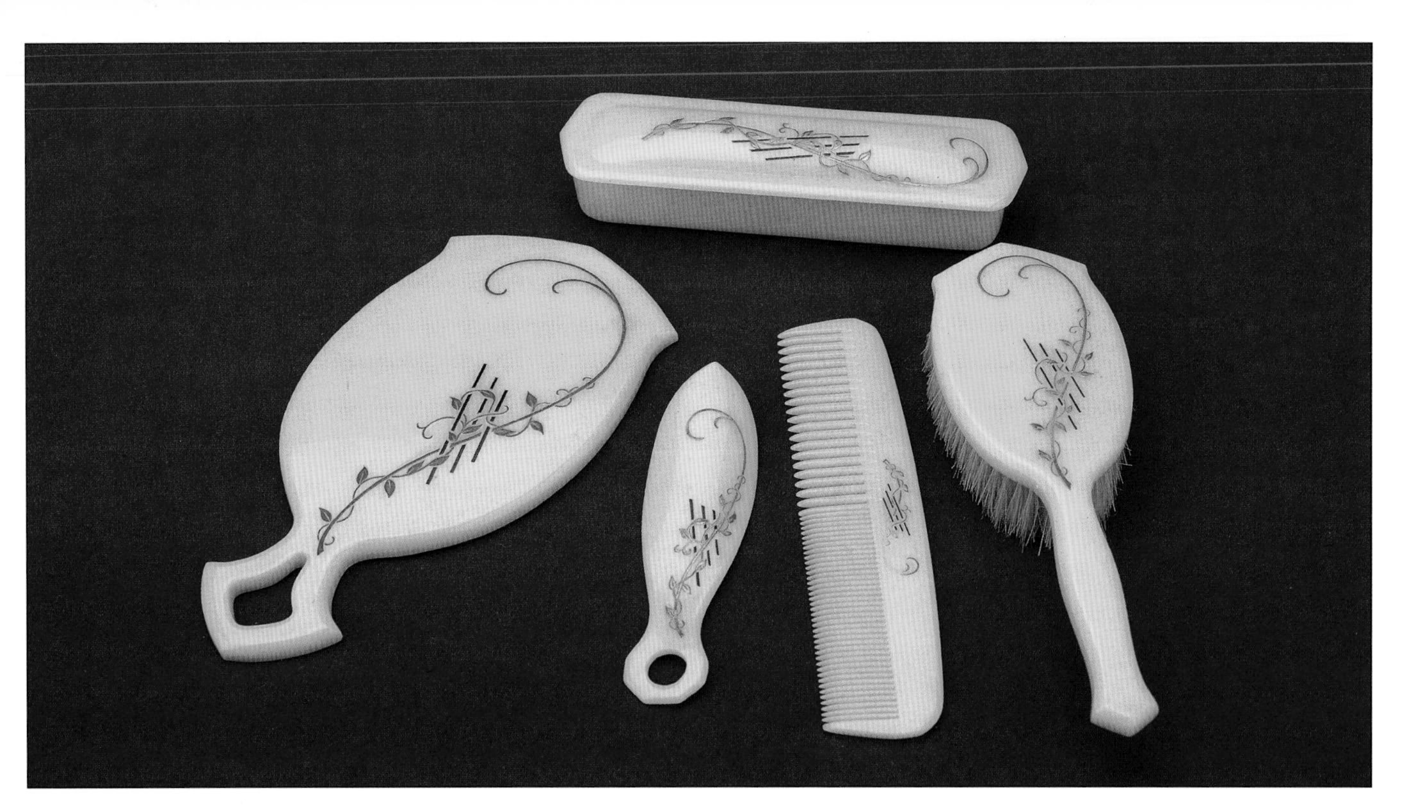

Unusual five-piece dresser set in a buttery yellow color with fancy engraved decorations highlighted in gold and black. The set is stamped *DuPont Lucite U.S.A.* $185-225 set.

Wonderful dark green dresser set made up of a mirror, brush, shoe buttoner, nail file, comb and toothbrush holder stamped *DuPont Lucite.* $135-185 set.

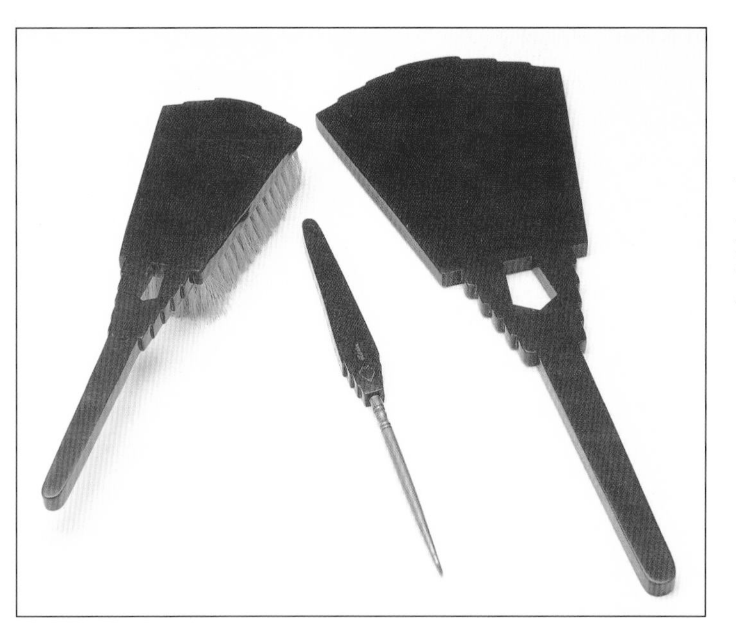

Vibrant red dresser set consisting of a large hand mirror, brush and shoe buttoner with a striking Art Deco design. This early plastic resembles Bakelite and is made of a solid piece instead of laminated layers like Pyralin and Arlton but it is stamped *La Futuriste Agalin Pat. July 17, 1928.* $245-295 set.

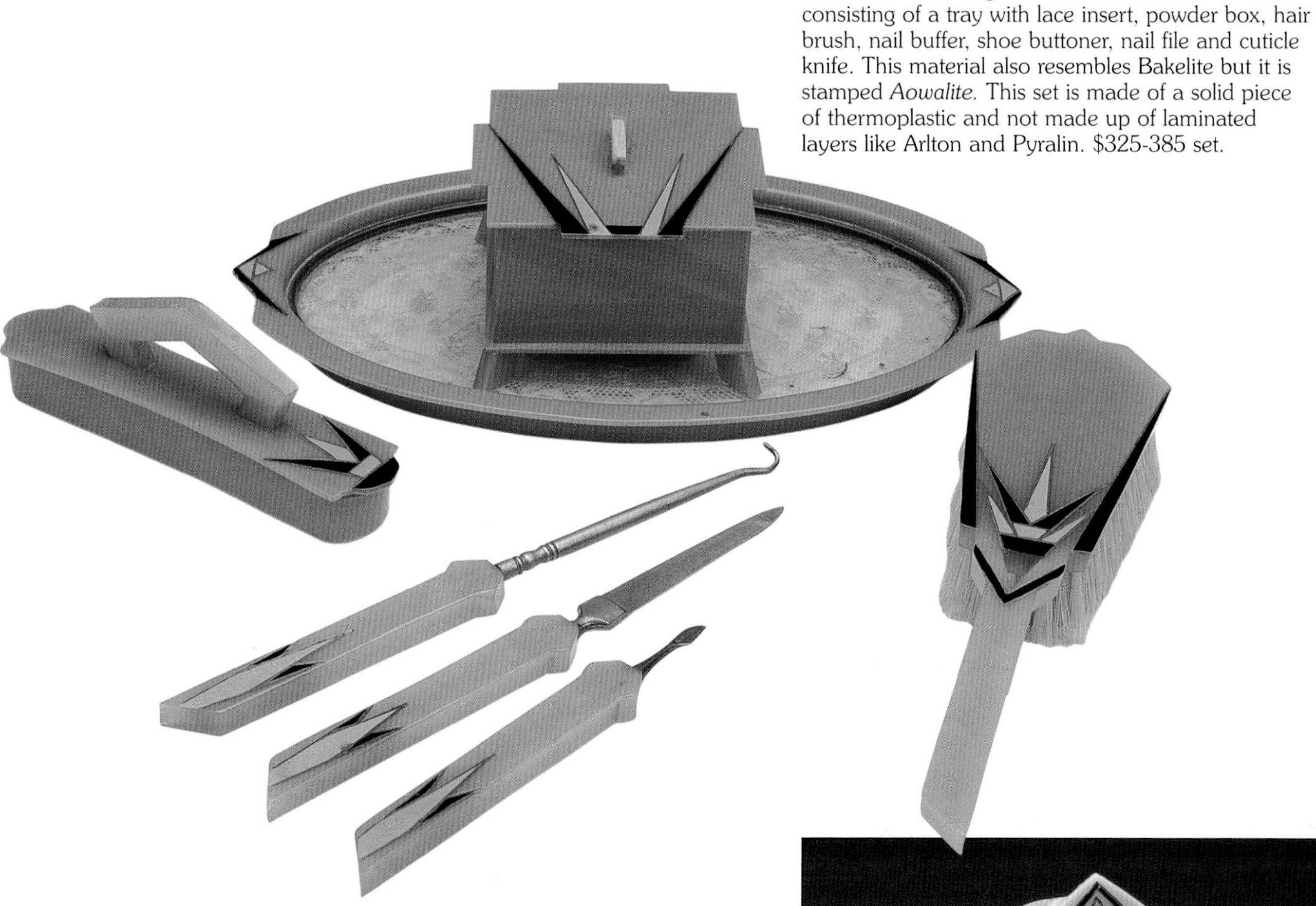

Modern and striking dresser set from the 1930s consisting of a tray with lace insert, powder box, hair brush, nail buffer, shoe buttoner, nail file and cuticle knife. This material also resembles Bakelite but it is stamped *Aowalite.* This set is made of a solid piece of thermoplastic and not made up of laminated layers like Arlton and Pyralin. $325-385 set.

Dresser clock made in the Gothic style and accented with engraved and hand-colored decorations. The clock movement is stamped *New Haven.* $85-115.

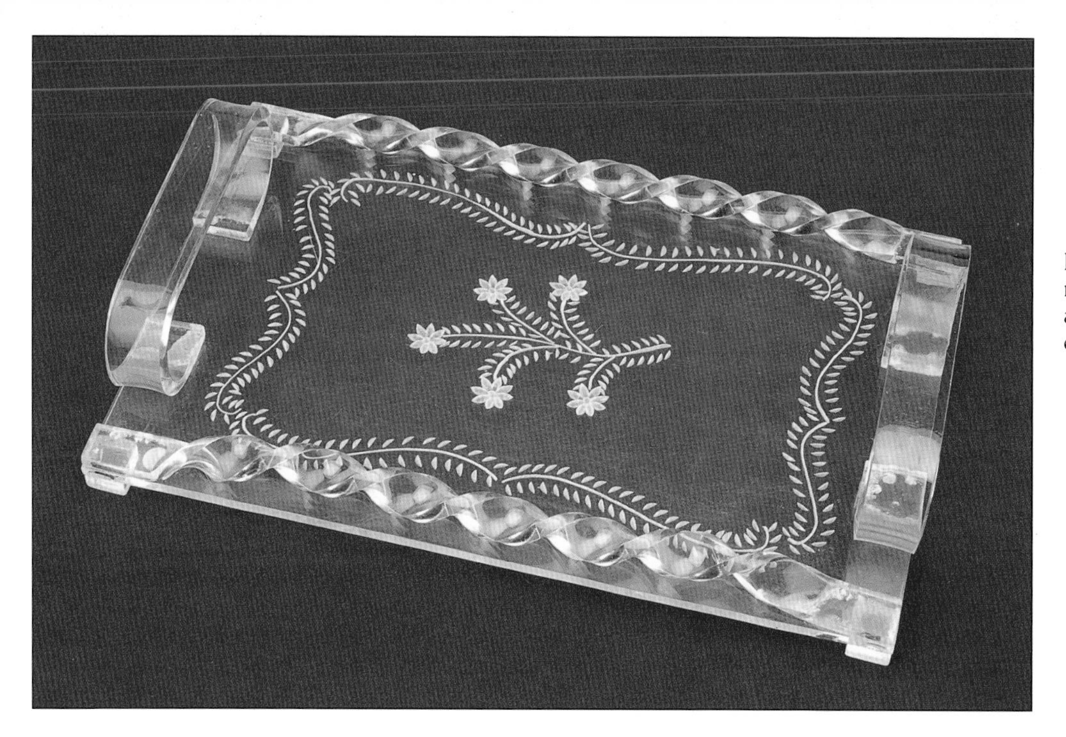

Large handled dresser tray made of clear Lucite with attractive twisted sides and engraved designs. $75-90.

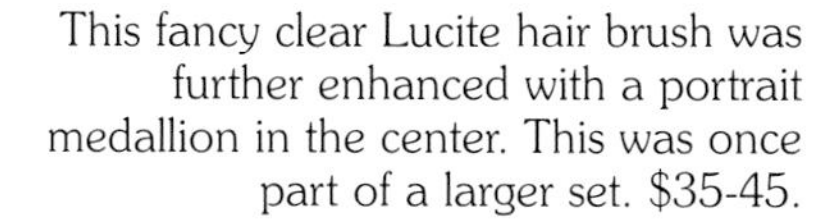

This fancy clear Lucite hair brush was further enhanced with a portrait medallion in the center. This was once part of a larger set. $35-45.

1950s clear Lucite dresser set consisting of a mirror, brush, shoe horn, mirrored tray, powder jar, two smaller jars and cuticle knife. The clear Lucite handles were molded to resemble earlier sets with cut glass handles. $140-180 set.

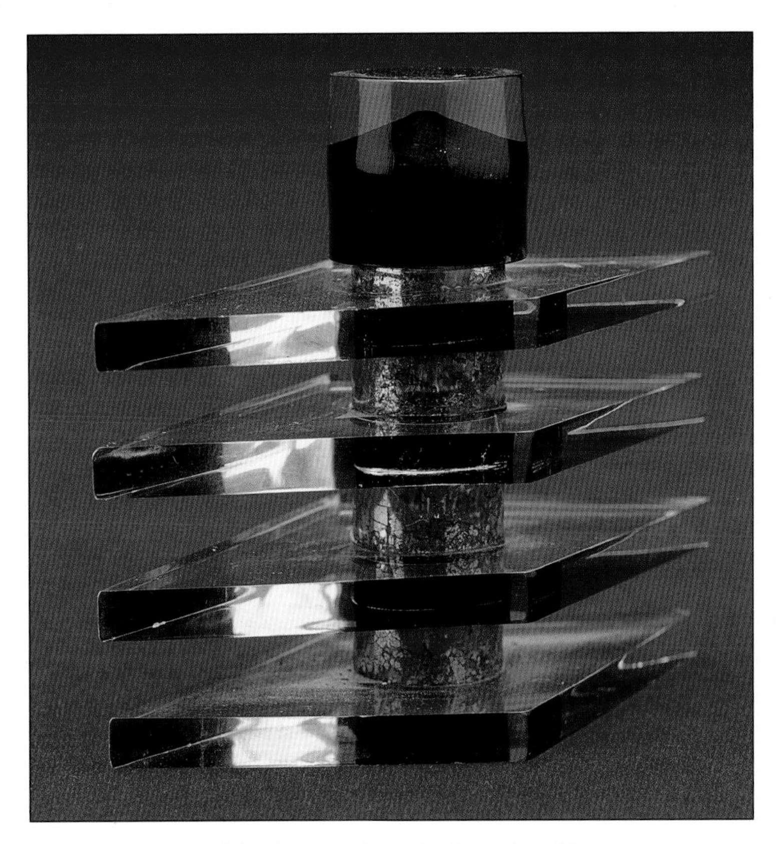

Very unusual Art Deco perfume bottle made of 4 layers of diamond-shaped clear Lucite wrapped around a glass bottle. $145-175.

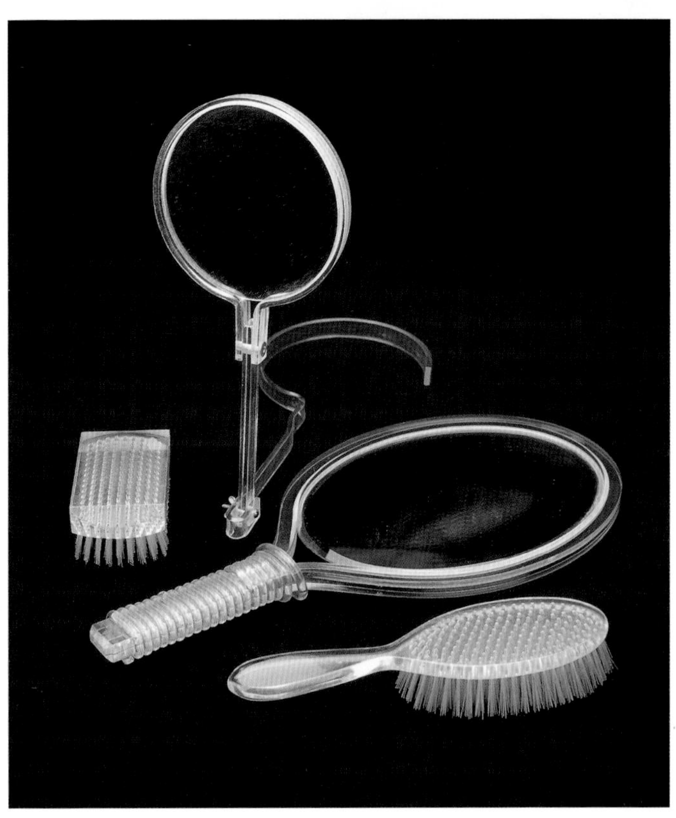

Clear Lucite dresser accessories became all the rage in the 1940s. Some were made fancy while others were very plain. Some were clear while others were made in jewel-like colors. They provided the consumer with an ultra modern look. $25-65 each.

Comb and mirror set for a little girl called *Prim 'n' Pretty* made of blue Lucite and still on its original presentation card. $20-35 set.

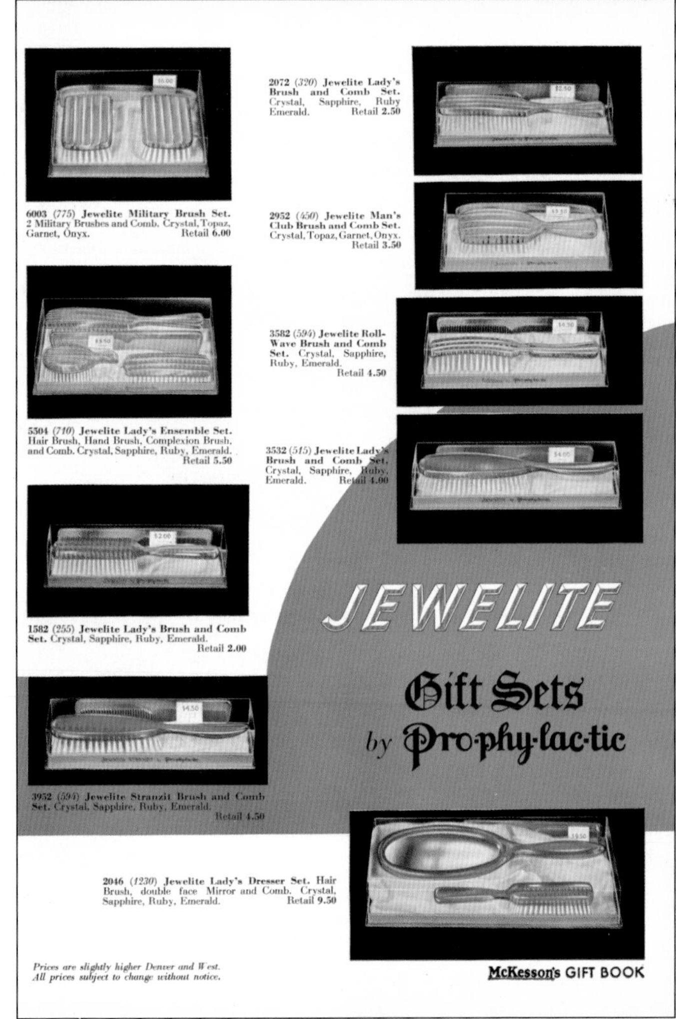

6003 (*775*) **Jewelite Military Brush Set.** 2 Military Brushes and Comb. Crystal, Topaz, Garnet, Onyx. Retail **6.00**

2072 (*320*) **Jewelite Lady's Brush and Comb Set.** Crystal, Sapphire, Ruby Emerald. Retail **2.50**

2952 (*450*) **Jewelite Man's Club Brush and Comb Set.** Crystal, Topaz, Garnet, Onyx. Retail **3.50**

5504 (*710*) **Jewelite Lady's Ensemble Set.** Hair Brush, Hand Brush, Complexion Brush, and Comb. Crystal, Sapphire, Ruby, Emerald. Retail **5.50**

3582 (*594*) **Jewelite Roll-Wave Brush and Comb Set.** Crystal, Sapphire, Ruby, Emerald. Retail **4.50**

3532 (*515*) **Jewelite Lady's Brush and Comb Set.** Crystal, Sapphire, Ruby, Emerald. Retail **4.00**

1582 (*255*) **Jewelite Lady's Brush and Comb Set.** Crystal, Sapphire, Ruby, Emerald. Retail **2.00**

JEWELITE

Gift Sets

by Pro-phy-lac-tic

3952 (*594*) **Jewelite Stranzit Brush and Comb Set.** Crystal, Sapphire, Ruby, Emerald. Retail **4.50**

2046 (*1230*) **Jewelite Lady's Dresser Set.** Hair Brush, double face Mirror and Comb. Crystal, Sapphire, Ruby, Emerald. Retail **9.50**

Prices are slightly higher Denver and West. All prices subject to change without notice.

McKesson's GIFT BOOK

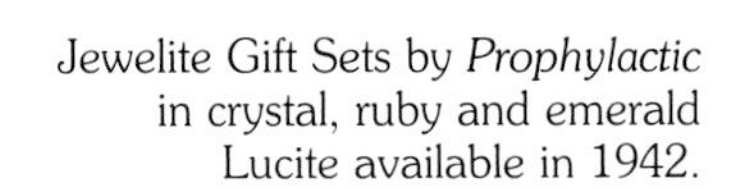

Jewelite Gift Sets by *Prophylactic* in crystal, ruby and emerald Lucite available in 1942.

A novelty in dresser accessories came in the mid-20th century. Perfume atomizers, cologne bottles, lipstick holders, ring holders, dresser trays, cigarette lighters and many other items were skillfully crafted with flowers embedded in clear Lucite. The flowers were actually hand carved and colored. Many items were sold as souvenirs in gift shops across the country. Three major manufacturers of these unique gifts were Bircraft of Huntingdon, Indiana, Jane Art of Elmhurst, New York and Evans Atomizers of Brooklyn, New York. Many of these wonderful Lucite items can still be found today. They make a wonderful collection.

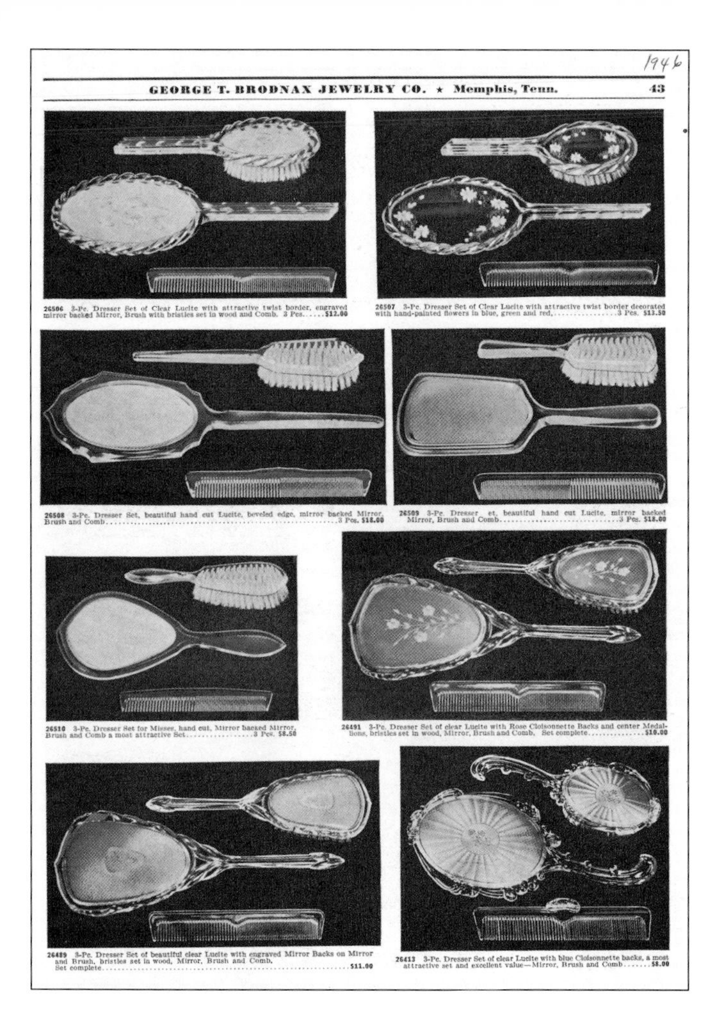

1946

GEORGE T. BRODNAX JEWELRY CO. ★ Memphis, Tenn. 43

26506 3-Pc. Dresser Set of Clear Lucite with attractive twist border, engraved mirror backed Mirror, Brush with bristles set in wood and Comb. 3 Pcs......$12.00

26507 3-Pc. Dresser Set of Clear Lucite with attractive twist border decorated with hand-painted flowers in blue, green and red,..................3 Pcs. $13.50

26508 3-Pc. Dresser Set, beautiful hand cut Lucite, beveled edge, mirror backed Mirror, Brush and Comb..3 Pcs. $18.00

26509 3-Pc. Dresser et, beautiful hand cut Lucite, mirror backed Mirror, Brush and Comb..............................3 Pcs. $18.00

26510 3-Pc. Dresser Set for Misses, hand cut, Mirror backed Mirror, Brush and Comb a most attractive Set..................3 Pcs. $8.50

26491 3-Pc. Dresser Set of clear Lucite with Rose Cloisonnette Backs and center Medallions, bristles set in wood, Mirror, Brush and Comb. Set complete...............$10.00

26489 3-Pc. Dresser Set of beautiful clear Lucite with engraved Mirror Backs on Mirror and Brush, bristles set in wood, Mirror, Brush and Comb. Set complete..$11.00

26413 3-Pc. Dresser Set of clear Lucite with blue Cloisonnette backs, a most attractive set and excellent value—Mirror, Brush and Comb.......$8.00

Clear Lucite dresser sets popular in 1946.

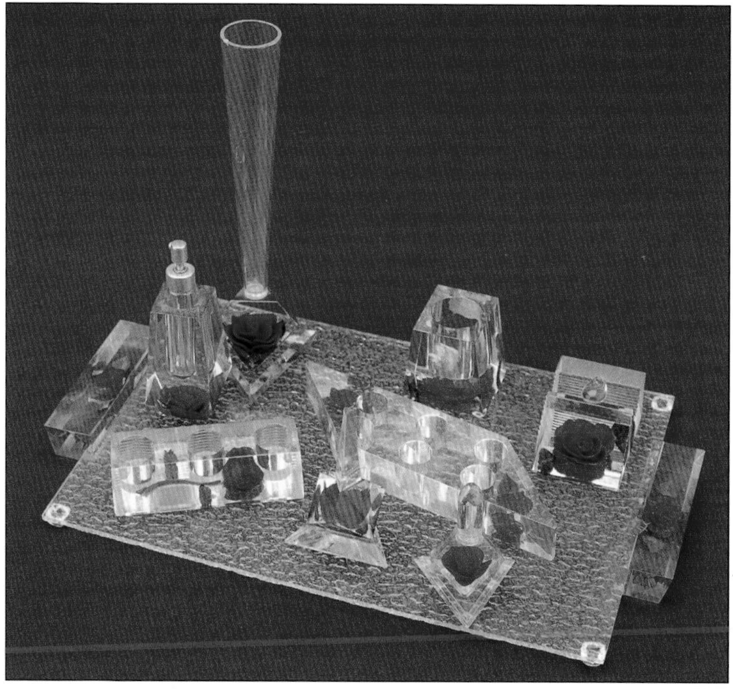

Wonderful assortment of clear Lucite dresser accessories with red roses embedded in each piece. The original labels remain on each piece which read: *Hand carved and colored by BIRCRAFT, Huntington, Indiana.* Dresser tray $85-100; vase $38-45; atomizer $70-95; all other pieces $25-45 each.

Lovely yellow flowers were embedded in these vintage clear Lucite dresser accessories. These pieces were fashioned by *Jane-Art, Elmhurst, New York.* $20-75 each.

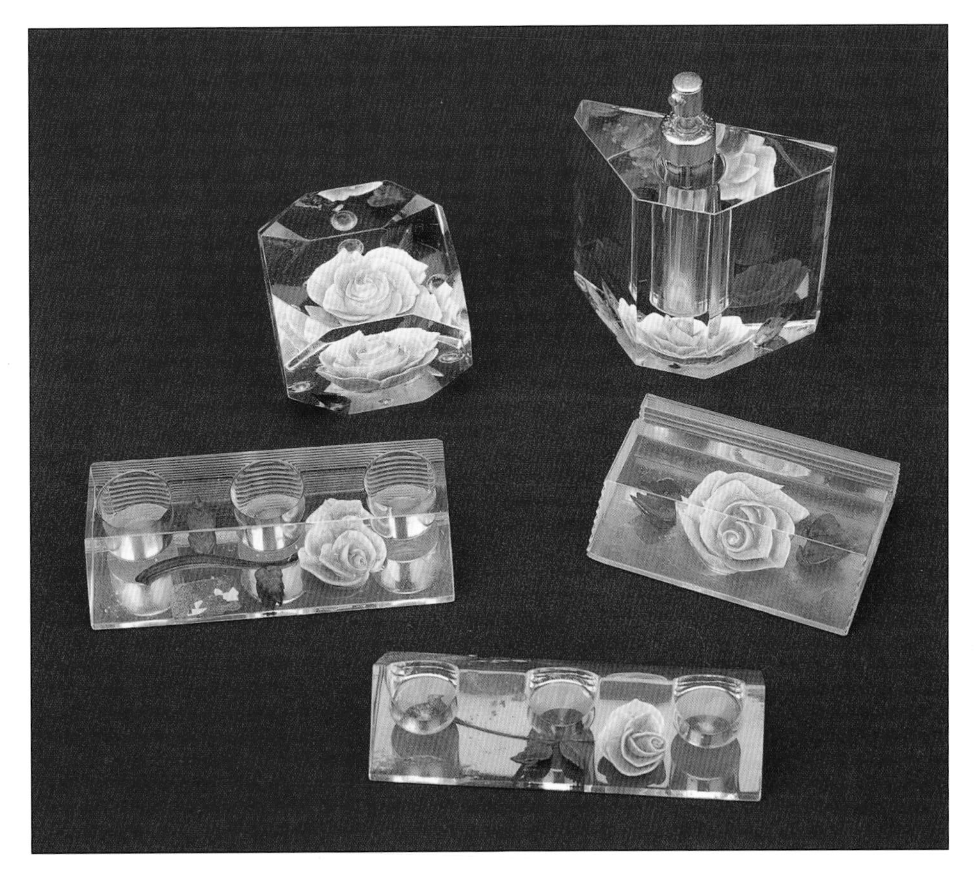

Perfume atomizer, paperweight, two lipstick holders and calling card holder made of clear Lucite with pink flowers hand carved into each piece. $25-75 each.

What makes these clear Lucite atomizers so striking is the added element of black. The diamond-shaped atomizer with the black bulb is labeled *Evans Atomizers Inc., Brooklyn, N.Y.* $85-115 each.

These two clear Lucite perfume atomizers are a little more elaborate than most. A large orchid is embedded into one while the other has a bouquet of red roses. $90-130 each.

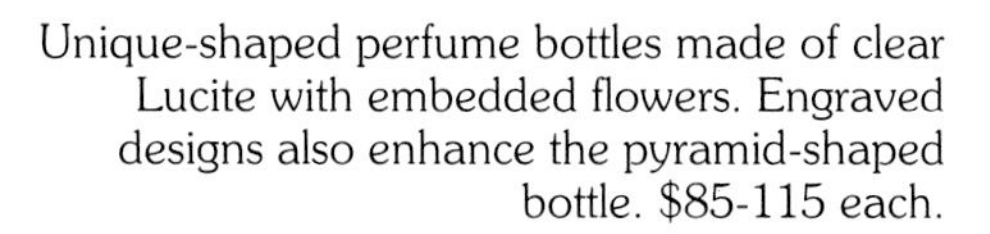

Unique-shaped perfume bottles made of clear Lucite with embedded flowers. Engraved designs also enhance the pyramid-shaped bottle. $85-115 each.

Trio of perfume atomizers made of clear Lucite with different flowers embedded in each one. Two of the atomizers are missing their original bulbs. $65-85 each.

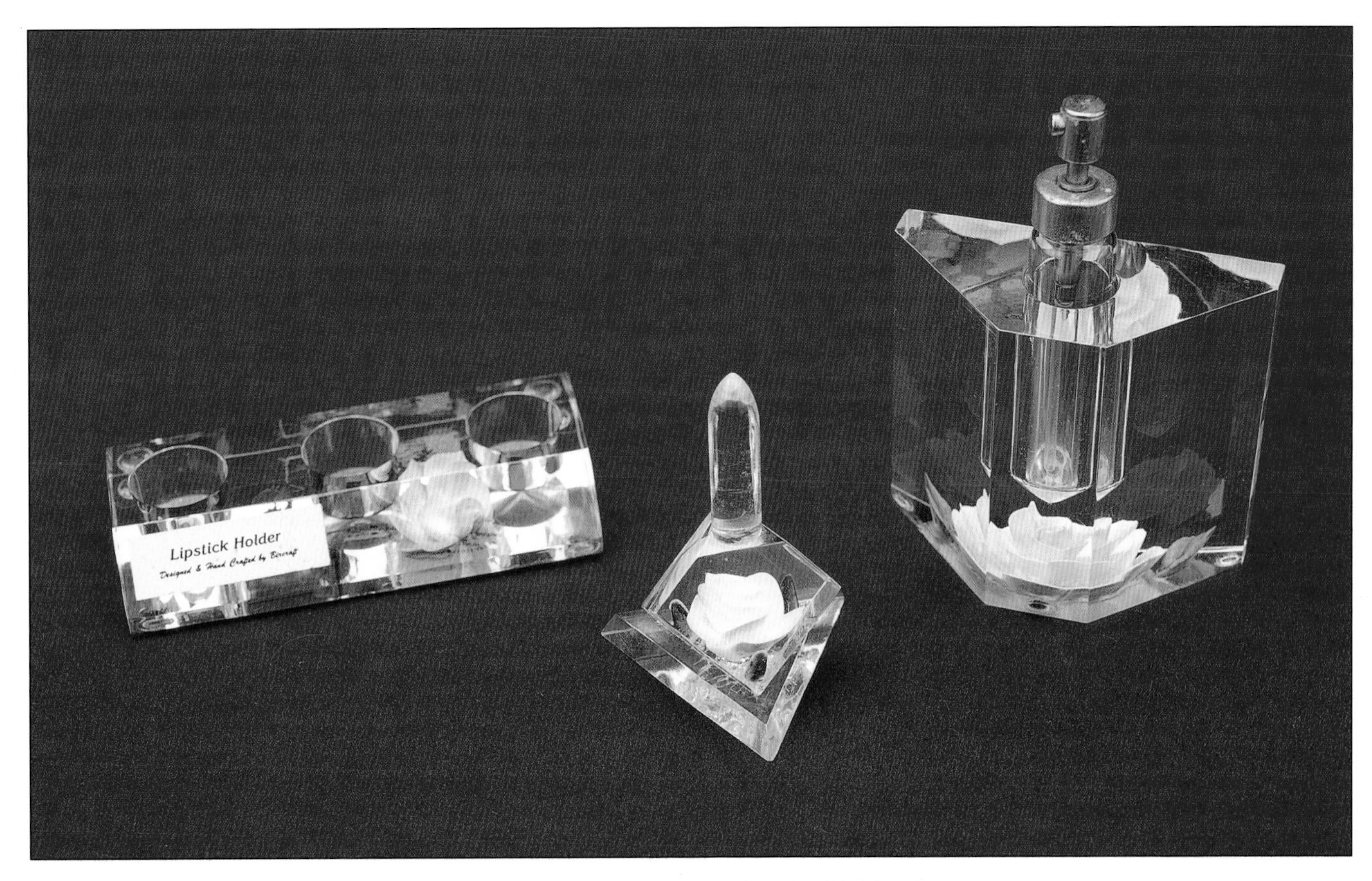

Lipstick holder, ring holder and perfume atomizer made of clear Lucite with embedded hand-carved flowers. The lipstick holder has an original label which reads: *Hand Carved and Hand Colored Genuine Bircraft Originals, Handmade in Huntington, Indiana, Every Flower Unique to Itself.* Ring holder $20-30; Lipstick holder $30-40; Atomizer $65-75.

These clear Lucite atomizers, made in different geometric shapes, were embedded with turquoise-colored exotic flowers and fashioned by *Jane-Art, Elmhurst, New York.* $70-95 each.

Two clear Lucite pincushions, both embedded with flowers and both hand carved and colored by *Bircraft*. $30-40 each.

Besides the usual dresser accessories, desk accessories such as letter openers and paperweights were also hand crafted from clear Lucite and decorated with embedded flowers. $25-60 each.

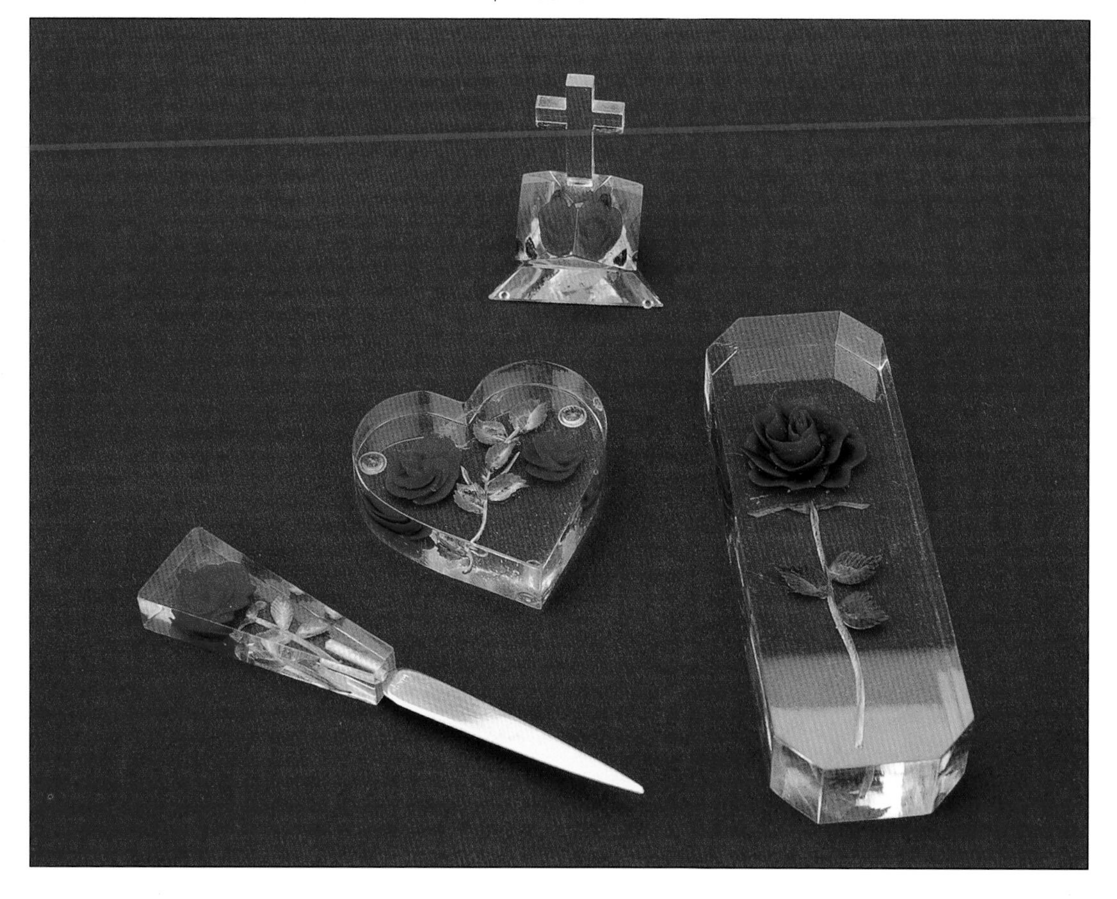

Chapter Three

Enamel Dresser Sets

Fine enamel toilet ware was extremely stylish in the Edwardian era and its popularity continuing into the Art Deco period. Inspiration was drawn from the exquisite Russian Imperial style of Faberge. *Guilloche* or engine turned enameling was a desirable technique and executed on dresser sets in addition to perfume bottles, picture frames, trinket boxes, cigarette cases and other *objets d'art.* Items were exquisitely crafted and sometimes encrusted with jewels.

Dresser clock and nylon bristle hair brush made of sapphire blue enamel with twisted metal borders. I am sure that these two pieces were originally part of a much larger and very elegant dresser set. Clock $85-125; brush $45-65.

The colors used for enamel dresserware were most often jewel tones like sapphire, ruby or emerald. This engine-turned enamel set, although missing some important pieces, is still impressive with its emerald green color. $145-185 set.

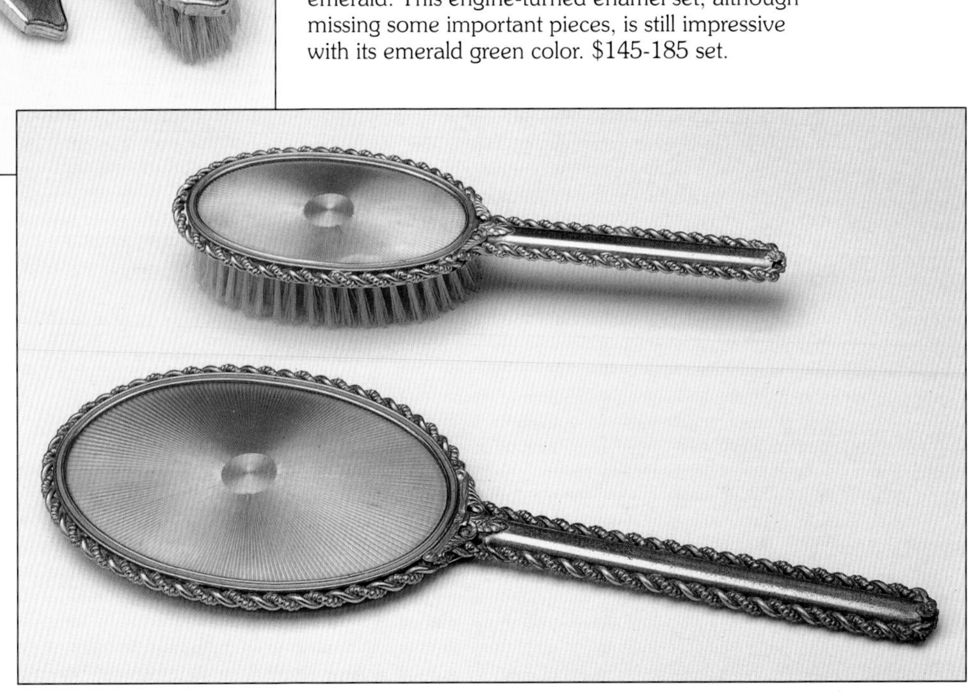

Light blue enamel mirror and brush set with twisted metal trim. This set is almost identical to the sapphire blue set except the handles are different. $135-185 set.

Beautiful three-piece jewel tone dresser set in its original presentation box. The bright green backs are trimmed in gilt metal and accented with glass handles. This style was referred to as *Jewel-Glo* and was especially popular around 1940. $175-195 set.

Lovely jade green enamel dresser set consisting of oval mirror and hair brush accented with gilt trim and fancy cut glass handles. This was a fashionable look in dresser ware in the late 1930s and early 1940s. $170-220 set.

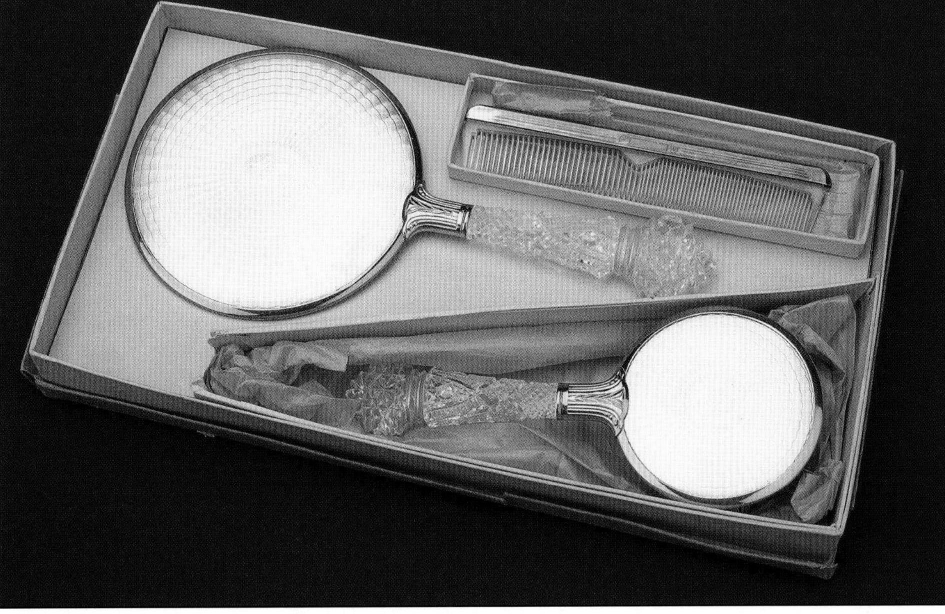

Almost identical to the green Jewel-Glo set, the silvery color backs on these dresser pieces are further enhanced with a flower basket medallion in the center of each piece. The backs of the mirror and brush are made of a thin laminate which resembles *guilloche* enamel. These sets are very similar to the ones made by Astorloid. $125-175 set.

Extravagant enameled dresser sets were manufactured by the Victor A. Picard Company of New York in the early 20th century. They were advertised as *exquisite beauty of contour and graceful restrained decoration.* Pastel shades of enamel were used in a pattern called *Lucienne*. The *Astree* pattern was made with translucent enamel in blue, green, mauve, rose, yellow and grey. The *Rosemarie* pattern was transparent enamel in pastel shades of blue, pink, rose, orchid, yellow and green. A novel design appeared in the *Dranda* pattern which sported a hand-carved centerpiece of either jade, carnelian, amethyst, lapis lazuli or rose quartz surrounded by delicate pastel enamel. The *Drusilla* pattern was described as being *enchantingly reminiscent of the days of the minuet with a hand painted floral ornament over cream-colored enamel.* These sets were exquisite in every detail.

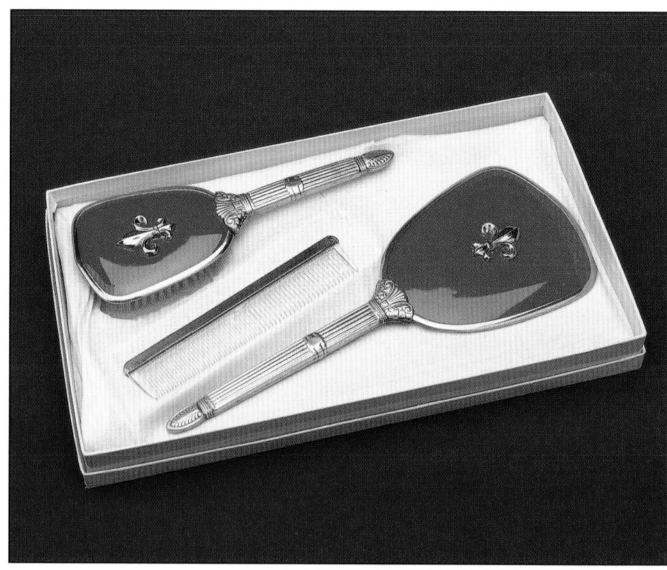

Three-piece dresser set consisting of a brush, mirror and comb made of gold-plated metal and red laminated backs which resemble enamel work. An applied gold-plated *fleur de les* adds the finishing touch. $70-95 set.

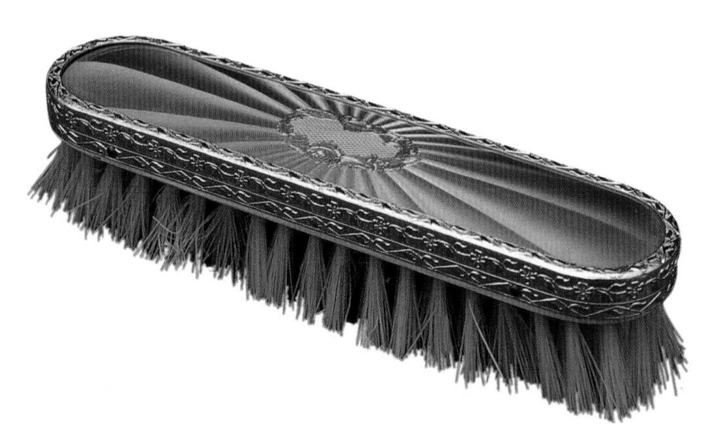

Cloth brush made to look like sterling silver with engine-turned enamel but actually made of aluminum with a plastic laminate. $40-50.

A very popular look is dresser ware in the 1940s were these pretty sets made with gold laminate backs topped with genuine *cloisonné* medallions. $50-75 set.

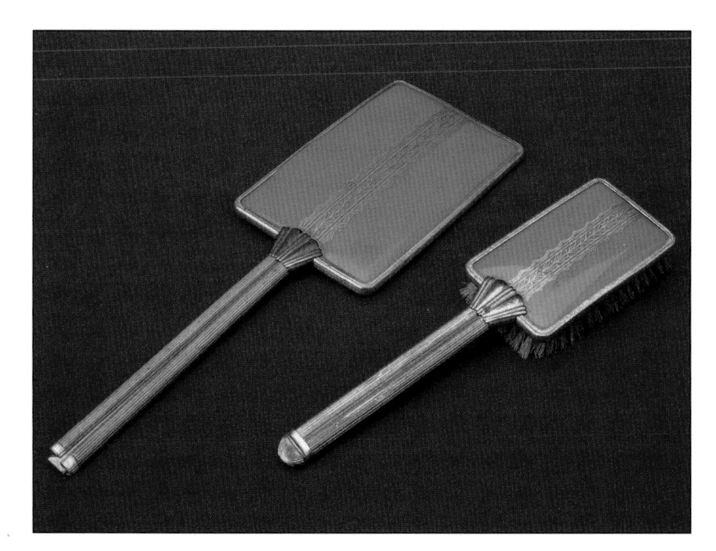

Rectangular-shaped two-piece dresser set made of gold plate with blue backs trimmed in gold. $60-85 set.

Not everyone was able to afford fancy enamel toiletware. Many manufacturers began marketing enamel dresser sets in assorted styles and price ranges to suit most budgets. Some sets were made of sterling silver while others were chromium-plated. In 1930, the May & Malone Company offered dresser sets made of genuine cloisonné on sterling silver. Floral decorations and ribbon and bow designs created a sumptuous look. This was called the *Princess Carol* pattern and it was offered in blue, green, white and maize enamel. The hand mirror alone listed for $58.80. The *Betsy Williams* pattern was also genuine cloisonné on sterling silver with a flower basket motif and a bow design. This style was available in light green, blue or a straw color and the flower basket was done in a harmonizing shade. The mirror for this set listed for $70.00. Although it was exceptionally made, it was still a substantial sum to pay in 1930. These sets were obviously designed for the upper classes. A similar look for less was toiletware plated with non-tarnishing chromium in two-toned enamel with a floral design in the center. A traditional three-piece set in a fancy gift box sold for $21.00 in 1932.

Flower basket motifs decorate this five-piece lavender and chromium-plated dresser set from the 1930s. $95-115 set.

Striking Art Deco three-piece black enamel and chromium-plated set designed with a shield for monogramming. $85-115 set.

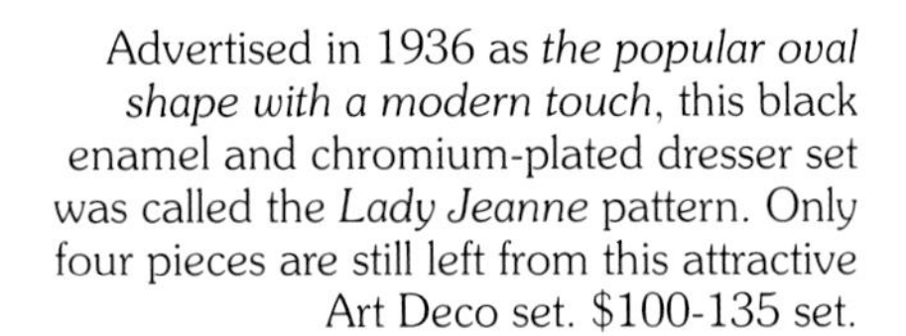

Advertised in 1936 as *the popular oval shape with a modern touch*, this black enamel and chromium-plated dresser set was called the *Lady Jeanne* pattern. Only four pieces are still left from this attractive Art Deco set. $100-135 set.

This two-piece dresser set in its original box was made of chrome and black enamel. Part of the *La Pierre* line of 1935, the set retains wonderful Art Deco appeal. $115-145 set.

	Model No. 900	Model No. 400	Model No. 400A	Model No. 400B	Model No. 1500	Model No. 800	Model No. 700	Model No. 800A	Model No. 1400
All Black	900/1	400/1	400A/1	400B/1	1500/1	800/1	700/1	800A/1	1400/1
Ivory with Black		400/5	400A/5	400B/5	1500/5	800/5	700/5	800A/5	1400/5
Pirate Green with Black		400/6	400A/6	400B/6	1500/6	800/6	700/6	800A/6	1400/6
All Pirate Green	900/7	400/7	400A/7	400B/7	1500/7	800/7	700/7	800A/7	1400/7
Shaded Black	900/9	400/9	400A/9	400B/9	1500/9	800/9	700/9	800A/9	1400/9
Shaded Green	900/10	400/10	400A/10	400B/10	1500/10	800/10	700/10	800A/10	1400/10
All Ivory	900/11								

French enamel dresser sets became desirable boudoir accessories in the mid 1930s in addition to chrome and enamel dresserware manufactured by the International Silver Company made in streamlined Art Deco designs. All black enamel with chrome trim was in vogue as well as ivory with black, Pirate green with black, all Pirate green, shaded black, shaded green and all ivory. Traditional three-piece sets listed for around $12.00 while a ten-piece set could cost in excess of $38.00. Chrome and enamel sets were also manufactured by the Evans Case Company. Evans was advertised as *the worlds' largest manufacturer of style accessories.* Solid black French enamel sets were available in addition to a unique enamel with a pearl moiré effect and another with a simulated gold veined lapis finish. Evans also offered fine sterling silver and enamel sets. A ten-piece set retailed for $64.50 in the early 1930s.

Above and below:
Chrome and enamel dresser ware by the *International Silver Company* of Meriden, Connecticut, which was offered for sale in 1935. Described as "*dresser ware of outstanding craftsmanship*" the enamel was available in assorted colors and combinations of color and the chrome was non-tarnishing.

INTERNATIONAL SILVER CO. MERIDEN, CONN.

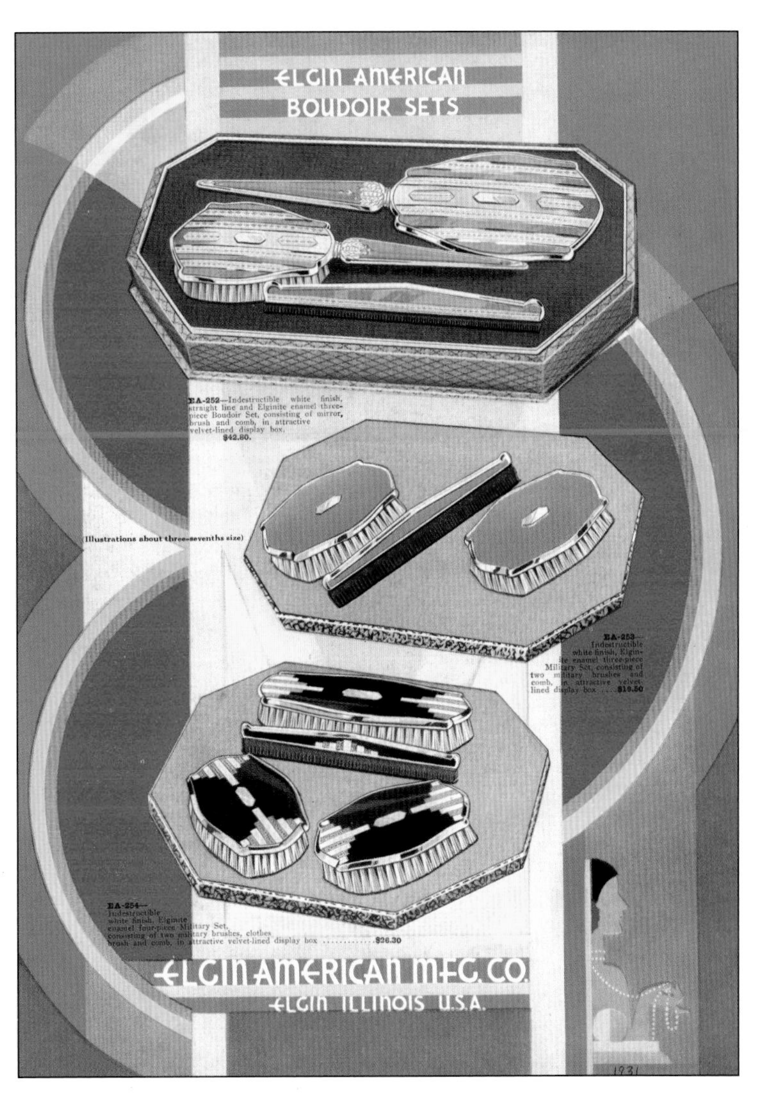

Elgin American Boudoir Sets fashionable in the early 1930s. Described as having an" *indestructible white finish and Elginite enamel,*" these sets had wonderful Art Deco appeal.

Dresser sets were advertised as perfect gifts for birthdays, weddings, anniversaries and graduations. The "Adoration Dresser Set" was non-tarnishing chromium-plate and French enamel set with a thin model mirror with beveled edge. It was offered in six color combinations. The brushes were fitted with the finest quality Russian boar bristles. A few different models were offered with the most popular being an oval shape and then a semi-oblong shape. Each complete set consisting of a brush, mirror and comb sold for $20.00.

Beautiful chrome and enamel dresser sets were also manufactured by Elgin American in the 1930s. Described as *indestructible, white finish Elginite enamel* an eleven-piece set in a velvet-lined presentation box listed for $91.50 in 1931.

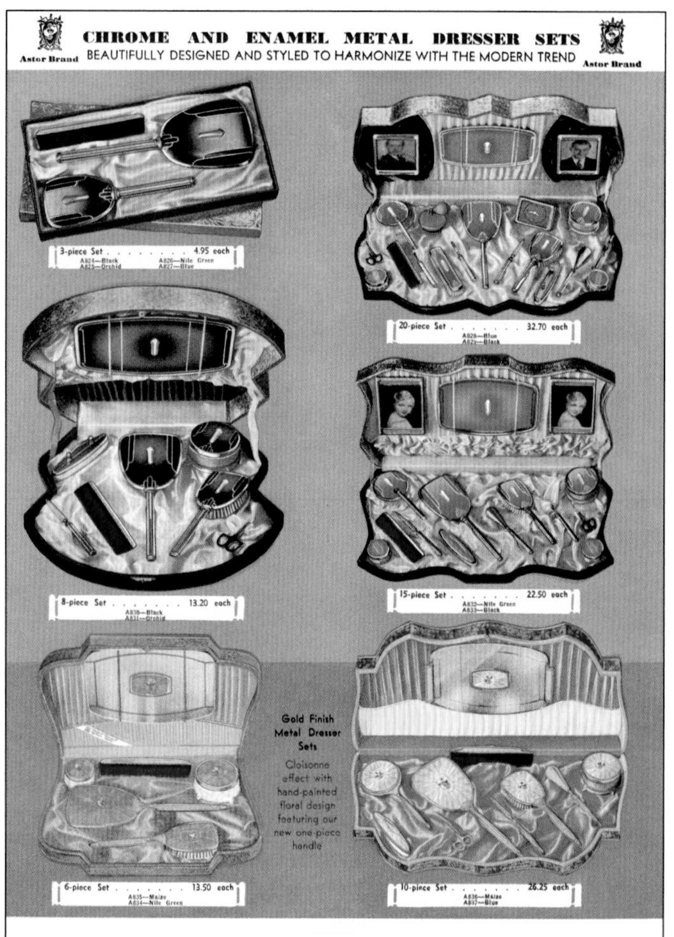

Wonderful assortment of 1938 models of chrome and enamel dresser ware by *Astor Brand* in beautiful fabric-lined presentation boxes. Blue, black, orchid and Nile green were popular colors.

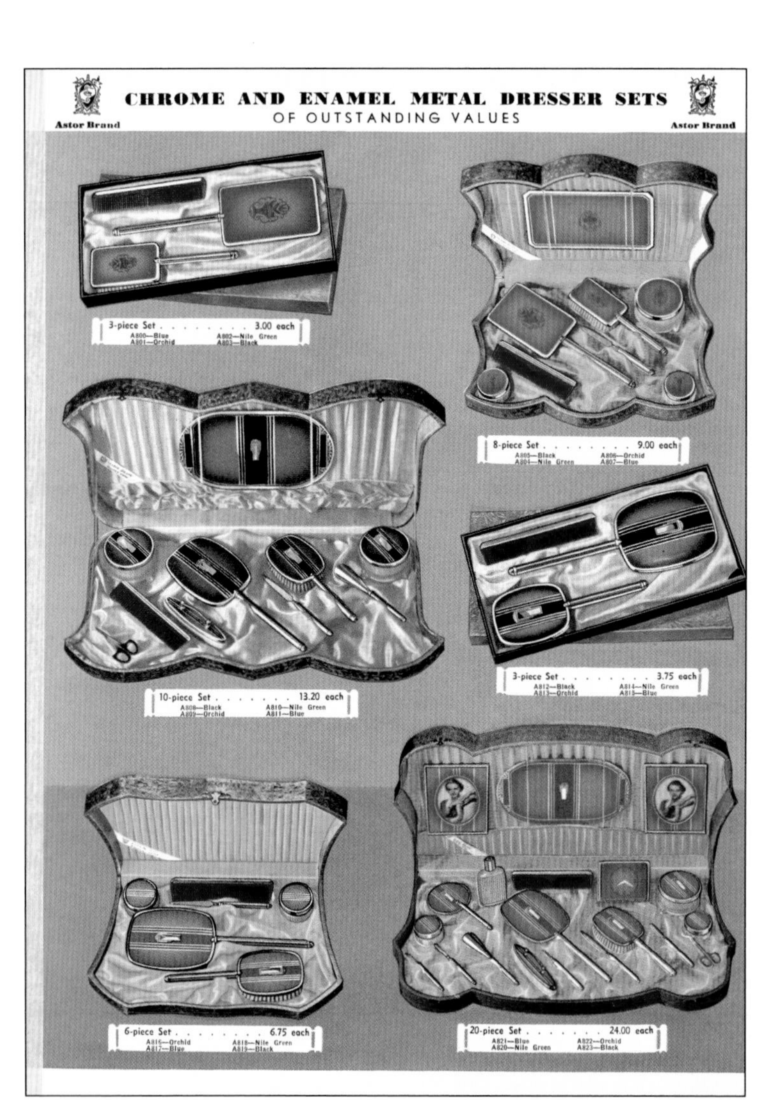

Right:
Complete boudoir sets by *Elgin American* made of chromium plate and enamel. An eleven-piece set in black enamel with a white finish listed for $91.50 in the 1930s.

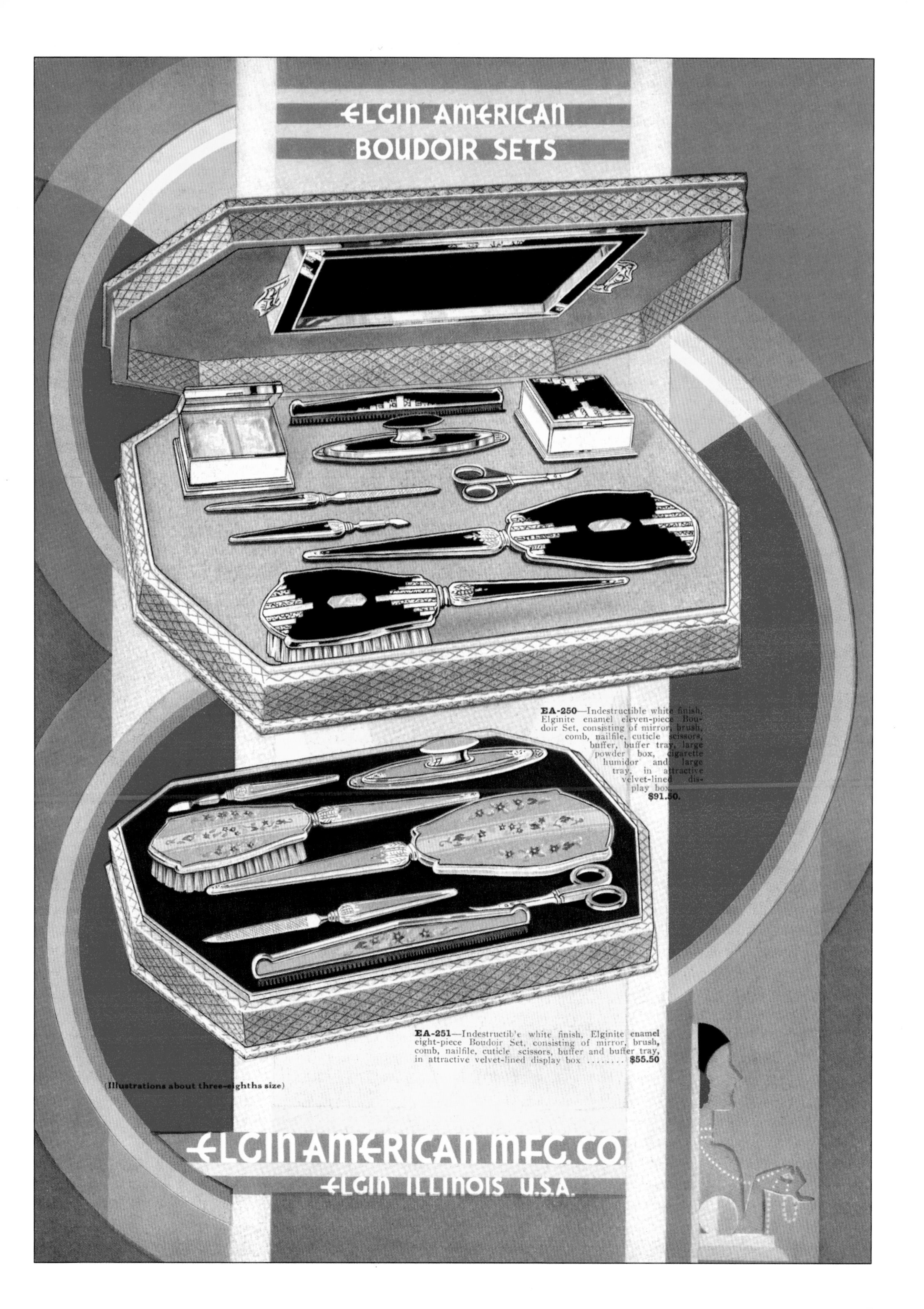

ELGIN AMERICAN
BOUDOIR SETS
EA-250—Indestructible white finish, Elginite enamel eleven-piece Boudoir Set, consisting of mirror, brush, comb, nailfile, cuticle scissors, buffer, buffer tray, large powder box, cigarette humidor and large tray, in attractive velvet-lined display box. $91.50.
EA-251—Indestructib'e white finish, Elginite enamel eight-piece Boudoir Set, consisting of mirror, brush, comb, nailfile, cuticle scissors, buffer and buffer tray, in attractive velvet-lined display box $55.50
(Illustrations about three-eighths size)
ELGIN AMERICAN MFG. CO.
ELGIN ILLINOIS U.S.A.

Dresser sets by Astorloid were common in the late 1930s. Blue, rose and black on ivory were fashionable colors and accented with genuine cloisonné enamel medallions. By 1939, dresser sets by Ramel were in vogue. The enamel on these sets was shaded in harmonizing colors and trimmed in 22K gold. This style was referred to as *Gloria*. The *Regina* model was chromium finished with pastel colors and accented with gold and black decorations. The *Marcelle* was also a 22K gold finished set with a cloisonné ornament. The *Madeline* pattern had a genuine cloisonné shield; the *Lorette* pattern was accented with an etched gold metal plaque and the *Rosalie* was 22K gold with shaded colors and a gold plaque for engraving. The variety was still overwhelming.

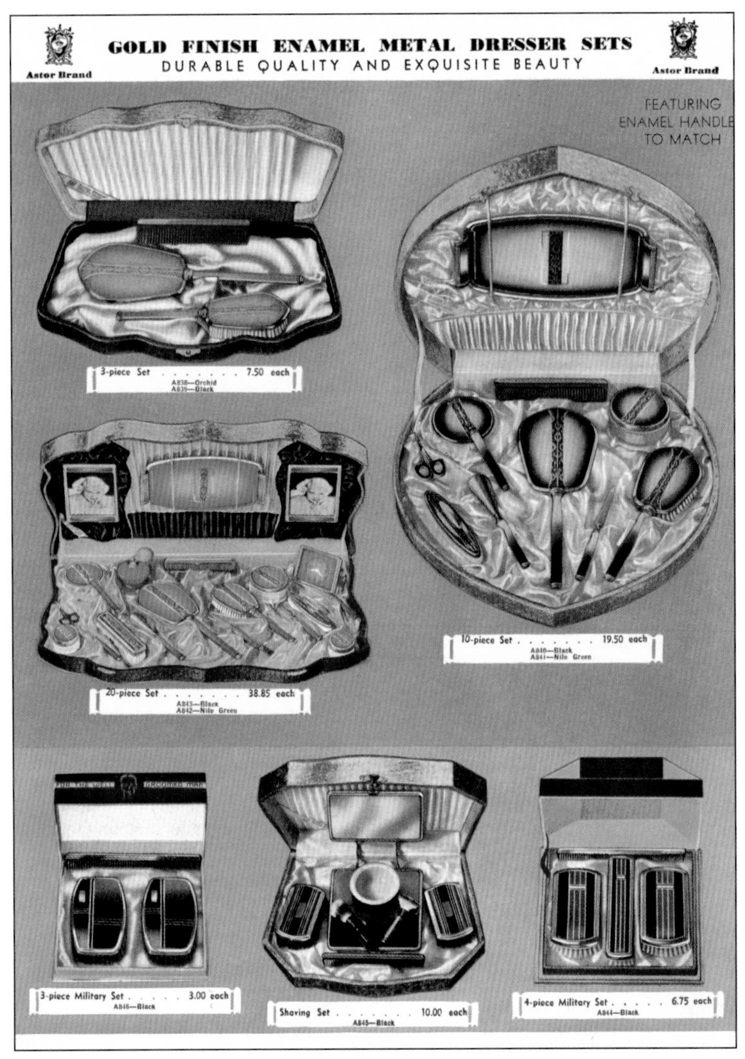

Gold finished enamel dresser sets for women and military and shaving sets for men by *Astor Brand* fashionable in 1938.

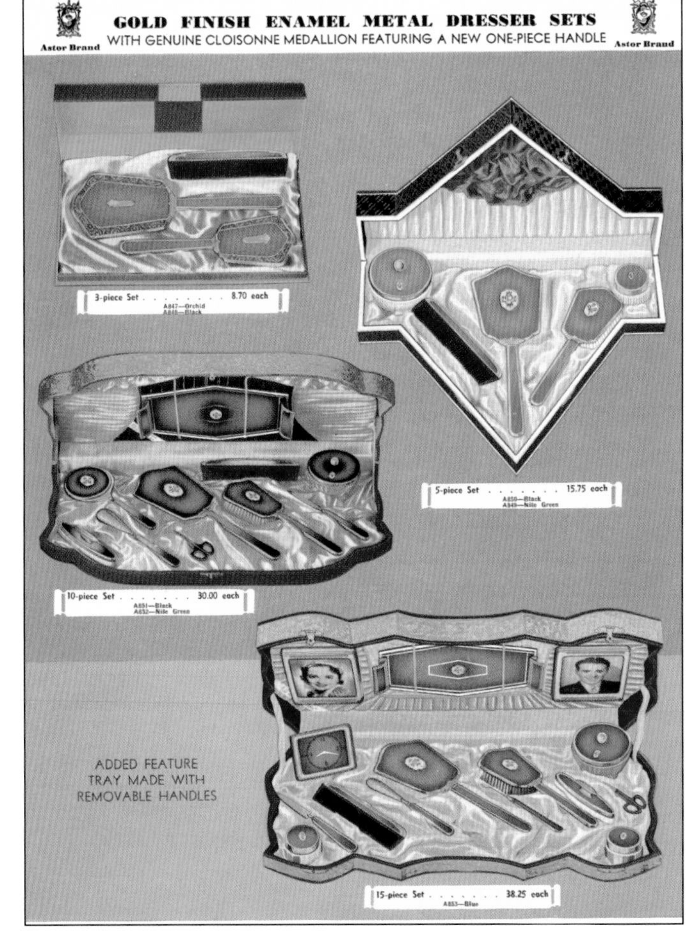

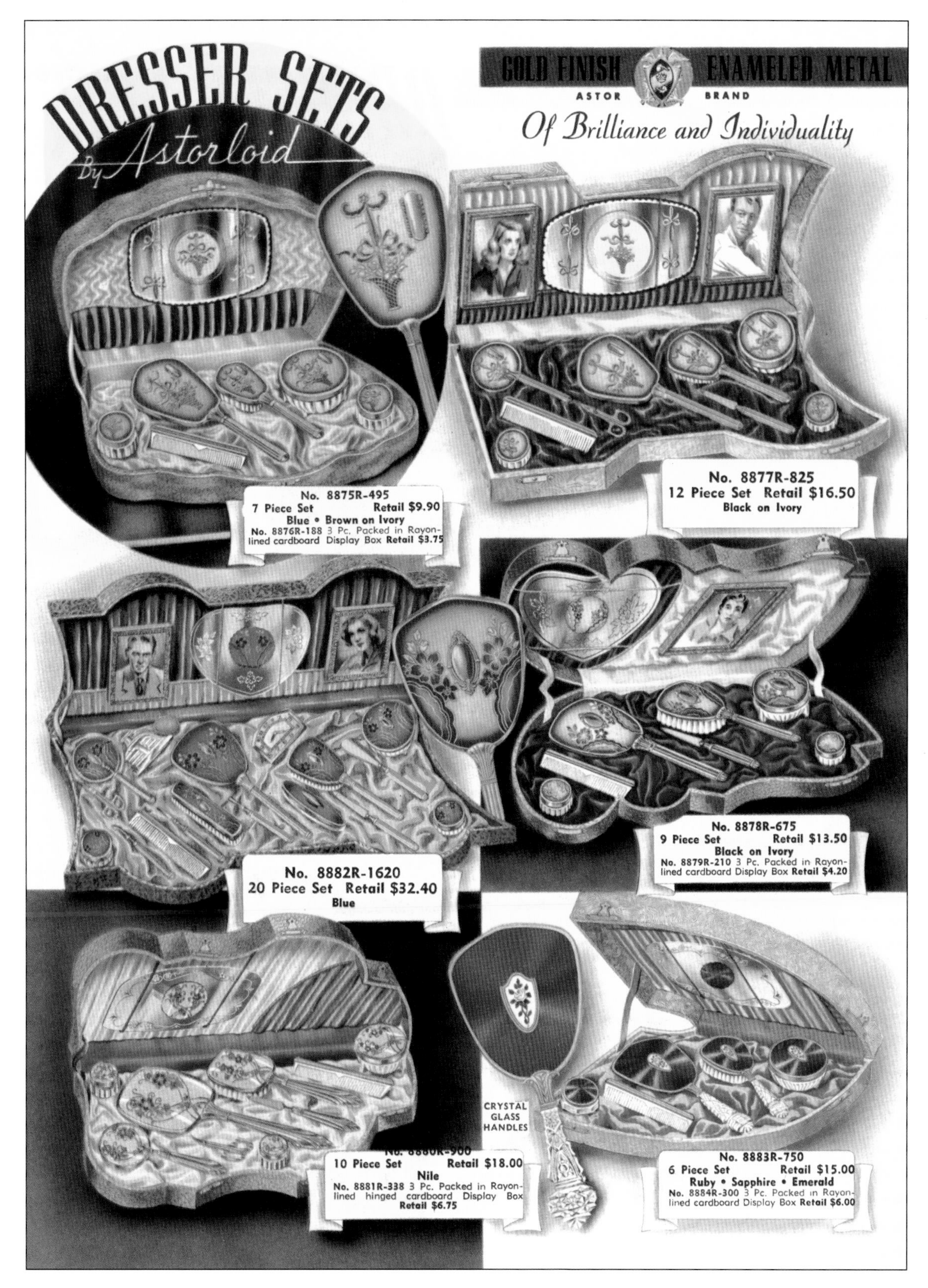

DRESSER SETS
By Astorloid
GOLD FINISH
ENAMELED METAL
ASTOR
BRAND
Of Brilliance and Individuality
No. 8875R-495
7 Piece Set Retail $9.90
Blue • Brown on Ivory
No. 8876R-188 3 Pc. Packed in Rayon-lined cardboard Display Box Retail $3.75
No. 8877R-825
12 Piece Set Retail $16.50
Black on Ivory
No. 8882R-1620
20 Piece Set Retail $32.40
Blue
No. 8878R-675
9 Piece Set Retail $13.50
Black on Ivory
No. 8879R-210 3 Pc. Packed in Rayon-lined cardboard Display Box Retail $4.20
10 Piece Set Retail $18.00
Nile
No. 8881R-338 3 Pc. Packed in Rayon-lined hinged cardboard Display Box Retail $6.75
CRYSTAL GLASS HANDLES
No. 8883R-750
6 Piece Set Retail $15.00
Ruby • Sapphire • Emerald
No. 8884R-300 3 Pc. Packed in Rayon-lined cardboard Display Box Retail $6.00

Gold finished *Astorloid* dresser sets offered for sale in 1942. A six-piece set with crystal glass handles and packaged in a rayon-lined cardboard display box retailed for $15.00.

In the early 1940s, "Jewel-Tone" enamel dresser sets were advertised for sale in mail order and wholesale catalogs. Most were gilt finished with gilt handles and a center shield for a monogram or a cloisonné medallion. Other patterns were referred to as jewel-glo backs with fancy crystal handles. Popular colors were emerald green, ruby red, and sapphire blue: hence the name, Jewel-glo.

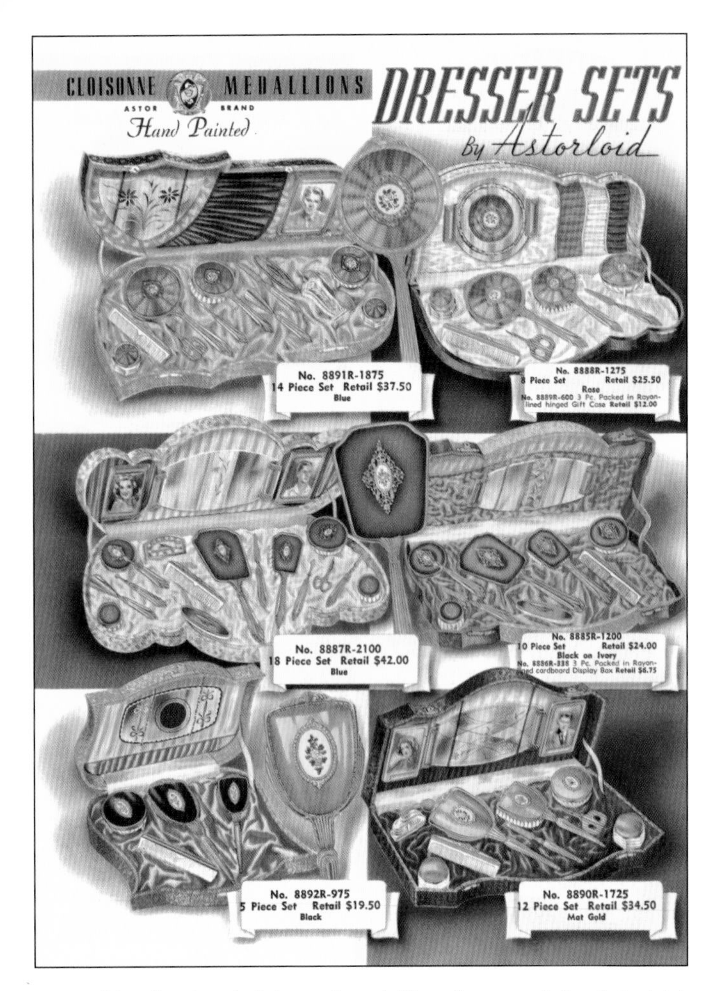

Hand painted cloisonné medallion dresser sets by *Astorloid* in 1942. Notice the touch of patriotism in the one dresser case lined with red, white and blue fabric.

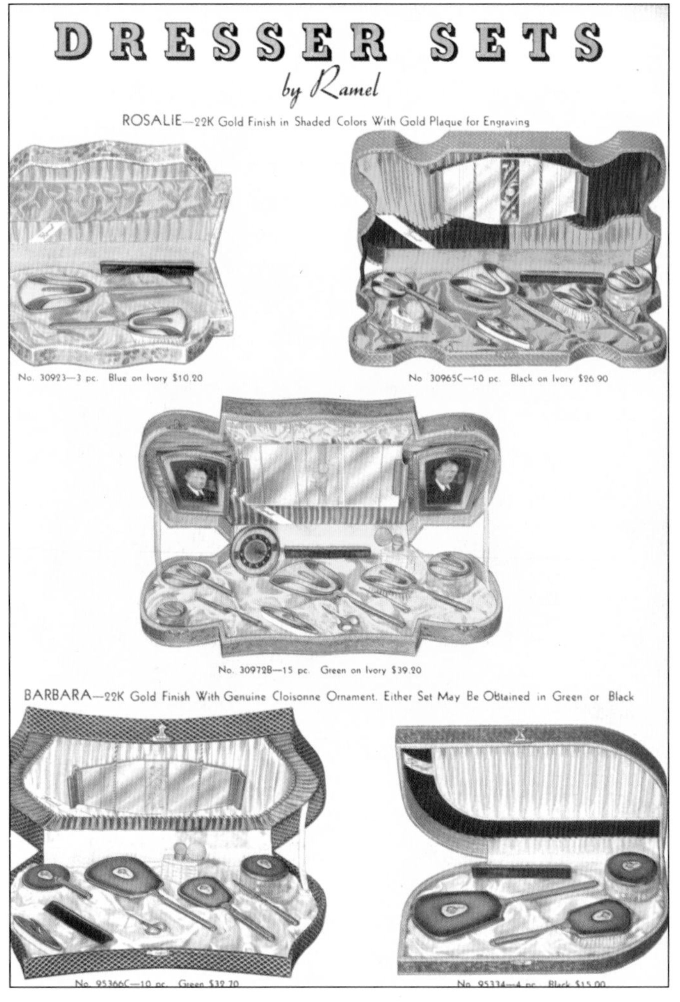

Dresser sets by *Ramel* were popular in 1939. The *Rosalie* pattern was 22K gold finished with shaded colors and a gold plaque for engraving. The *Barbara* pattern was decorated with a genuine cloisonné ornament. Other pattern names included the *Madeline* which was 22K brush gold with a genuine cloisonné shield; the *Lorette* pattern which was 22K gold finished with etched gold metal plaques and many others.

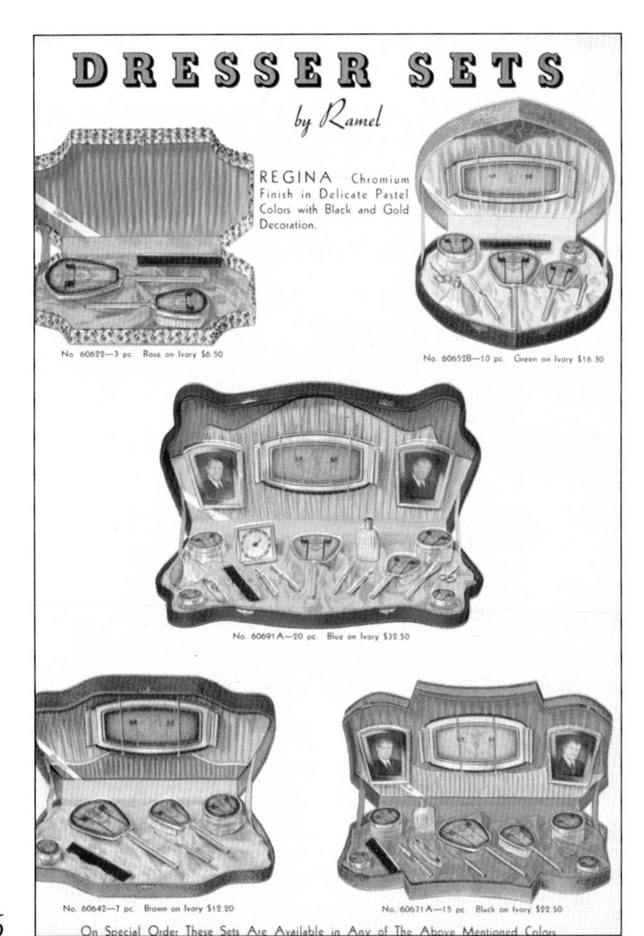

The *Regina* pattern was designed with a chromium finish and pastel enamel trimmed in black and gold. This was another popular look in the late 1930s.

Chapter Four

Glass Dresser Sets

Many books have already been written about the varieties of glass used in the manufacture of toilet ware, dresser items and particularly perfume and scent bottles. This book, however, primarily focuses on metals and the early plastics used for making boudoir accessories. A few glass sets are pictured to show some diversity of what was available in the late 19th and early 20th centuries but books specifically devoted to glass are plentiful and many of them picture wonderful dresser sets made of all types of glass and porcelain. Many American glass companies, especially in the depression years produced endless varieties of ladies toilet items to include powder jars, perfume bottles, trays and trinket boxes. The glass was made clear, frosted, satin, transparent and opaque. It came in all colors and textures. Companies like Fostoria, Westmoreland, Tiffin, Paden City, New Martinsville, Heisey, and Fenton were a few of the more common ones. DeVilbiss, Steuben and Lalique were companies noted for creating exquisite perfume and scent bottles and atomizers that are highly sought after today. Wonderful examples of glass and porcelain were also made by many European glass manufacturers. Bristol glass from England, Bohemian glass from Czechoslovakia, Venetian glass from Italy, Limoges from France and Nippon from Japan have delighted collectors for decades.

Two large Victorian scent bottles with raised and gilded decorations. This type of glass, referred to as opal glass or white opaline glass, was very popular in the late 19th and early 20th centuries. $165-195 pair.

Two unusual-shaped dresser trays made of opal glass with raised and beaded decorations. In the 1890s, trays like this were advertised as being *novelty-shaped*. $135-150 Pair.

Matched dresser set consisting of fancy-shaped tray, puff box and two trinket boxes made of Victorian opal glass with raised and gilded decorations. Notice how most of the gilding is still on one of the trinket boxes but only remnants remain on the other pieces. All four pieces have a floral theme. $195-225 set.

Three wonderful opal glass comb and brush trays with raised and gilded decorations. The tray in the middle is also tinted with a bluish green color. The large tray was made with an ornate lion and scroll pattern. Lion tray $85-100; small trays $60-90 each.

Puff box, trinket box and small handled tray made of opal glass with raised and heavily gilded decorations. Each piece was also hand painted with floral motifs although much of the paint has worn off through the years. tray $50-70; puff box $55-75; trinket box $55-75.

Fancy opal glass collar box and three trinket boxes made in different shapes with raised decorations and gilt trim. Collar box $95-135; trinket boxes $55-75 each.

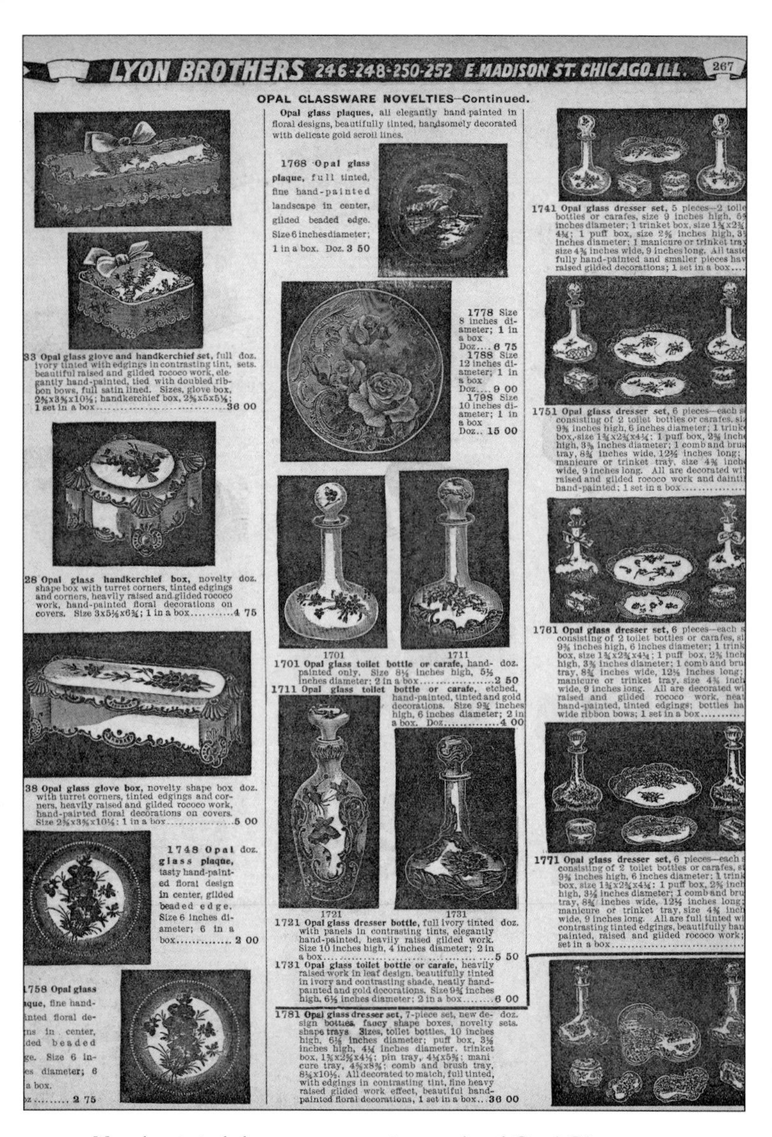

LYON BROTHERS 246-248-250-252 E. MADISON ST. CHICAGO, ILL. 267

OPAL GLASSWARE NOVELTIES—Continued.

Opal glass plaques, all elegantly hand painted in floral designs, beautifully tinted, handsomely decorated with delicate gold scroll lines.

1768 **Opal glass plaque,** full tinted, fine hand-painted landscape in center, gilded beaded edge. Size 6 inches diameter; 1 in a box. Doz. 3 50

1778 Size 8 inches diameter; 1 in a box Doz.... 6 75

1788 Size 12 inches diameter; 1 in a box Doz.... 9 00

1798 Size 10 inches diameter; 1 in a box Doz.. 15 00

33 **Opal glass glove and handkerchief set,** full ivory tinted with edgings in contrasting tint, beautiful raised and gilded rococo work, elegantly hand-painted, tied with doubled ribbon bows, full satin lined. Sizes, glove box, 2⅝x3⅝x10½; handkerchief box, 2⅝x5x5⅛; 1 set in a box. doz. sets. 36 00

28 **Opal glass handkerchief box,** novelty shape box with turret corners, tinted edgings and corners, heavily raised and gilded rococo work, hand-painted floral decorations on covers. Size 3x5⅛x6¾; 1 in a box. doz. 4 75

38 **Opal glass glove box,** novelty shape box with turret corners, tinted edgings and corners, heavily raised and gilded rococo work, hand-pairted floral decorations on covers. Size 2⅜x3⅝x10¼; 1 in a box. doz. 5 00

1748 **Opal glass plaque,** tasty hand-painted floral design in center, gilded beaded edge. Size 6 inches diameter; 6 in a box. doz. 2 00

758 **Opal glass** ...que, fine hand-...inted floral de-...ns in center, ...ded beaded ...ge. Size 6 in-...es diameter; 6 ...a box. ...z 2 75

1701 1711

1701 **Opal glass toilet bottle or carafe,** hand-painted only. Size 8¼ inches high, 5½ inches diameter; 2 in a box. doz. 2 50

1711 **Opal glass toilet bottle or carafe,** etched, hand-painted, tinted and gold decorations. Size 9¾ inches high, 6 inches diameter; 2 in a box. Doz. 4 00

1721 1731

1721 **Opal glass dresser bottle,** full ivory tinted with panels in contrasting tints, elegantly hand-painted, heavily raised gilded work. Size 10 inches high, 4 inches diameter; 2 in a box. doz. 5 50

1731 **Opal glass toilet bottle or carafe,** heavily raised work in leaf design, beautifully tinted in ivory and contrasting shade, neatly hand-painted and gold decorations. Size 9¾ inches high, 6½ inches diameter; 2 in a box. 6 00

1781 **Opal glass dresser set,** 7-piece set, new design bottles, fancy shape boxes, novelty shape trays. Sizes, toilet bottles, 10 inches high, 6⅛ inches diameter; puff box, 3⅛ inches high, 4¼ inches diameter, trinket box, 1⅜x2⅝x4½; pin tray, 4¼x5⅝; manicure tray, 4⅝x8⅜; comb and brush tray, 8¼x10½. All decorated to match, full tinted, with edgings in contrasting tint, fine heavy raised gilded work effect, beautiful hand-painted floral decorations, 1 set in a box. doz. sets. 36 00

1741 **Opal glass dresser set,** 5 pieces—2 toile[t] bottles or carafes, size 9 inches high, 6... inches diameter; 1 trinket box, size 1¾x2¾... 4¼; 1 puff box, size 2⅝ inches high, 3... inches diameter; 1 manicure or trinket tray size 4⅝ inches wide, 9 inches long. All taste[fully] hand-painted and smaller pieces hav[e] raised gilded decorations; 1 set in a box....

1751 **Opal glass dresser set,** 6 pieces—each s[et] consisting of 2 toilet bottles or carafes, si[ze] 9¾ inches high, 6 inches diameter; 1 trinke[t] box, size 1¾x2¾x4¼; 1 puff box, 2⅝ inche[s] high, 3⅝ inches diameter; 1 comb and brus[h] tray, 8¾ inches wide, 12¼ inches long; manicure or trinket tray, size 4⅝ inche[s] wide, 9 inches long. All are decorated wit[h] raised and gilded rococo work and dainti[ly] hand-painted; 1 set in a box.......

1761 **Opal glass dresser set,** 6 pieces—each s[et] consisting of 2 toilet bottles or carafes, si[ze] 9¾ inches high, 6 inches diameter; 1 trink[et] box, size 1¾x2¾x4¼; 1 puff box, 2⅝ inch[es] high, 3⅝ inches diameter; 1 comb and bru[sh] tray, 8¾ inches wide, 12¼ inches long; manicure or trinket tray, size 4⅝ inch[es] wide, 9 inches long. All are decorated wi[th] raised and gilded rococo work, neat[ly] hand-painted, tinted edgings; bottles ha[ve] wide ribbon bows; 1 set in a box..........

1771 **Opal glass dresser set,** 6 pieces—each s[et] consisting of 2 toilet bottles or carafes, s[ize] 9¾ inches high, 6 inches diameter; 1 trink[et] box, size 1¾x2¾x4¼; 1 puff box, 2⅝ inch[es] high, 3½ inches diameter; 1 comb and bru[sh] tray, 8¾ inches wide, 12¼ inches long; manicure or trinket tray, size 4⅝ inch[es] wide, 9 inches long. All are full tinted wi[th] contrasting tinted edgings, beautifully han[d-] painted, raised and gilded rococo work; set in a box..............

Hand painted dresser accessories made of *Opal Glassware* advertised for sale in 1899. Toilet bottles or carafes, trays of all shapes and sizes and boxes with lids were popular for a multitude of uses. Individual pieces were sold in addition to complete sets.

Fancy milk glass toilet ware with artistic finishes and referred to as "Opal Glassware" was very popular in the late 19th and early 20th centuries. Toilet sets, consisting of two large cologne or scent bottles with glass stoppers, unusual-shaped trays, powder jars, trinket and jewel boxes were lavishly made of opal glassware and occasionally gilded, handsomely tinted or hand painted with dainty flowers. Special pieces to hold gloves, handkerchiefs, collar and cuffs were also available. Smoker's sets consisting of tobacco and cigar jars, an ash receiver, match holder and a tray were popular gifts for men. A four-piece set listed for $18.00 in 1899.

Opal glass trays were made in a variety of unusual shapes. They were designed for many uses. Some were manicure trays, some brush and comb trays and some were designed to hold hairpins or even match sticks. Trays were often enameled, hand painted, tinted and gilded. Edges were notched and designs were raised. The glass was hand decorated and borders were fancy. Some trays were described as being Rococo styled. Opal glass puff boxes with lids were usually round; pin boxes with lids were rectangular while trinket boxes were oddly-shaped. All were hand decorated and each one was a little different. The overall look of Victorian and Edwardian opal glassware was quite appealing.

This three-piece 19th century hand painted Bohemian glass dresser set looks similar to Bristol glass. Consisting of two large scent bottles and a powder jar, the set is very attractive with its hand painted designs. The bottles are mainly frosted pink with large fancy stoppers and a similar lid for the powder jar. The bottoms are clear glass. $375-425 set.

Four Victorian revival toilet bottles labeled *Astringent, Cologne, Salts and Cotton* made of hand painted china and studded with clear rhinestones marked *Geo. Lefton 1956*. The original foil labels are still attached to the bottoms of the bottles. The label reads: *Lefton's Exclusives Japan*. $175-225 set.

Two Victorian revival toilet bottles made of hand painted china with violet decorations, trimmed in gold and studded with rhinestones. This set is marked *Norcrest Fine China Japan*. $85-135 pair.

Two perfume bottles and a matching powder jar from the 1930s made of an embossed diamond pattern in clear glass and accented with pink glass stoppers and a lid. $115-165 set.

Four-piece Art Deco dresser set consisting of two perfume bottles, powder jar and a matching tray made of jadeite and black glass by the *New Martinsville Glass Manufacturing Company*. Because of the stark geometric design of this fantastic set, it is very hard to find these pieces in mint condition. This was a very popular look in 1928 and 1929 and very desirable today. $295-345 set.

Clear glass powder jar with lid and fan-shaped finial plus three small trinket boxes made by the *New Martinsville Glass Company* in 1930. $100-150 set.

An unusually large assortment of atomizers, scent bottles, powder jars and a tray made of pink and white porcelain dating back to the 1960s. Scent bottles $75-95 each; powder jars $50-75 each; atomizers $85-125 each; tray $35-50.

Gorgeous three-piece 20th century Bohemian glass dresser set consisting of two large scent bottles and a powder jar made of pink glass ornately decorated with gold and enamel. $275-325 set.

Hand painted decorations and seashell-shaped ornaments enhance this lovely two-piece dresser set from the 1950s marked *Arnart Creation Japan*. $85-125 set.

Pressed glass powder jar with jade pearl-on-amber pyralin lid accented with inlaid designs. $40-55.

Three crystal powder jars with fancy embossed silver-plated, gold-plated and aluminum lids. The aluminum and silver-plated lids display ornate Art Nouveau motifs; one is monogrammed. $85-145 each.

Lovely hand-painted flowers decorate this two-piece porcelain dresser set made up of a powder jar and hair receiver. Both pieces are stamped *Made in Japan*. $80-100 pair.

Unfortunately, the bottom half of this Art Deco powder jar is missing but the top half depicts a lovely flapper and was well worth keeping. $70-90.

Unique china doll powder jar designed with the top half of the body as the lid and the bottom half are the legs The jar is marked *Bavaria*. $150-200.

1920s ceramic lady head powder jar with exaggerated facial features and a bouffant hairdo with a hairnet. The ruffled collar almost looks clown-like. $95-135.

Large powder jar with china doll finial. The lid is designed to look like the dolls dress. Gold accents add the finishing touches. The bottom is stamped *Taube China Hand painted.* $135-165.

Five-piece *Nippon* dresser set consisting of two perfume bottles, large scent or astringent bottle, powder jar and trinket box made of hand-painted porcelain. This set was made in the 1980s. $150-200 set.

Notice how the two perfume bottles fit nicely on the tray and the tray turns into the powder jar. A beautiful angel finial sits on top of the lid. This lovely angelic ceramic set is labeled *Royal Sealy Japan.* $55-75 set.

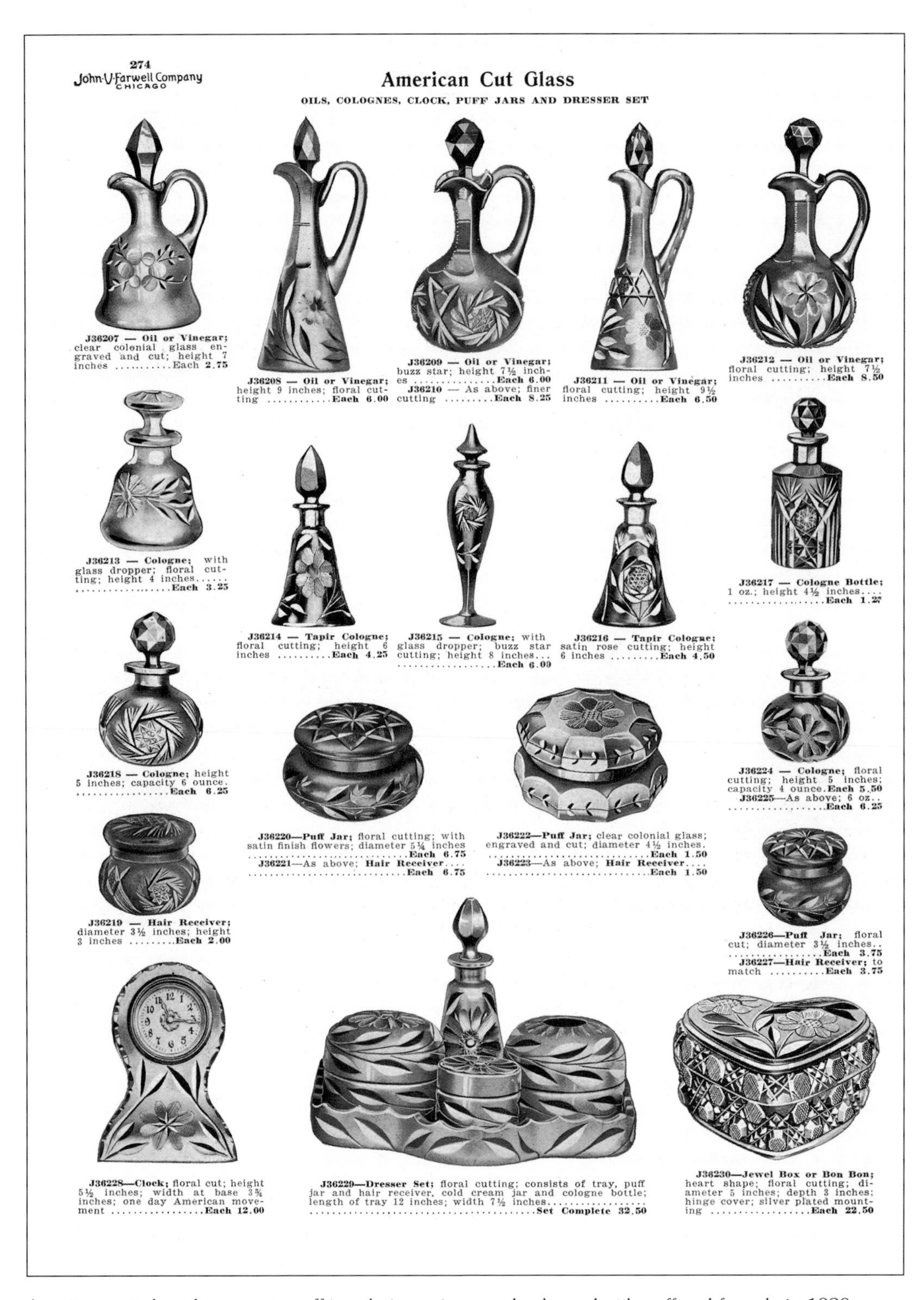

274
John-V-Farwell Company CHICAGO
American Cut Glass
OILS, COLOGNES, CLOCK, PUFF JARS AND DRESSER SET
J36207 — Oil or Vinegar; clear colonial glass engraved and cut; height 7 inches ....Each 2.75
J36208 — Oil or Vinegar; height 9 inches; floral cutting ....Each 6.00
J36209 — Oil or Vinegar; buzz star; height 7½ inches ....Each 6.00
J36210 — As above; finer cutting ....Each 8.25
J36211 — Oil or Vinegar; floral cutting; height 9½ inches ....Each 6.50
J36212 — Oil or Vinegar; floral cutting; height 7½ inches ....Each 8.50
J36213 — Cologne; with glass dropper; floral cutting; height 4 inches ....Each 3.25
J36214 — Tapir Cologne; floral cutting; height 6 inches ....Each 4.25
J36215 — Cologne; with glass dropper; buzz star cutting; height 8 inches ....Each 6.00
J36216 — Tapir Cologne; satin rose cutting; height 6 inches ....Each 4.50
J36217 — Cologne Bottle; 1 oz.; height 4½ inches ....Each 1.25
J36218 — Cologne; height 5 inches; capacity 6 ounce. ....Each 6.25
J36220—Puff Jar; floral cutting; with satin finish flowers; diameter 5¼ inches ....Each 6.75
J36221—As above; Hair Receiver ....Each 6.75
J36222—Puff Jar; clear colonial glass; engraved and cut; diameter 4½ inches. ....Each 1.50
J36223—As above; Hair Receiver ....Each 1.50
J36224 — Cologne; floral cutting; height 5 inches; capacity 4 ounce. Each 5.50
J36225—As above; 6 oz. ....Each 6.25
J36219 — Hair Receiver; diameter 3½ inches; height 3 inches ....Each 2.00
J36226—Puff Jar; floral cut; diameter 3½ inches ....Each 3.75
J36227—Hair Receiver; to match ....Each 3.75
J36228—Clock; floral cut; height 5½ inches; width at base 3¾ inches; one day American movement ....Each 12.00
J36229—Dresser Set; floral cutting; consists of tray, puff jar and hair receiver, cold cream jar and cologne bottle; length of tray 12 inches; width 7½ inches ....Set Complete 32.50
J36230—Jewel Box or Bon Bon; heart shape; floral cutting; diameter 5 inches; depth 3 inches; hinge cover; silver plated mounting ....Each 22.50

American cut glass dresser sets, puff jars, hair receivers and cologne bottles offered for sale in 1920.

American cut glass was also very desirable in the late 19th and early 20th centuries. Complete dresser sets consisting of a tray, puff jar, hair receiver, cold cream jar and cologne bottle with a fancy floral cutting sold for $32.50 in 1920. Diamond cuts and star cuts were also available. Individual pieces could be purchased and added to the basic set. In 1923, a four-piece dresser set made of the highest quality American cut glass was described as a *beautiful design which adds grace to any dresser setting. Particularly in the evening when its remarkable rainbow hues emanate their elusive rays, it has an individuality of warmth and splendor.* Cut glass was also advertised to harmonize with silver toiletware as well as French Ivory Pyralin.

341

Blue-White Diamond Radiant Rich American Cut Glass Puff, Handkerchief and Glove Boxes, Hair Receivers, Jewel Cases and Colognes

Best Quality Leaded Blanks   Mastership Cutting and Expert Workmanship

Women adore such lovely cut glass dresser articles as these two complete sets shown. Harmonize splendidly with silver or ivory toilet pieces. Make charming Christmas or Anniversary Gifts.

PUFF BOXES
No. S244 4 inch, each.......$6.75
No. S245 5 inch, each....... 7.50

HAIR RECEIVERS
No. S246 4 inch, each.......$6.75
No. S247 5 inch, each....... 7.50

JEWEL CASES
White Silk Lining
No. S248 6 inch, each.........$11.25
No. S249 7 inch, each......... 14.25
No. S250 8 inch, each......... 18.00

COLOGNE
Drop Stopper
No. Each
S251 2 oz..$6.75

HANDKERCHIEF BOX
White Silk Lining
No. S252 6½ inches square, each.....$30.00

GLOVE BOX
White Silk Lining
No. S253 Length 10½ inches, width 4½ inches, each........$30.00

COLOGNE
Each
No. S254 4 oz..$ 7.50
No. S255 6 oz.. 9.00
No. S256 8 oz.. 11.25
No. S257 12 oz.. 12.75

A New Creation That is Most Exquisitely Cut in a Mastership Fashion

PUFF BOX
No. S258 5 inch, each....$10.50

HAIR RECEIVER
No. S259 5 inch, each....$10.50

JEWEL CASES
White Silk Lining
No. S260 6 inch, each............$11.25
No. S261 7 inch, each............ 15.00
No. S262 8 inch, each............ 19.50

HANDKERCHIEF BOX
White Silk Lining
No. S263 6½ inches square, each........$31.50

GLOVE BOX
White Silk Lining
No. S264 Length 10½ inches, width 4½ inches, each.......$31.50

COLOGNE
Drop Stopper
No. Each
S265 4 oz.....$7.50
S266 6 oz..... 9.00

Beautiful cut glass boxes, hair receivers and cologne bottles in assorted shapes and sizes offered for sale in 1923. Cut glass was being promoted as a perfect item to use in harmony with silver or ivory dresser pieces.

PERFUME BOTTLE GIFT SETS

Excellent gift suggestions in inexpensive items appreciated by every woman.
Our own importations make these low prices possible.

Prices Subject to Wholesale Discounts. See Page 1.

EL6960 4 Piece Gift Set in Box............................ **$10.00**
High grade heavy cut crystal. Two cut perfume bottles with ground glass stoppers, 5½ inches high, powder jar 4 inches in diameter, plateau mirror 9 x 6 inches. Decorated covered wood hinged gift box, lined with fine celanese in attractive harmonizing colors. Box size 12¼ x 8¾ x 4 inches. Colors: White, Blue, Rose. Splendid gift set.

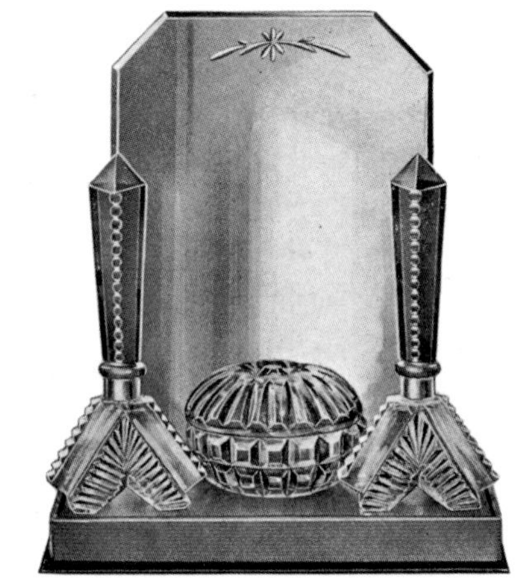

EL6961 4 Piece Vanity Set.................................. **$9.00**
Fine quality heavy cut crystal. Two exquisite perfume bottles with ground glass stoppers, 6¼ inches high. Powder jar 3½ inches in diameter. Large plateau mirror sets in back, 6 x 9 inches. Supplied in cut out gift box. Colors: Crystal in combination with maize, blue or rose.

EL6962 Crystal Perfume and Powder Jar Set.......................... **$8.00**
Finest quality heavy cut crystal. Two crystal perfume bottles, 4¾ inches high, ground glass stoppers. Round powder jar, 2½ inches high. Colors: White, Rose, Blue.

EL6963 Crystal Perfume Set, 4 Pieces............. .... **$6.50**
Finest quality heavy cut crystal. Three square clear crystal bottles, 5 inches high, each with a top of a different color; ground glass stoppers. Nested into an attractive tray, 6¾ inches long. Splendid gift set. Colors: White, Rose, Blue.

EL6964 3 Piece Vanity Set.. **$5.00**
Fine quality heavy cut crystal perfume bottle with ground glass stopper, 4¾ inches high; atomizer 3⅛ inches high. Mirror tray 7 x 4 inches. Supplied in cut out gift box. Colors: White, Rose or Blue.

EL6965 2 Piece Crystal Gift Set.. **$5.00**
Fine quality heavy cut crystal perfume bottle with ground glass stopper, 6½ inches high. Double vanity mirror, one side magnifying, with crystal handle to match. Attractive hinged decorated box with fine celanese lining to harmonize. Size 7¼ x 5¼ inches.

Prices Subject to Wholesale Discounts.

L. & C. MAYERS CO. FIFTH AVE., NEW YORK

[329]

Three and four-piece perfume bottle sets became great gift items in the 1930s. Numerous glass companies began marketing sets similar to these for years.

Chapter Five

Boxed Sets

Fancy boxes designed to house dresser sets reached their peak in the 1930s. Huge unusual-shaped presentation boxes, either velvet, satin, silk, rayon or sateen-lined created the perfect backdrop for the wonderful creations that saturated the market in the early 20th century. These sets created *real eye appeal* and made perfect gift items for women for all occasions. Sets made of sterling silver, silver plate, Celluloid, Pyralin, Lucite, enamel and more were placed within the folds of the sumptuous fabric. The way the fabric was applied to the interior of the box was also done with skill and expertise. Tucks, folds, gathers and draped fabric created luxurious interiors. The exterior of the boxes were sometimes covered in embossed foil in rich jewel colors while other boxes were covered in floral paper or paper with geometric designs similar to wallpaper.

Sterling silver dresser sets offered for sale in 1935 from the Benjamin Allen Company of Chicago, were housed in either silk moiré-lined two-toned suede crackled cases or velvet-lined gray moreen covered cases. Large sets of fourteen or more pieces were housed in a large case with a drawer. Each piece of the set had its own designated area when it snugly fit into place.

Magnificent ten-piece dresser set in its original presentation box made of translucent green and solid jet black Pyralin. Each piece is engraved and hand colored in yellow, green and silver. There are no manufacturers' marks anywhere on the set or the box but it is most likely an *ART-Y-ZAN* dresser set from the 1930s. $500-750.

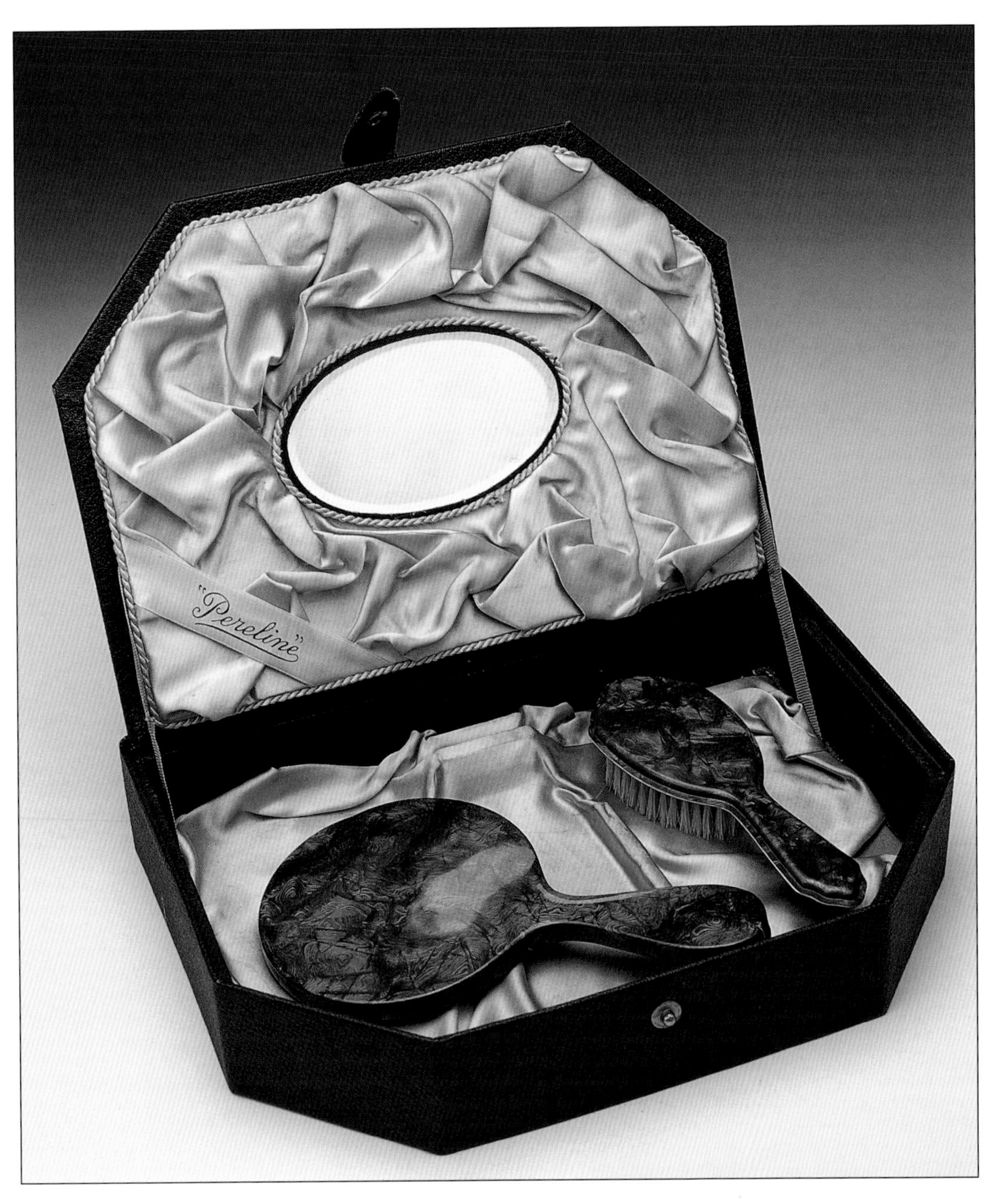

This two-piece dresser set of deep blue plastic with a pearl effect is called *Pereline.* It is still in its original satin-lined presentation box with oval mirror. This set was originally designed for a child. It is missing the comb. $65-85 set.

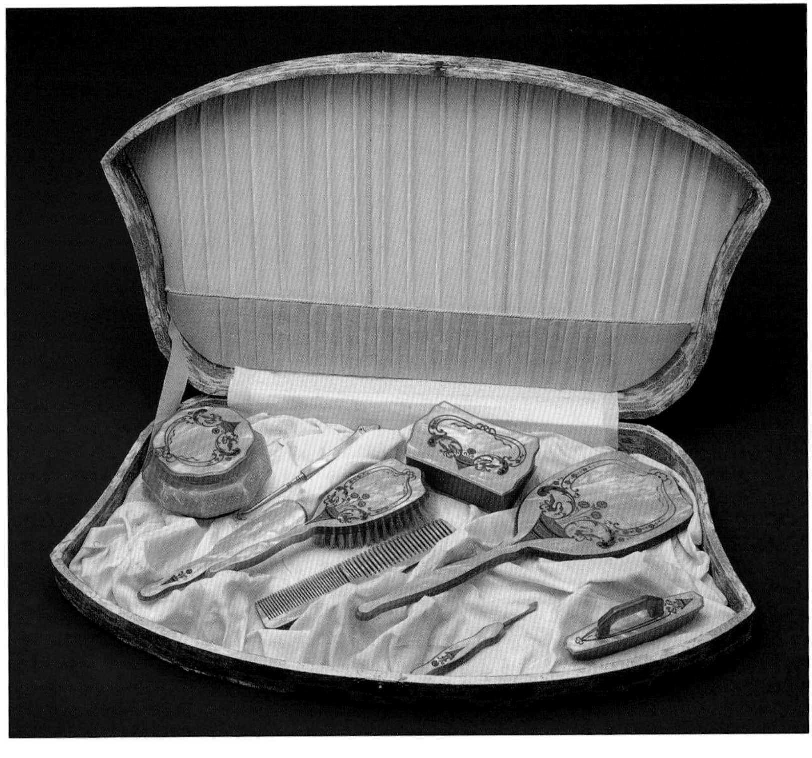

Nine-piece dresser set made of pastel green pearl-on-amber Pyralin with inlaid decorations in gold and black. The set is still housed in the original sateen-lined display box. This set dates back to the 1930s. $350-400 set.

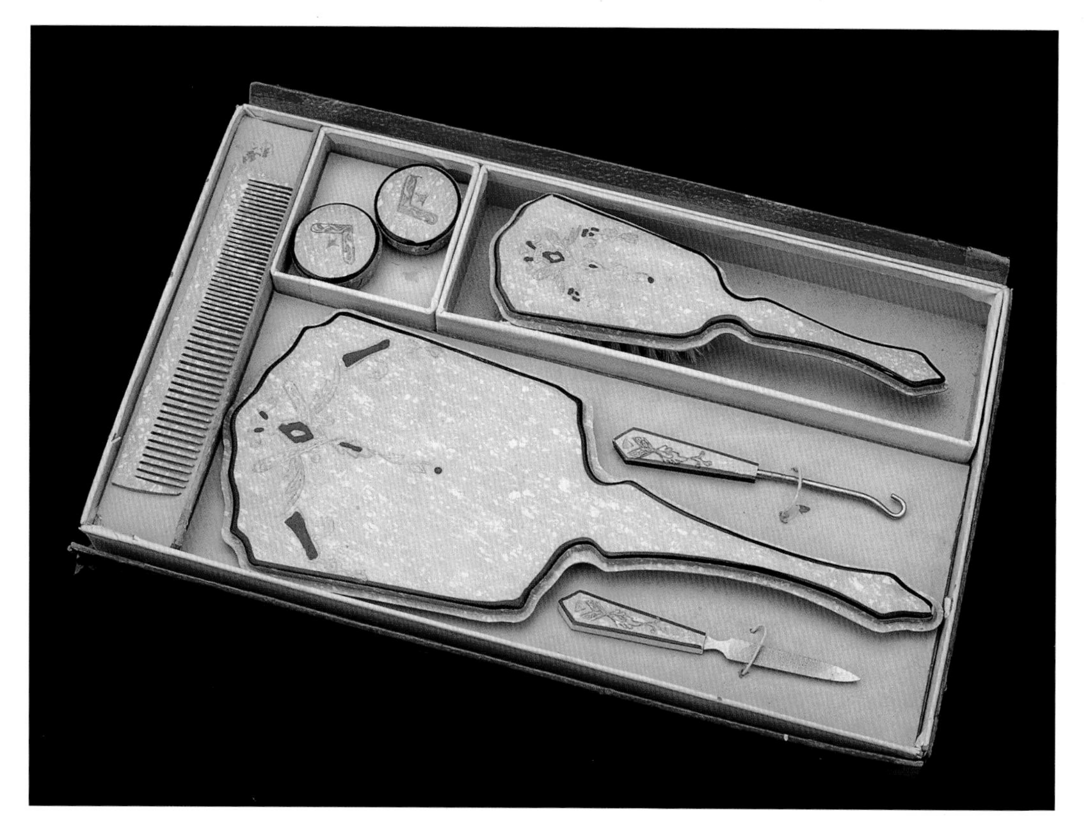

Striking seven-piece celluloid dresser set made with a marbled finish and accented with fancy inlaid designs that were colored in gold and black. This was probably an inexpensive set offered through a mail order company in the 1920s and still in its original cardboard box. $115-145 set.

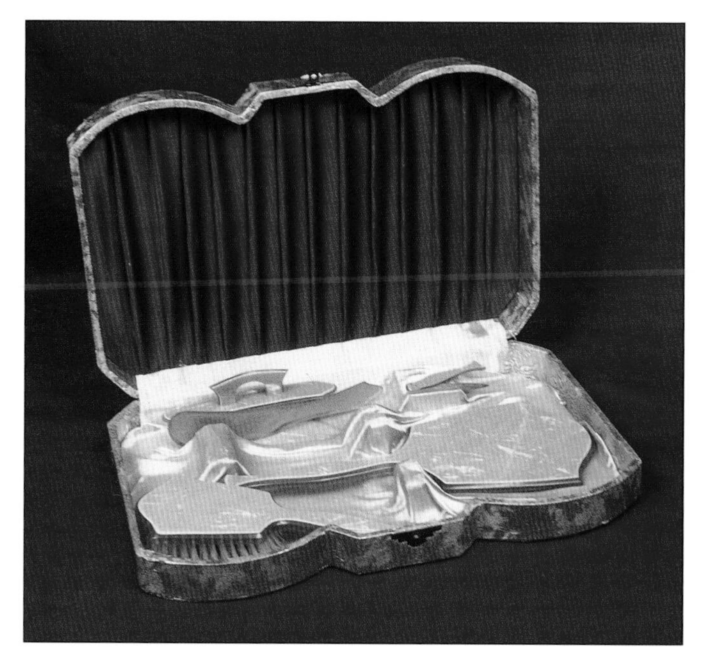

Lovely five-piece set of rose pearl-on-amber Pyralin by Dupont. The set is still housed in its original presentation box. $150-200 set.

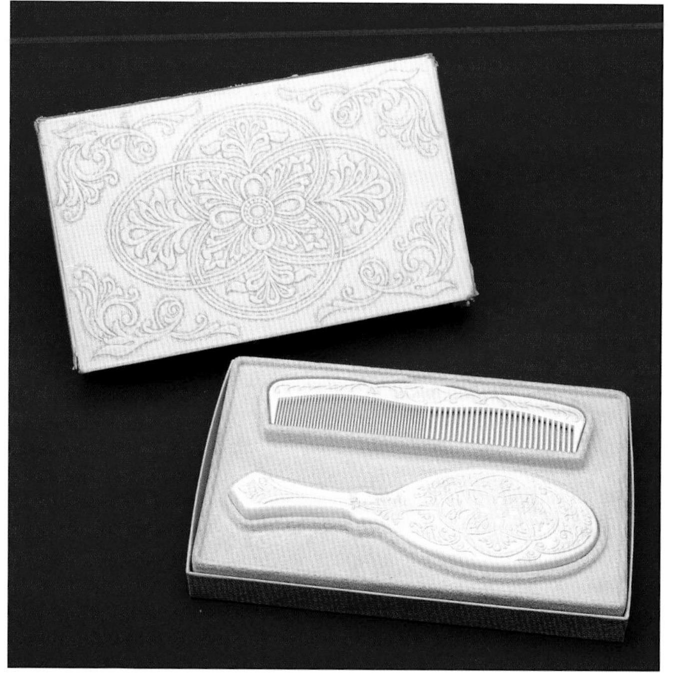

This 1970s two-piece dresser set by *Avon* called "Past & Present" was designed to look like earlier French Ivory celluloid sets from the late 19th century. $40-50 set.

A sumptuous gold and silver brocade fabric, under a thin sheet of clear plastic, makes this vintage vanity set by *Deltah* really unique. Its heart-shaped box adds more charm to this already wonderful set. $200-250 set.

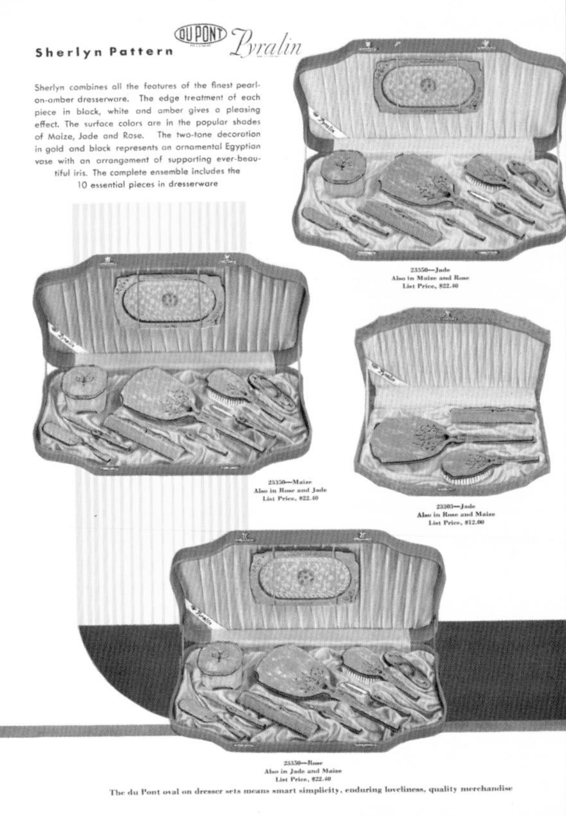

Lovely boxed dresser sets with the *Sherlyn* pattern of DuPont's Pearl-on-Amber Pyralin fittings popular in 1932. The two-toned decoration on the tops of each piece is a representation of an Egyptian iris in a vase.

Boxed gift sets of DuPont Lucite dresser ware in the *Sweet Pea* pattern of 1932.

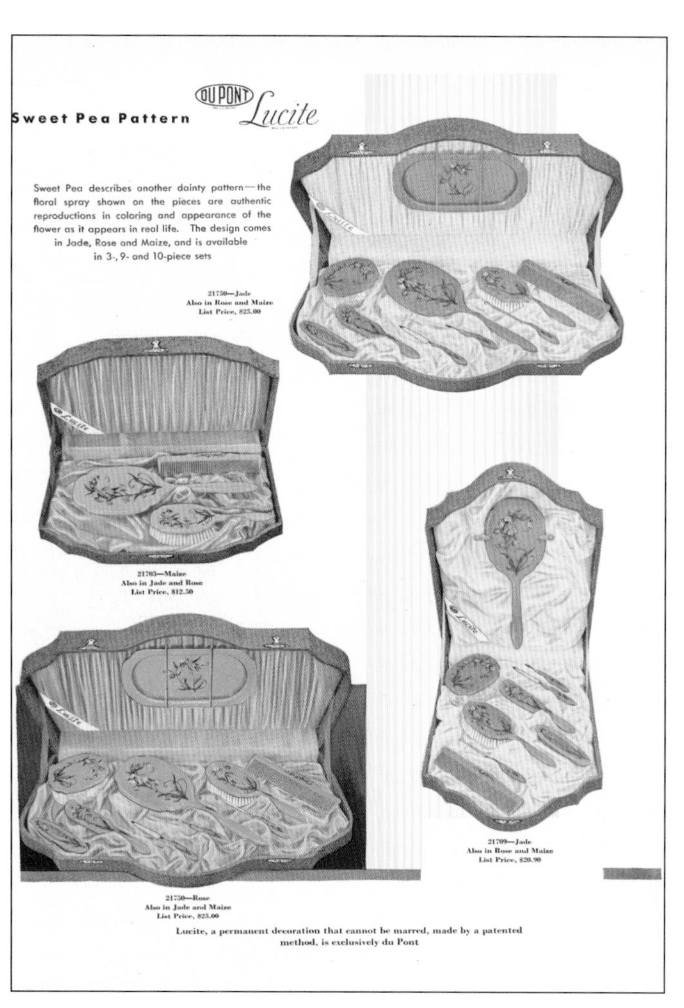

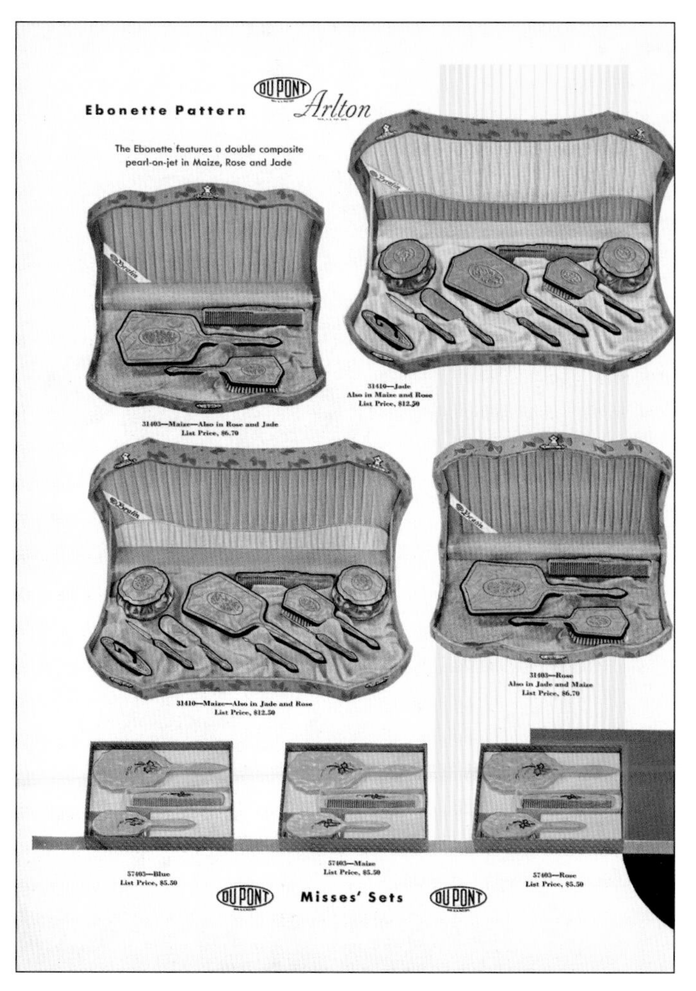

Misses' boxed sets of the *Ebonette* pattern of DuPont Arlton dresser ware available in 1932. This pattern featured a double composite of pearl-on-jet in assorted colors.

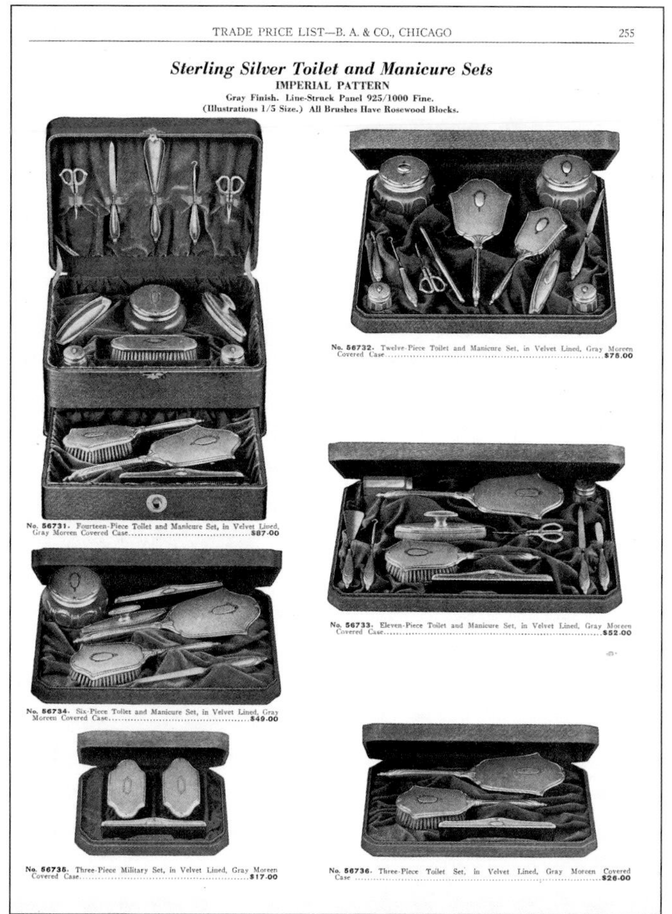

Boxed sets from 1935 fitted with the *Imperial* pattern of sterling silver toilet and manicure implements.

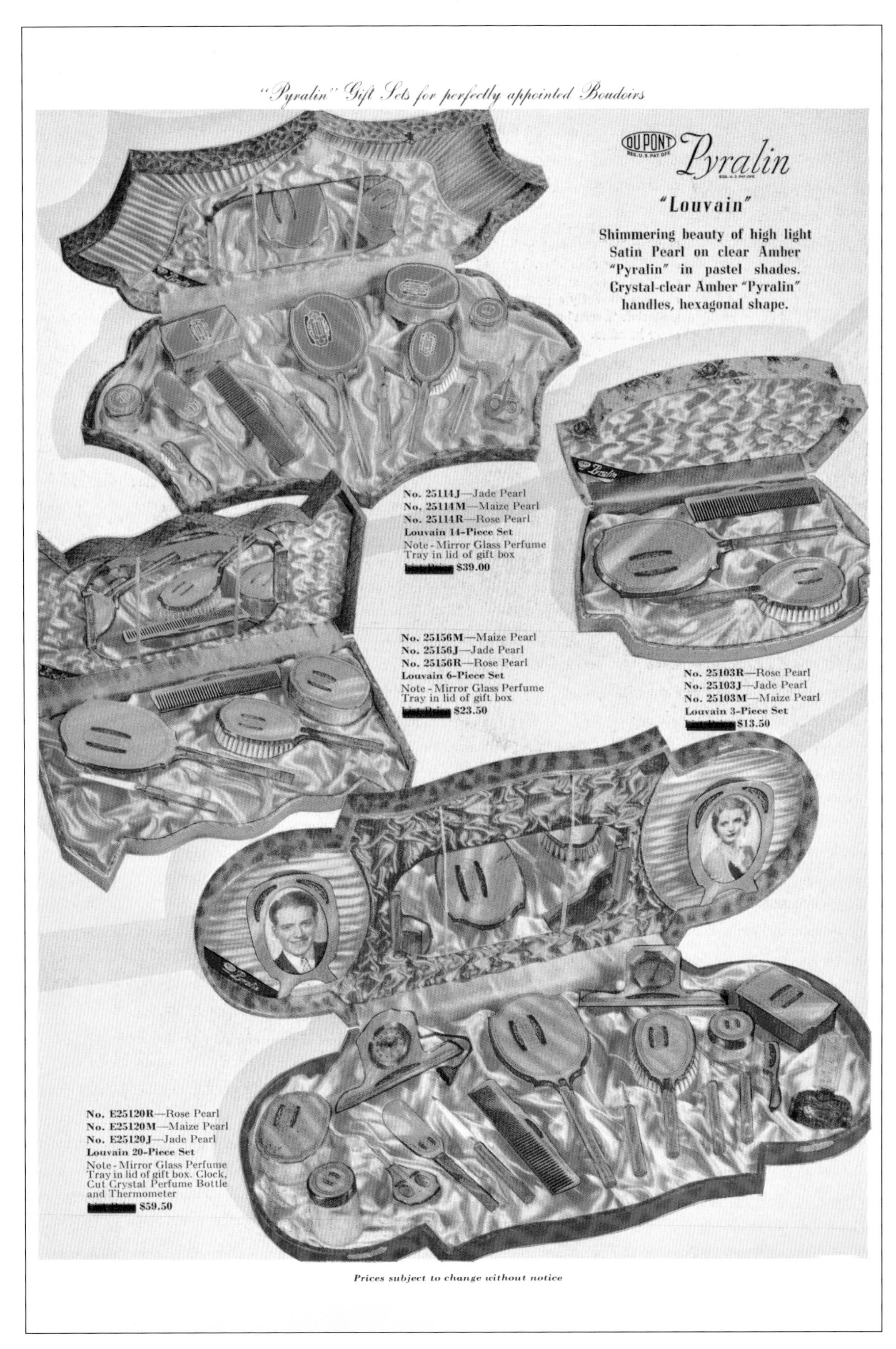

"Pyralin" Gift Sets for perfectly appointed Boudoirs
DU PONT Pyralin
"Louvain"
Shimmering beauty of high light Satin Pearl on clear Amber "Pyralin" in pastel shades. Crystal-clear Amber "Pyralin" handles, hexagonal shape.
No. 25114J—Jade Pearl
No. 25114M—Maize Pearl
No. 25114R—Rose Pearl
Louvain 14-Piece Set
Note - Mirror Glass Perfume Tray in lid of gift box
$39.00
No. 25156M—Maize Pearl
No. 25156J—Jade Pearl
No. 25156R—Rose Pearl
Louvain 6-Piece Set
Note - Mirror Glass Perfume Tray in lid of gift box
$23.50
No. 25103R—Rose Pearl
No. 25103J—Jade Pearl
No. 25103M—Maize Pearl
Louvain 3-Piece Set
$13.50
No. E25120R—Rose Pearl
No. E25120M—Maize Pearl
No. E25120J—Jade Pearl
Louvain 20-Piece Set
Note - Mirror Glass Perfume Tray in lid of gift box. Clock, Cut Crystal Perfume Bottle and Thermometer
$59.50
Prices subject to change without notice

Extremely elaborate as well as very impressive boxed sets offered from DuPont in 1939. Pictured here is the *Louvain* pattern of satin pearl on clear amber Pyralin. A twenty-piece set listed for $59.50.

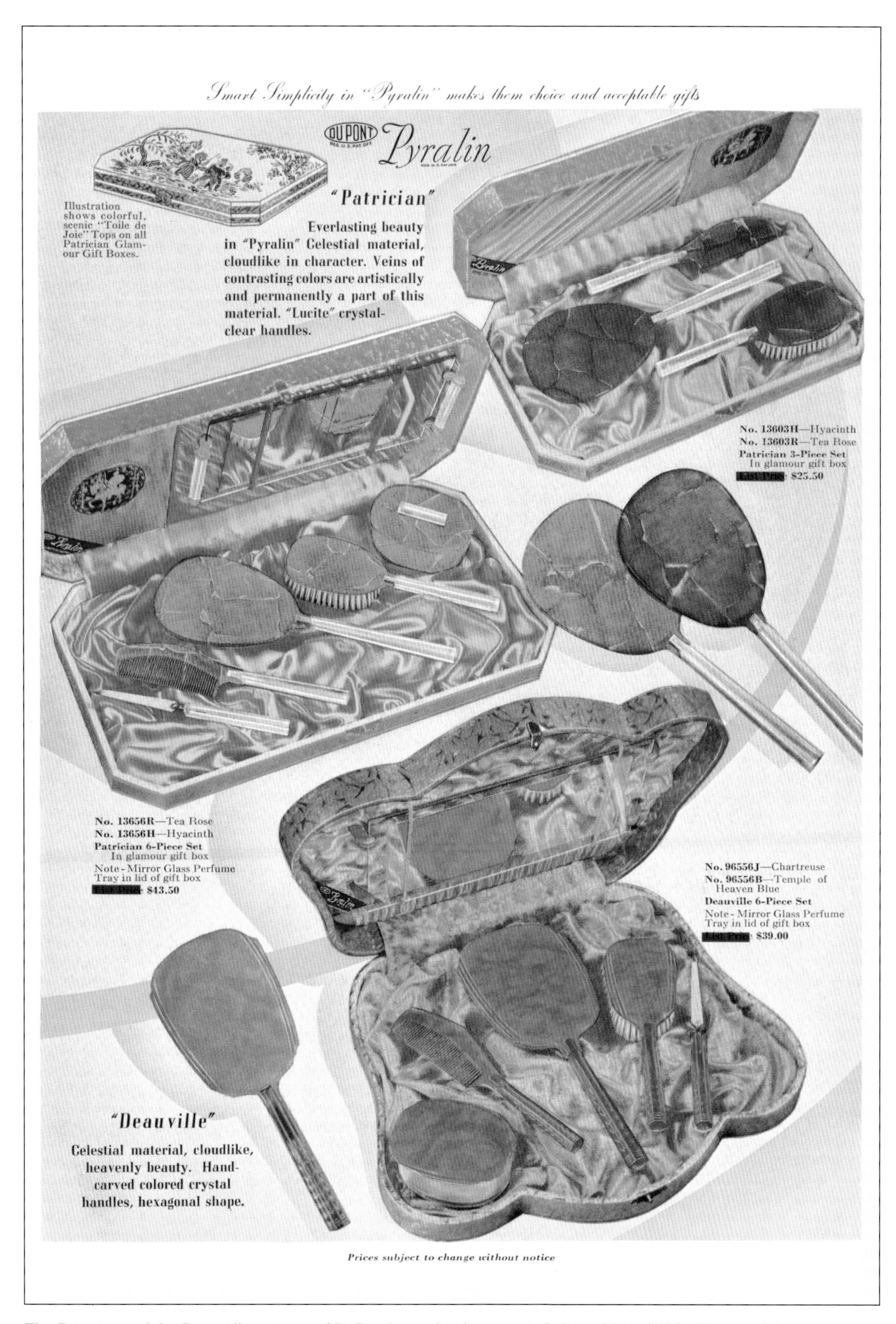

The *Patrician* and the *Deauville* patterns of DuPont's pyralin dresser sets fashionable in 1939. The top of the Patrician glamour box featured a scenic *Toile de Joie* pattern. The Patrician had clear crystal handles while the Deauville pattern had carved, crystal handles in a hexagon shape.

The Lucite and Pyralin dresser sets of the 1930s were housed in the largest and most glamorous boxes. They were even sometimes referred to as "glamour boxes". Impetus could have come from Hollywood and the glamorous lifestyles of the movie stars of the period. The larger the set, the better it seemed to be. Some sets, consisting of more than twenty pieces contained picture frames and even a glass mirrored dresser tray usually suspended in the lid of the box. Occasionally a picture frame would come complete with a stock picture of a movie star. This only enticed the consumer even more. But the luxurious fabric and the way it was draped in the box created a product so desirable that it appealed to women of all walks of life and all age groups. They became very desirable items for gift giving.

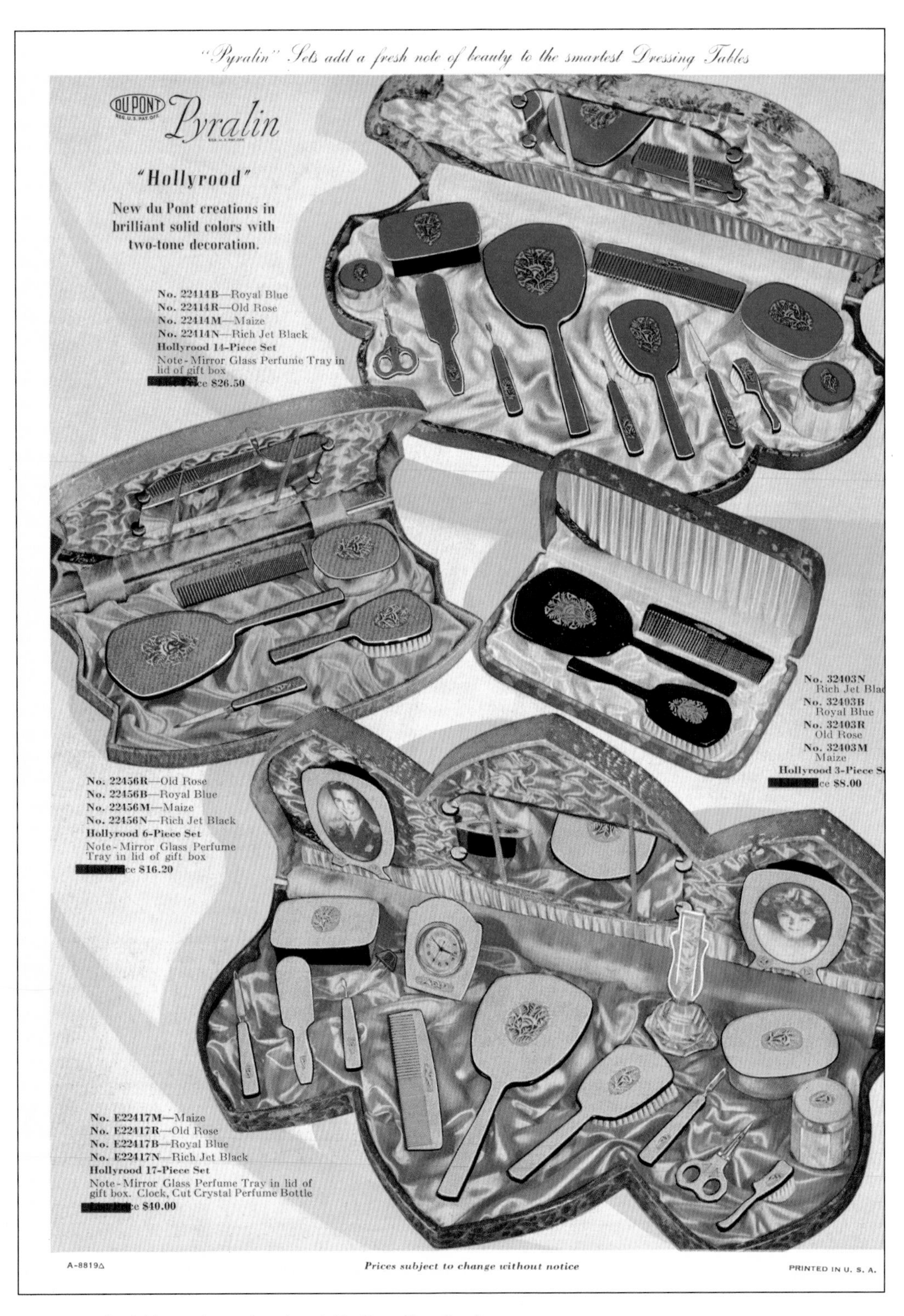

More wonderful boxed sets fitted with DuPont Pyralin dresser sets and manicure implements and offered for sale in 1939.

Chapter Six

Travel Sets

Fitted Cases

Travel sets, overnight cases, tray cases, outfitted cases and fitted luggage were only a few of the many terms used to describe a bag made of real or imitation leather designed for either a man or a woman and fitted with complete toilet and manicure sets. Advertised as being *...Ideal for weekend or long trips...* these sturdy well-made cases were very practical. They were designed in many shapes and sizes and offered in different qualities depending on how exotic the exterior leather was. The two main styles were the fitted overnight cases and the removable folding tray case. The fitted overnight case had the fittings strapped right into the lid of the suitcase. The removable folding tray case was a separate small case with the fittings inside which fit right into the larger suitcase. Another type of fitted case had a drop lid which opened up like a small dressing table. Still others were designed with a zippered case, fitted with toiletware, which fit snugly into a larger suitcase.

In 1879, *Ehrich's Fashion Quarterly* offered a traveling valise called "The Furnished Pellisier". This bag, made in the Gladstone style, was constructed of Morocco leather and lined with muslin. It was fitted with nickel-covered bottles for soap, brushes and perfume. It also came equipped with a hairbrush, comb case and whisk broom. This fitted valise listed for $8.50.

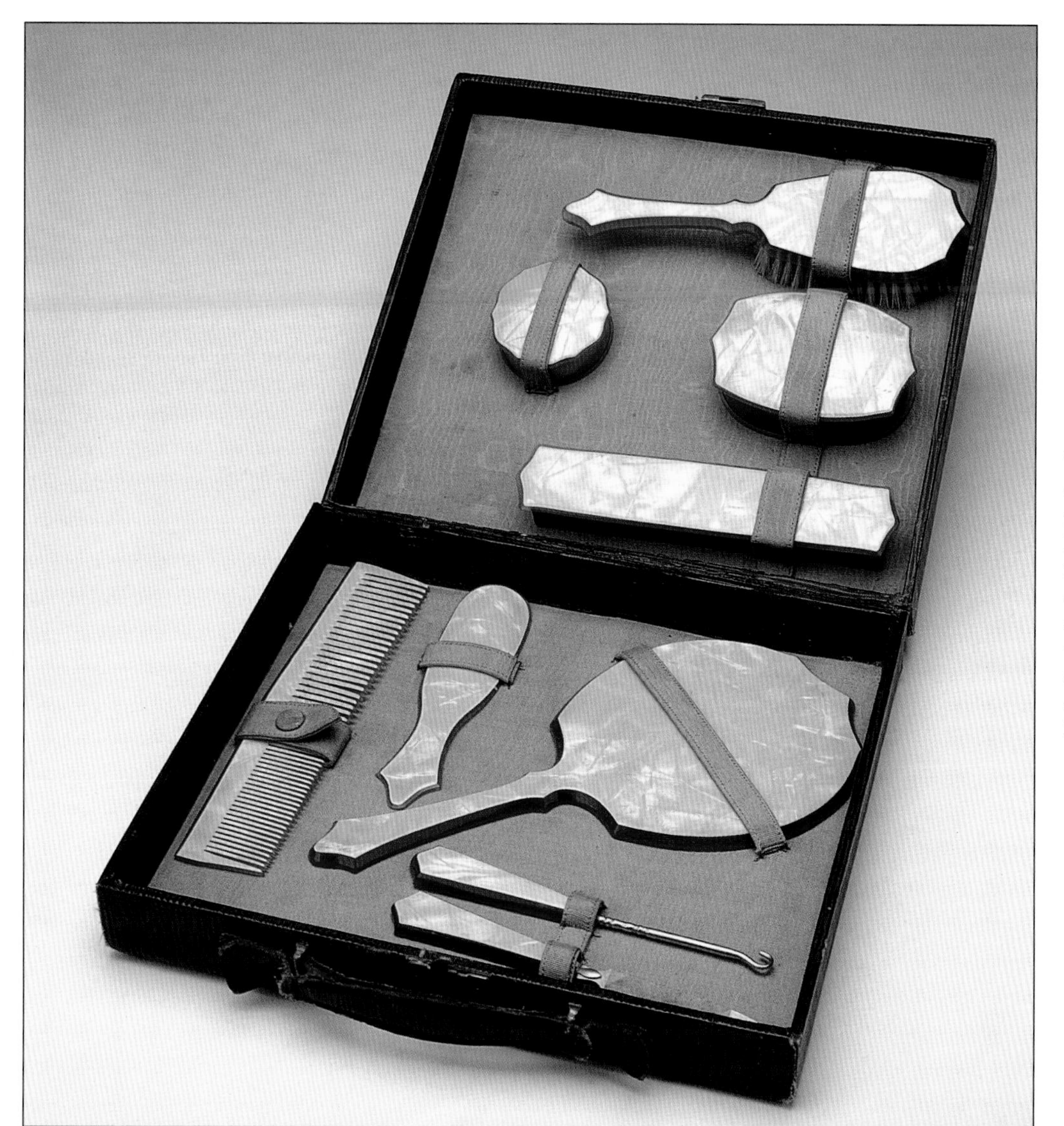

This ladies' traveling case, fitted with nine maize pearl-on-amber Pyralin toilet accessories, fits into a larger top grain cowhide suitcase. Called a *Ladies' Tray Case*, these unique cases were extremely popular in the early 20th century. $245-295.

Ladies tray case fitted with eight toilet implements, most of which are clear glass jars or bottles. This cowhide case fits into a larger traveling case. $195-245.

Ladies fitted tray case made of top grain cowhide and equipped with an eight piece toilet set consisting of mirror, brush, comb, nail file, two glass bottles and two glass jars. Again, this small case fits into a larger case and was the perfect traveling accessory in the 1930s. $200-250.

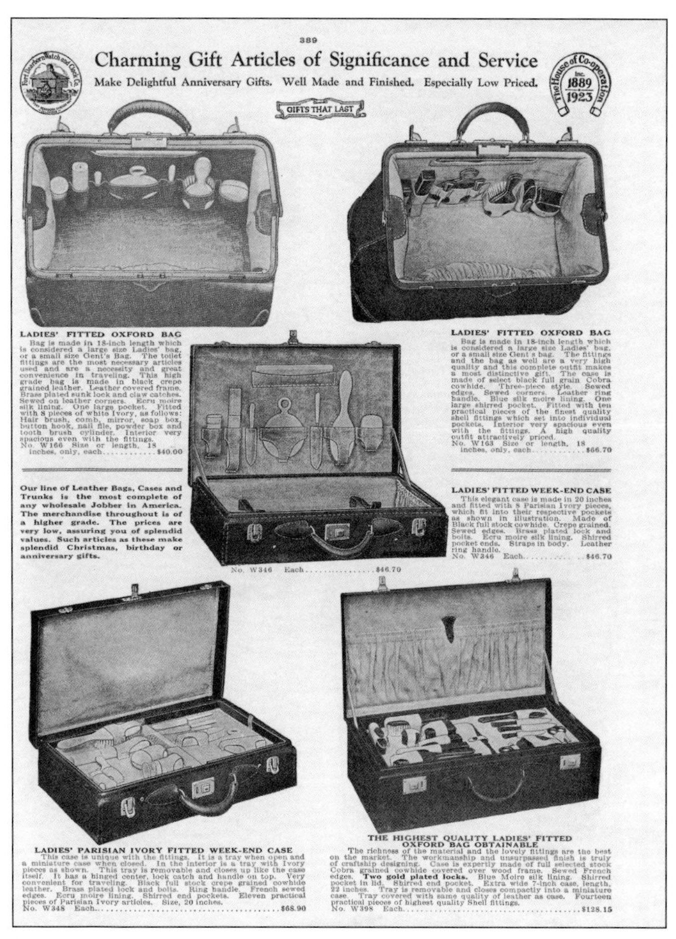

389

Charming Gift Articles of Significance and Service

Make Delightful Anniversary Gifts. Well Made and Finished. Especially Low Priced.

GIFTS THAT LAST

The House of Co-operation Inc. 1889 1923

LADIES' FITTED OXFORD BAG

Bag is made in 18-inch length which is considered a large size Ladies' bag, or a small size Gent's Bag. The toilet fittings are the most necessary articles used and are a necessity and great convenience in traveling. This high grade bag is made in black crepe grained leather. Leather covered frame. Brass plated sunk lock and claw catches. Sewed on leather corners. Ecru moire silk lining. One large pocket. Fitted with 8 pieces of white Ivory, as follows: Hair brush, comb, mirror, soap box, button hook, nail file, powder box and tooth brush cylinder. Interior very spacious even with the fittings.
No. W166 Size or length, 18 inches, only, each.............$40.00

LADIES' FITTED OXFORD BAG

Bag is made in 18-inch length which is considered a large size Ladies' bag, or a small size Gent's bag. The fittings and the bag as well are a very high quality and this complete outfit makes a most distinctive gift. The case is made of select black full grain Cobra cowhide. Three-piece style. Sewed edges. Sewed corners. Leather ring handle. Blue silk moire lining. One large shirred pocket. Fitted with ten practical pieces of the finest quality shell fittings which set into individual pockets. Interior very spacious even with the fittings. A high quality outfit attractively priced.
No. W163 Size or length, 18 inches, only, each.............$66.70

Our line of Leather Bags, Cases and Trunks is the most complete of any wholesale Jobber in America. The merchandise throughout is of a higher grade. The prices are very low, assuring you of splendid values. Such articles as these make splendid Christmas, birthday or anniversary gifts.

LADIES' FITTED WEEK-END CASE

This elegant case is made in 20 inches and fitted with 8 Parisian Ivory pieces, which fit into their respective pockets as shown in illustration. Made of Black full stock cowhide. Crepe grained. Sewed edges. Brass plated lock and bolts. Ecru moire silk lining. Shirred pocket ends. Straps in body. Leather ring handle.
No. W346 Each.............$46.70

No. W346 Each.................$46.70

LADIES' PARISIAN IVORY FITTED WEEK-END CASE

This case is unique with the fittings. It is a tray when open and a miniature case when closed. In the interior is a tray with Ivory pieces as shown. This tray is removable and closes up like the case itself. It has a hinged center, lock catch and handle on top. Very convenient for traveling. Black full stock crepe grained cowhide leather. Brass plated lock and bolts. Ring handle. French sewed edges. Ecru moire lining. Shirred end pockets. Eleven practical pieces of Parisian Ivory articles. Size, 20 inches.
No. W348 Each.......................................$68.90

THE HIGHEST QUALITY LADIES' FITTED OXFORD BAG OBTAINABLE

The richness of the material and the lovely fittings are the best on the market. The workmanship and unsurpassed finish is truly of craftship designing. Case is expertly made of full selected stock Cobra grained cowhide covered over wood frame. Sewed French edges. **Two gold plated locks.** Blue Moire silk lining. Shirred pocket in lid. Shirred end pocket. Extra wide 7-inch case, length, 22 inches. Tray is removable and closes compactly into a miniature case. Tray covered with same quality of leather as case. Fourteen practical pieces of highest quality Shell fittings.
No. W398 Each.......................................$128.15

Ladies' Fitted Oxford Bags and weekend cases equipped with French Ivory fittings fashionable in 1923.

326
John V. Farwell Company CHICAGO

Ladies' High Grade Outfitted Cases

Department 50—Fifth Floor

J445/50—Outfitted Case: black cobra grain cowhide; white fittings, light-weight case; full-length sewed leather hinge; flush lid; anchor handle; two gold plated side hasp locks; moire silk lining; two shirred pockets; furnished with nine toilet articles; size 22 inches.............Each 103.50

J446/50—Outfitted Case: black cobra grain cowhide; shell fittings, light-weight case; sewed full-length leather hinge; anchor handle; two gold-plated side hasp locks; moire silk lining; shirred pockets; nine pieces of shell fittings and steel scissors in lid; size 22 inches.............Each 150.00

J447/50—Outfitted Case: black panther grain cowhide; white fittings; a very unusual case; sewed full-length leather hinge; flush-fitting lid; French edges; anchor handle; two gold plated side hasp locks; silk striped foulard lining; shirred pockets and long shirred pocket in lid; fittings are carried in a box formed by the side of the case and a drop lid held by two button-lock flaps. (See illustration). A very compact way of carrying fittings entirely separated from the contents of the case; furnished with fourteen pieces of white fittings; size 20 inches.............Each 156.00

J448/50—Outfitted Case: black panther grain cowhide; white fittings, light-weight case; French edges; sewed full-length leather hinge; anchor handle; two gold-plated side hasp locks; crepe moire silk lining; shirred pockets; fitted with removable toilet case which forms separate, complete unit holding fifteen white toilet articles; size 20 inches.............Each 170.00

J449/50—Outfitted Case: black cobra grain cowhide; hand-decorated white fittings; light-weight case; full-length sewed leather hinge; flush lid; anchor handle; two gold plated side hasp locks; lined with crepe moire silk lining; shirred pockets on ends; hand-decorated white fittings; borders matching the lining of the case; fourteen karat gold leaf is used on these decorations; size 22 inches..........Each 180.00

J450/50—Outfitted Case: black panther grain cowhide; white fittings; light-weight French edge case; full-length sewed leather hinge; flush lid; anchor handle; two gold plated side hasp locks; moire silk lining; shirred pockets; removable toilet case with envelope pocket for handkerchief and gloves; this case also forms a separate unit containing fifteen white toilet articles; size 22 inches...............Each 210.00

Ladies' High Grade Outfitted Cases with French Ivory fittings. The black panther grain cowhide case with the removable toilet case listed for $210.00 in 1920. The black cobra grain case with nine fittings listed for $103.50. These unique cases were rather costly when they were first introduced but extremely well-made and very practical.

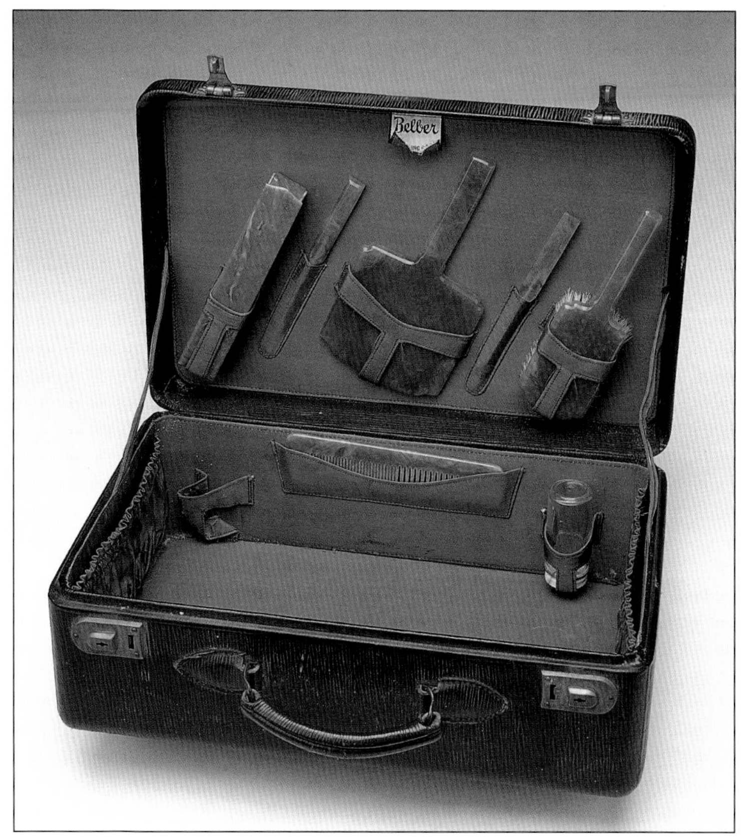

Belber Traveling Goods created this top grain overnight case fitted with seven (at one time-eight) toilet accessories. The brush, mirror, comb, file, shoe buttoner and toothbrush holder are made of jade green Lucite. $225-275.

Ladies' traveling bags, also called Oxford bags, were made like doctor's bags or satchels. These bags were also carried by men. A ladies' Oxford bag was eighteen inches long and the gent's bag was slightly smaller. These bags were extremely fashionable in 1920. They were made of many exotic grains of cowhide, particularly bison grain, walrus grain, seal grain and cobra grain. A fine walrus grain leather bag, complete with nine French Ivory fittings, listed for $57.00. A fine seal grain leather bag fitted with eleven French Ivory fittings sold for $106.00. Outfitted cases made of black cobra grain cowhide and black panther grain cowhide were also extravagantly priced at $103.50 with French Ivory fittings and $150.00 with tortoiseshell fittings. A more elaborate tray case in black panther grain, with French Ivory fittings listed for $210.00. They were substantial sums to pay in 1920.

The cases were always described as being lightweight due to the construction of the leather over a basswood frame. Rounded corners, padded tops and smartly lined interiors were attractive features. Bronze, brass or nickel hardware sometimes harmonized with the interior fittings. The fittings, however, were the most desirable feature. French Ivory, Pyralin, Lucite, sterling silver and chrome and enamel were the most popular choices.

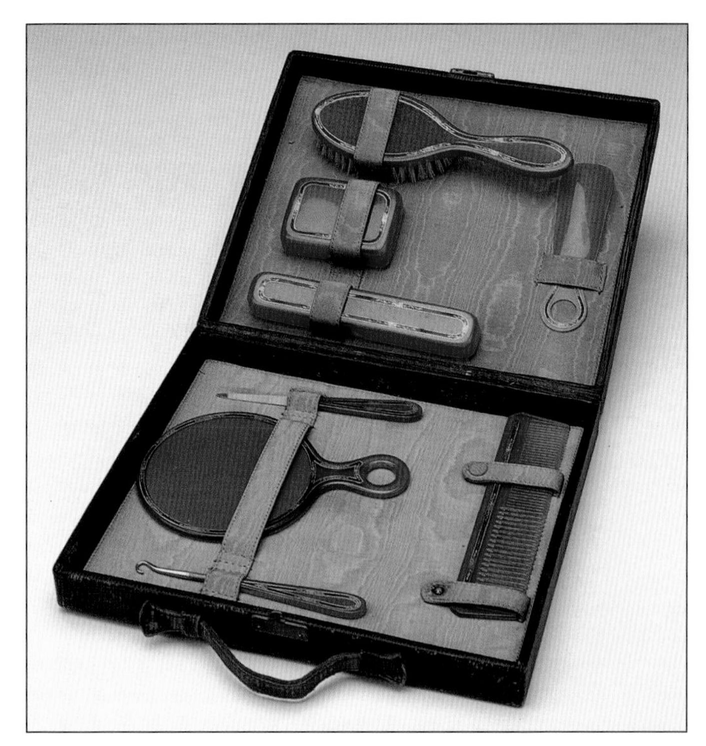

An amber Pyralin dresser set with inlaid designs accented in gold and black is fitted nicely into this tray case made of top grain cowhide. The set includes a round mirror with ring handle, comb, brush, nail file, shoe buttoner, soap holder, toothbrush holder and shoe horn. The case has a nice silvery moiré lining. $255-295.

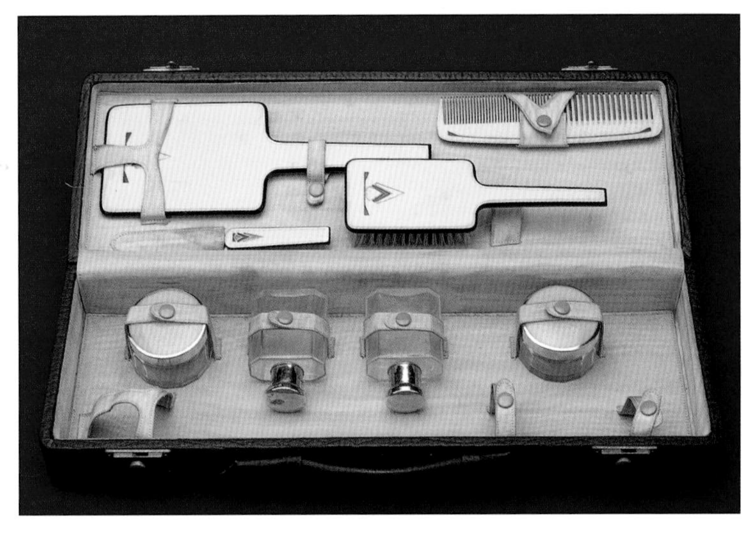

A larger top grain cowhide traveling case by *Knapp* housed this smaller fitted tray case with yellow Art Deco toilet accessories stamped *DuPont.* Inlaid designs in black and gold create a striking modernistic set. Unfortunately the set is missing two pieces. $200-245.

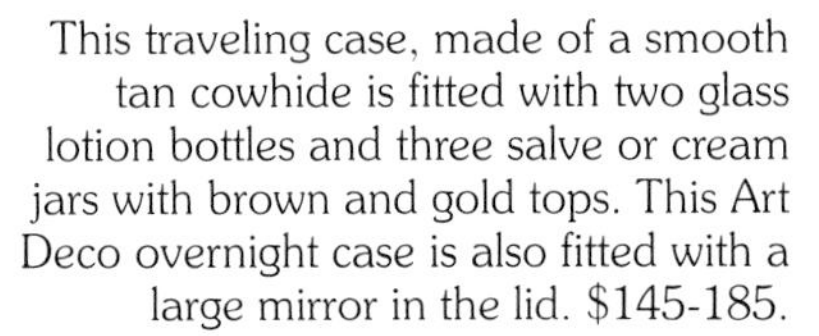

This traveling case, made of a smooth tan cowhide is fitted with two glass lotion bottles and three salve or cream jars with brown and gold tops. This Art Deco overnight case is also fitted with a large mirror in the lid. $145-185.

Sleek and sophisticated fitted overnight case equipped with gold-plated necessary requisites for travel. The geometric designs and fluted appearance create a very streamlined and modern set. This Art Deco traveling set was made by *Evans*. The top grain cowhide case is lined in moiré fabric. The lid has two compartments for jewelry and other items. Unfortunately one bottle is missing. $275-325.

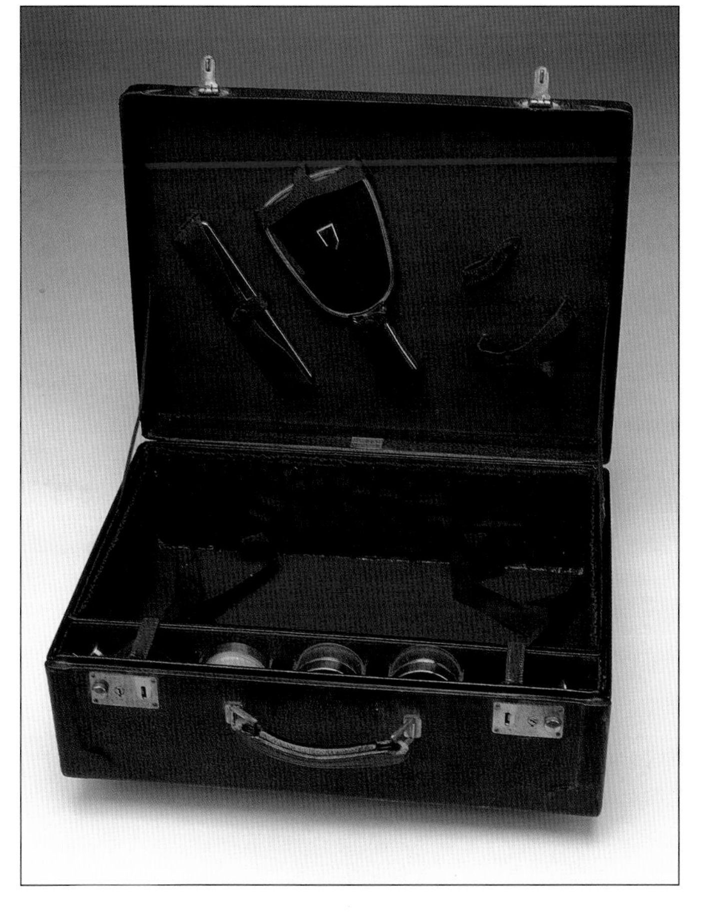

Ladies overnight case made of black top grain cowhide by *Lido Luggage*. The inside is fitted with seven toilet requisites made of gold plate and blue enamel. The brush, however, is missing. The inside of the case also has shirred pockets to hold lingerie and a generous section for holding clothing. $195-245.

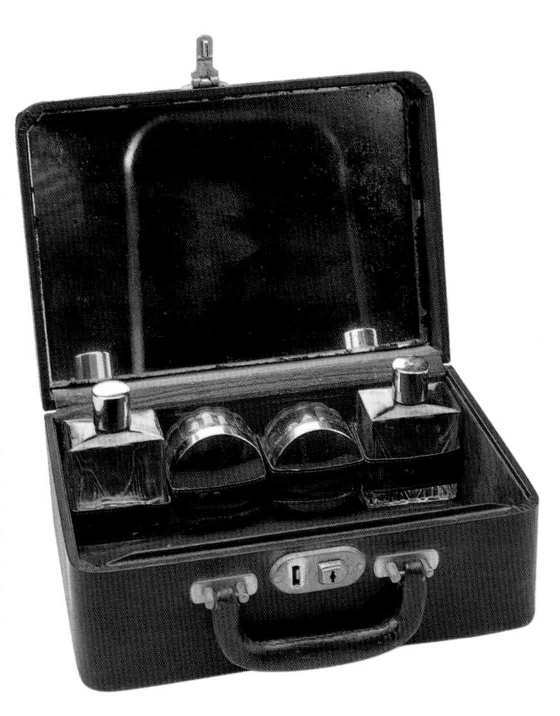

Small leather overnight case fitted with two glass lotion bottles and two cream jars. The mirror is secured in the lid. $115-145.

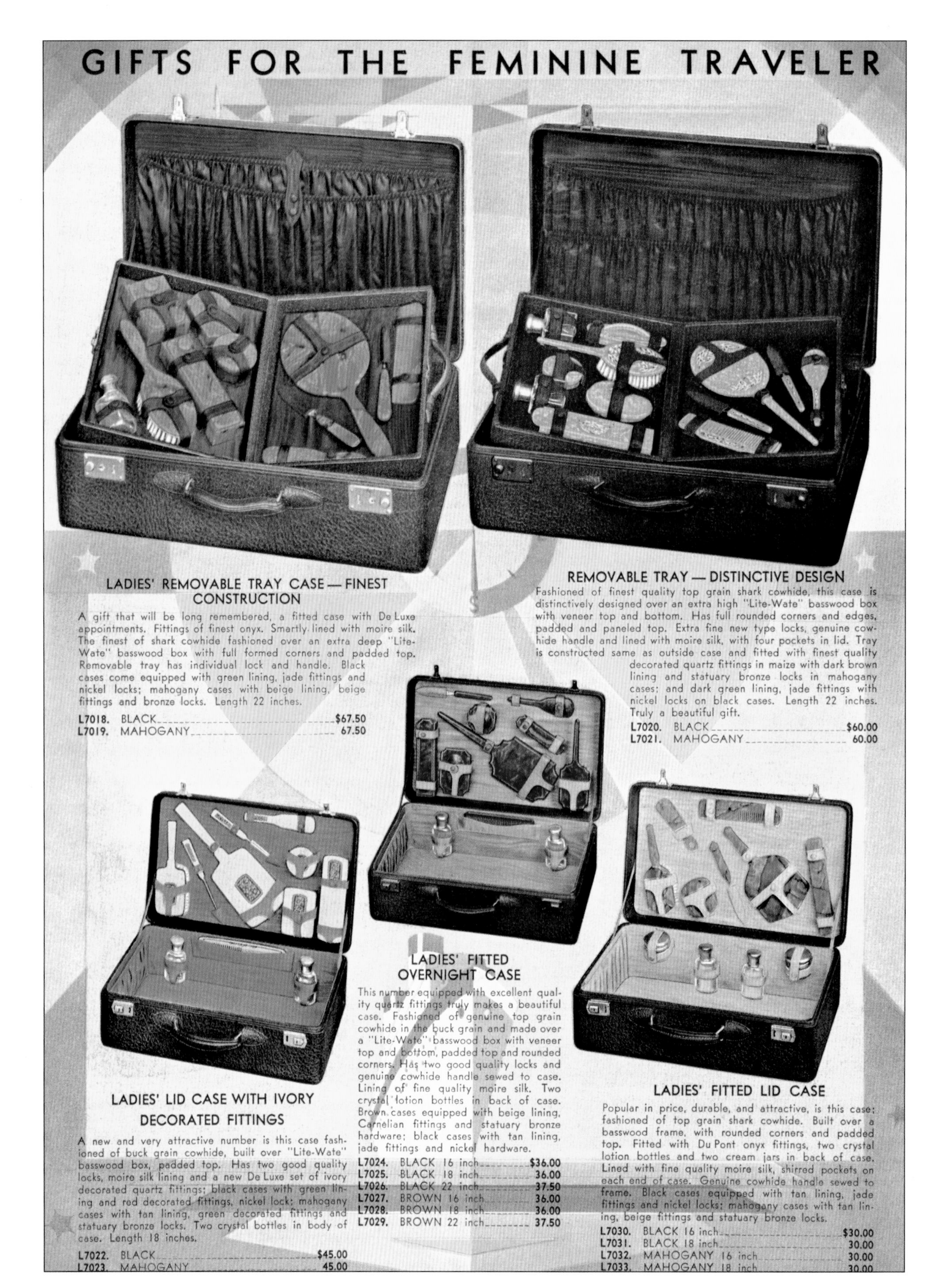

GIFTS FOR THE FEMININE TRAVELER

LADIES' REMOVABLE TRAY CASE — FINEST CONSTRUCTION

A gift that will be long remembered, a fitted case with De Luxe appointments. Fittings of finest onyx. Smartly lined with moire silk. The finest of shark cowhide fashioned over an extra deep "Lite-Wate" basswood box with full formed corners and padded top. Removable tray has individual lock and handle. Black cases come equipped with green lining, jade fittings and nickel locks; mahogany cases with beige lining, beige fittings and bronze locks. Length 22 inches.

L7018. BLACK $67.50
L7019. MAHOGANY 67.50

REMOVABLE TRAY — DISTINCTIVE DESIGN

Fashioned of finest quality top grain shark cowhide, this case is distinctively designed over an extra high "Lite-Wate" basswood box with veneer top and bottom. Has full rounded corners and edges, padded and paneled top. Extra fine new type locks, genuine cowhide handle and lined with moire silk, with four pockets in lid. Tray is constructed same as outside case and fitted with finest quality decorated quartz fittings in maize with dark brown lining and statuary bronze locks in mahogany cases; and dark green lining, jade fittings with nickel locks on black cases. Length 22 inches. Truly a beautiful gift.

L7020. BLACK $60.00
L7021. MAHOGANY 60.00

LADIES' FITTED OVERNIGHT CASE

This number equipped with excellent quality quartz fittings truly makes a beautiful case. Fashioned of genuine top grain cowhide in the buck grain and made over a "Lite-Wate" basswood box with veneer top and bottom, padded top and rounded corners. Has two good quality locks and genuine cowhide handle sewed to case. Lining of fine quality moire silk. Two crystal lotion bottles in back of case. Brown cases equipped with beige lining, Carnelian fittings and statuary bronze hardware; black cases with tan lining, jade fittings and nickel hardware.

L7024. BLACK 16 inch $36.00
L7025. BLACK 18 inch 36.00
L7026. BLACK 22 inch 37.50
L7027. BROWN 16 inch 36.00
L7028. BROWN 18 inch 36.00
L7029. BROWN 22 inch 37.50

LADIES' LID CASE WITH IVORY DECORATED FITTINGS

A new and very attractive number is this case fashioned of buck grain cowhide, built over "Lite-Wate" basswood box, padded top. Has two good quality locks, moire silk lining and a new De Luxe set of ivory decorated quartz fittings; black cases with green lining and red decorated fittings, nickel lock; mahogany cases with tan lining, green decorated fittings and statuary bronze locks. Two crystal bottles in body of case. Length 18 inches.

L7022. BLACK $45.00
L7023. MAHOGANY 45.00

LADIES' FITTED LID CASE

Popular in price, durable, and attractive, is this case; fashioned of top grain shark cowhide. Built over a basswood frame, with rounded corners and padded top. Fitted with Du Pont onyx fittings, two crystal lotion bottles and two cream jars in back of case. Lined with fine quality moire silk, shirred pockets on each end of case. Genuine cowhide handle sewed to frame. Black cases equipped with tan lining, jade fittings and nickel locks; mahogany cases with tan lining, beige fittings and statuary bronze locks.

L7030. BLACK 16 inch $30.00
L7031. BLACK 18 inch 30.00
L7032. MAHOGANY 16 inch 30.00
L7033. MAHOGANY 18 inch 30.00

Ladies' removable tray cases and overnight cases fitted with DuPont Pyralin and Lucite dresser accessories offered for sale in 1932 from the A.C.Becken catalog.

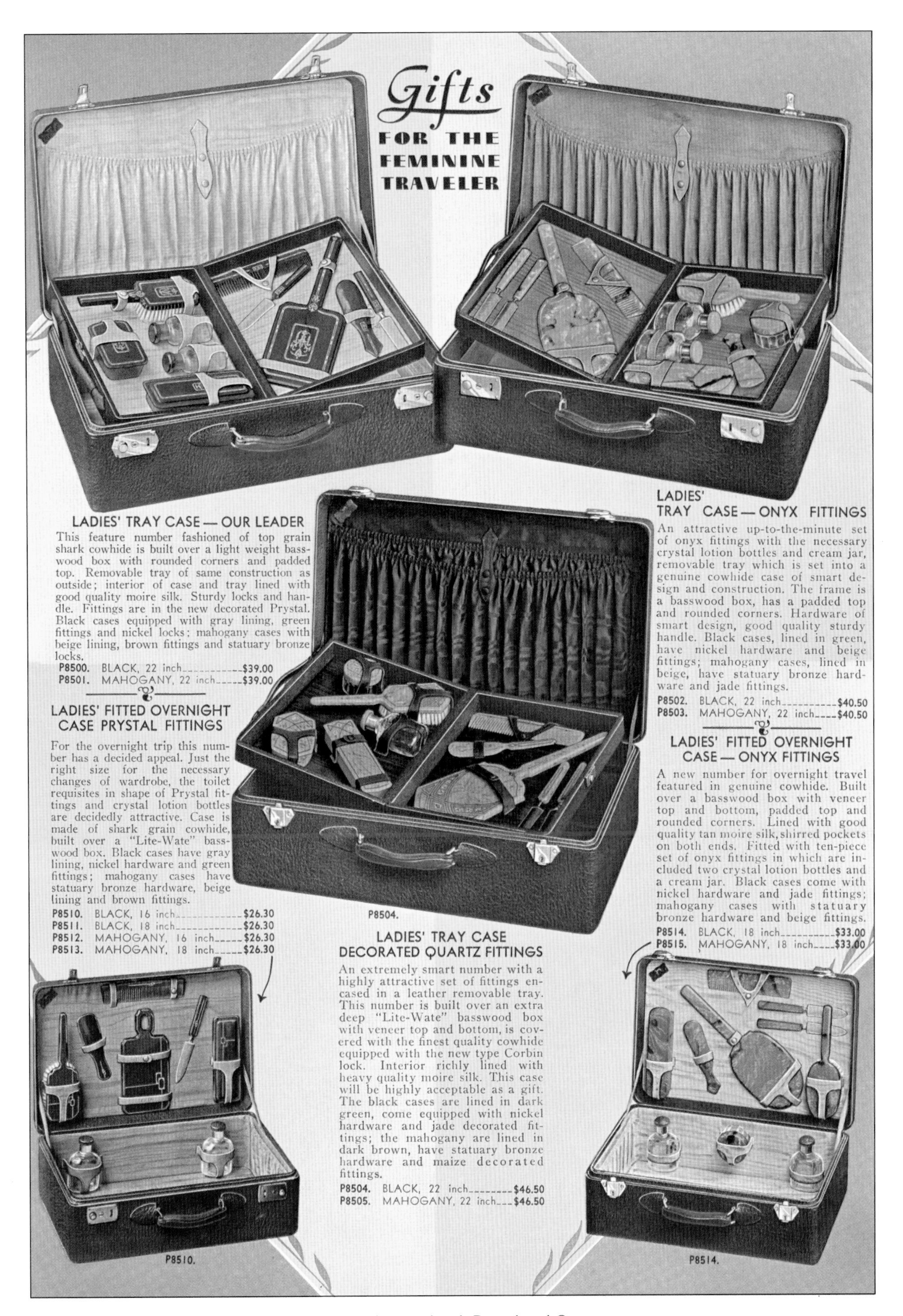

Gifts
FOR THE
FEMININE
TRAVELER
LADIES' TRAY CASE — OUR LEADER
This feature number fashioned of top grain shark cowhide is built over a light weight basswood box with rounded corners and padded top. Removable tray of same construction as outside; interior of case and tray lined with good quality moire silk. Sturdy locks and handle. Fittings are in the new decorated Prystal. Black cases equipped with gray lining, green fittings and nickel locks; mahogany cases with beige lining, brown fittings and statuary bronze locks.
P8500. BLACK, 22 inch $39.00
P8501. MAHOGANY, 22 inch $39.00
LADIES' FITTED OVERNIGHT CASE PRYSTAL FITTINGS
For the overnight trip this number has a decided appeal. Just the right size for the necessary changes of wardrobe, the toilet requisites in shape of Prystal fittings and crystal lotion bottles are decidedly attractive. Case is made of shark grain cowhide, built over a "Lite-Wate" basswood box. Black cases have gray lining, nickel hardware and green fittings; mahogany cases have statuary bronze hardware, beige lining and brown fittings.
P8510. BLACK, 16 inch $26.30
P8511. BLACK, 18 inch $26.30
P8512. MAHOGANY, 16 inch $26.30
P8513. MAHOGANY, 18 inch $26.30
LADIES' TRAY CASE — ONYX FITTINGS
An attractive up-to-the-minute set of onyx fittings with the necessary crystal lotion bottles and cream jar, removable tray which is set into a genuine cowhide case of smart design and construction. The frame is a basswood box, has a padded top and rounded corners. Hardware of smart design, good quality sturdy handle. Black cases, lined in green, have nickel hardware and beige fittings; mahogany cases, lined in beige, have statuary bronze hardware and jade fittings.
P8502. BLACK, 22 inch $40.50
P8503. MAHOGANY, 22 inch $40.50
LADIES' FITTED OVERNIGHT CASE — ONYX FITTINGS
A new number for overnight travel featured in genuine cowhide. Built over a basswood box with veneer top and bottom, padded top and rounded corners. Lined with good quality tan moire silk, shirred pockets on both ends. Fitted with ten-piece set of onyx fittings in which are included two crystal lotion bottles and a cream jar. Black cases come with nickel hardware and jade fittings; mahogany cases with statuary bronze hardware and beige fittings.
P8514. BLACK, 18 inch $33.00
P8515. MAHOGANY, 18 inch $33.00
P8504.
LADIES' TRAY CASE DECORATED QUARTZ FITTINGS
An extremely smart number with a highly attractive set of fittings encased in a leather removable tray. This number is built over an extra deep "Lite-Wate" basswood box with veneer top and bottom, is covered with the finest quality cowhide equipped with the new type Corbin lock. Interior richly lined with heavy quality moire silk. This case will be highly acceptable as a gift. The black cases are lined in dark green, come equipped with nickel hardware and jade decorated fittings; the mahogany are lined in dark brown, have statuary bronze hardware and maize decorated fittings.
P8504. BLACK, 22 inch $46.50
P8505. MAHOGANY, 22 inch $46.50
P8510.
P8514.

These early 1930s fitted cases for women were advertised with Prystal and Quartz fittings which were generic terms used to describe Pyralin and Lucite fittings.

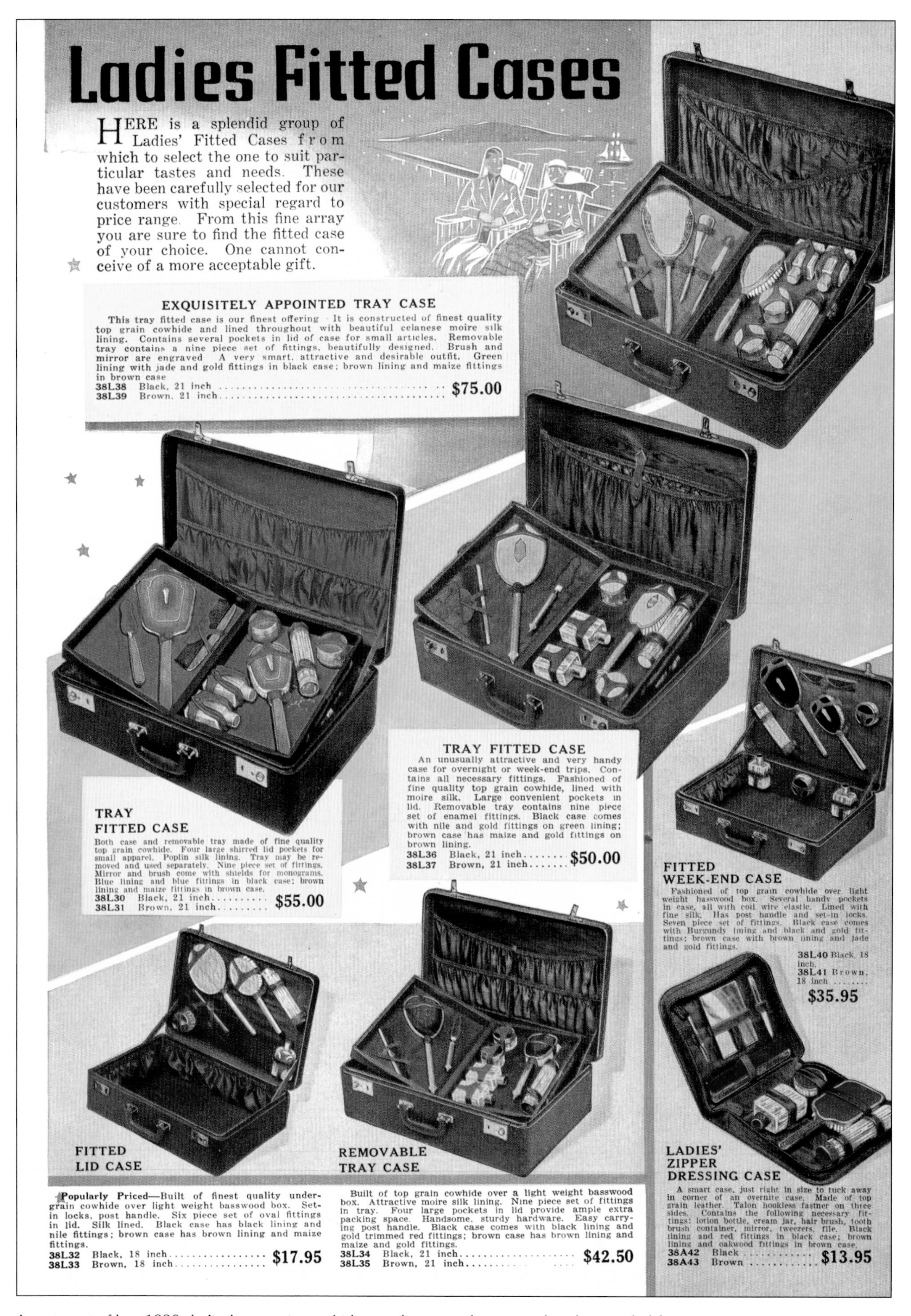

Ladies Fitted Cases
HERE is a splendid group of Ladies' Fitted Cases from which to select the one to suit particular tastes and needs. These have been carefully selected for our customers with special regard to price range. From this fine array you are sure to find the fitted case of your choice. One cannot conceive of a more acceptable gift.
EXQUISITELY APPOINTED TRAY CASE
This tray fitted case is our finest offering. It is constructed of finest quality top grain cowhide and lined throughout with beautiful celanese moire silk lining. Contains several pockets in lid of case for small articles. Removable tray contains a nine piece set of fittings, beautifully designed. Brush and mirror are engraved. A very smart, attractive and desirable outfit. Green lining with jade and gold fittings in black case; brown lining and maize fittings in brown case
38L38 Black, 21 inch
38L39 Brown, 21 inch $75.00
TRAY FITTED CASE
Both case and removable tray made of fine quality top grain cowhide. Four large shirred lid pockets for small apparel. Poplin silk lining. Tray may be removed and used separately. Nine piece set of fittings. Mirror and brush come with shields for monograms. Blue lining and blue fittings in black case; brown lining and maize fittings in brown case.
38L30 Black, 21 inch
38L31 Brown, 21 inch $55.00
TRAY FITTED CASE
An unusually attractive and very handy case for overnight or week-end trips. Contains all necessary fittings. Fashioned of fine quality top grain cowhide, lined with moire silk. Large convenient pockets in lid. Removable tray contains nine piece set of enamel fittings. Black case comes with nile and gold fittings on green lining; brown case has maize and gold fittings on brown lining.
38L36 Black, 21 inch
38L37 Brown, 21 inch $50.00
FITTED WEEK-END CASE
Fashioned of top grain cowhide over light weight basswood box. Several handy pockets in case, all with coil wire elastic. Lined with fine silk. Has post handle and set-in locks. Seven piece set of fittings. Black case comes with Burgundy lining and black and gold fittings; brown case with brown lining and jade and gold fittings.
38L40 Black, 18 inch.
38L41 Brown, 18 inch ....
$35.95
FITTED LID CASE
Popularly Priced—Built of finest quality undergrain cowhide over light weight basswood box. Set-in locks, post handle. Six piece set of oval fittings in lid. Silk lined. Black case has black lining and nile fittings; brown case has brown lining and maize fittings.
38L32 Black, 18 inch
38L33 Brown, 18 inch $17.95
REMOVABLE TRAY CASE
Built of top grain cowhide over a light weight basswood box. Attractive moire silk lining. Nine piece set of fittings in tray. Four large pockets in lid provide ample extra packing space. Handsome, sturdy hardware. Easy carrying post handle. Black case comes with black lining and gold trimmed red fittings; brown case has brown lining and maize and gold fittings.
38L34 Black, 21 inch
38L35 Brown, 21 inch $42.50
LADIES' ZIPPER DRESSING CASE
A smart case, just right in size to tuck away in corner of an overnite case. Made of top grain leather. Talon hookless fastner on three sides. Contains the following necessary fittings: lotion bottle, cream jar, hair brush, tooth brush container, mirror, tweezers, file. Black lining and red fittings in black case; brown lining and oakwood fittings in brown case.
38A42 Black
38A43 Brown $13.95

Assortment of late 1930s ladies' top grain cowhide travel cases with engraved and enameled fittings.

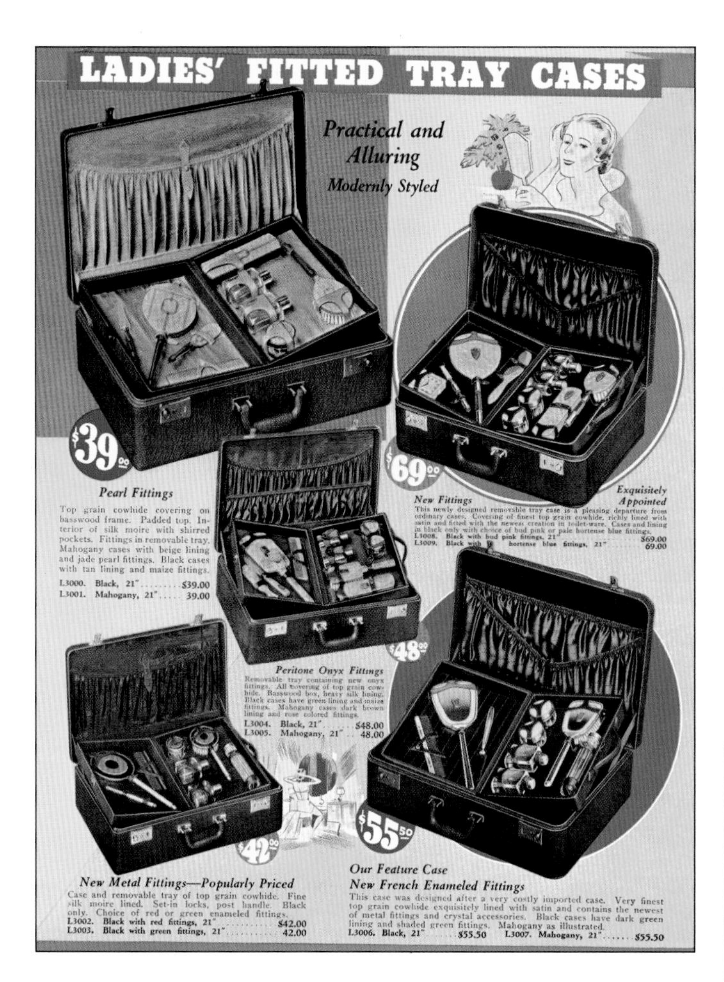

Lovely ladies' fitted tray cases made of the finest top grain cowhide and colorful Pyralin and French enameled fittings popular in 1938.

An unusual case was offered for sale in 1932 from the A.C.Becken catalog. Advertised as *smartly styled and fit for gift giving for any occasion.* A Ladies' English Style Dressing Case, expertly made of selected top grain cowhide, was designed with a "baboon grain". The case was fitted with eight hand-engraved Lucite fittings. The inside was lined with brocade. The case listed for $64.50. Another unusual case that was offered for sale in 1936 from the L & C Mayers Company catalog was a genuine top grain cowhide case with a smart shark grain and fitted with unique Birdseye Maple fittings. The fittings, however, were actually made of Celluloid, trimmed in chrome, and made to imitate the Birdseye Maple. This case listed for $55.00.

By the late 1940s and early 1950s, outfitted cases were beginning to disappear from wholesale and mail order catalogs. Regular luggage was beginning to take its place. Three-piece airplane ensembles, consisting of a wardrobe, weekend and Pullman cases were perfect for the mid-20th century jetsetter. When fitted cases were offered for sale, they were simpler that those of the previous decades. By the late 1950s, small fitted cases, now called vanity cases, were used to hold cosmetics, lotions and creams.

By the 1940s, fitted cases were changing slightly in style while the price was dramatically lower than in the previous decades.

Dress Kits for Men

Advertised as "travel necessities," toilet sets designed specifically for men, had a variety of names. Traveling sets, dress kits, dressing cases, tourist sets and military sets were a few of the most common. In 1920, roll-up traveling cases made of walrus grain leather and fitted with genuine black ebony military brushes, hard rubber comb, soap and utility boxes, nickel mirror, tooth brush box, shaving soap box, box with shaving brush, tweezers, nail file and glass tooth powder jar listed for $20.00. A brown cowhide traveling set with a smooth finish fitted with the same toilet articles, plus an extra area used for holding a safety razor listed for $27.00. Cases designed to hold genuine ebony fittings were usually made of exotic grained leather. Morocco was very common in addition to brown elephant grain, black seal grain and fine goat leather with a walrus grain. The elephant grain was the most expensive at $50.00. Similar roll cases, designed for women, were fitted with French Ivory fittings.

Thirteen-piece men's dress kit from the 1930s, consisting of ebony military brush, ebony hat brush, soap dish, toothbrush holder, razor box, shoe horn, two cuticle tools, nail file, cloth brush, easel-backed mirror, lotion bottle and shaving brush cylinder. $135-165.

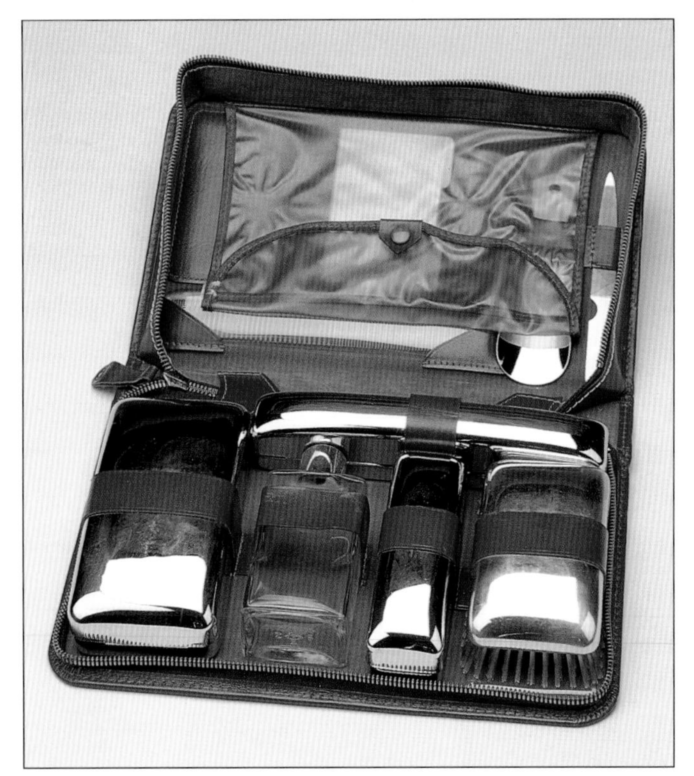

Toilet case for men by *Rumpp* called "Old Briar". The tan leather case zips open to reveal wonderful Art Deco chrome fittings. This set was never used and still in mint condition. $100-145.

Rumpp toilet case closed and with the original box.

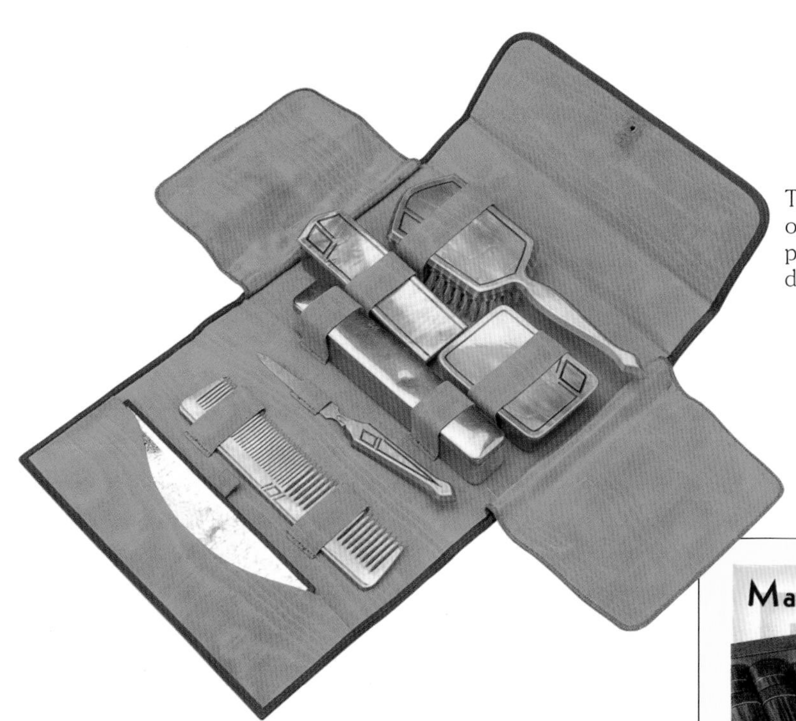

Traveling case designed for a woman made of top grain cowhide and fitted with maize pearl-on-amber Pyralin fittings with inlaid designs by *ART-Y-ZAN*. $145-175.

Mastercraft Dressing Cases

A Travel Necessity

MC443. Brown shark grain genuine seal, swirl grain leather lining. Turned edge wings. Fourteen fittings in natural ebony finish include—pair (2) military brushes, razor box, twin giant tubes for paste tubes, tooth brush and shaving brush tube with genuine Badger hair brush, cloth brush, leather covered mirror, nail clippers, shoe horn, file and comb. Extra silk lined pocket—sliding strap handle. Bronze Snap-Tite lock. Size closed 8x12 inches. Each **$49.50**

MC442. Brown shark grain cowhide with wine colored military striped leather lining. Turned edge wings and leather lined fitting straps. Fifteen de luxe fittings in natural ebony finish. Sliding strap handle. Closes with Snap-Tite lock. Size closed 7½x11¾ inches. Each . . **$37.50**

MC441—Black shark grain cowhide, brown trout grain leather lining; turned edge wings. Fifteen fittings in black ebony finish. Shaving brush tube equipped with genuine Badger hair brush. Sliding strap handle and Snap-Tite lock. Size closed 8x12 inches. Each **$30.00**

MC440. Brown shark grain cowhide, swirl grain leather lining. Twelve fittings in natural ebony finish—pair military brushes, utility box, shaving brush tube, tooth brush tube, razor box, lotion bottle in metal container, leather covered mirror, hat brush, comb, scissors and file. Sliding strap handle. Snap-Tite lock. Size closed 7⅛x11½ inches. Each **$27.00**

MC439. Brown shark grain cowhide with tan levant leather lining. Fitting loops leather lined. Fittings are in natural ebony finish and consist of military brush, utility box, shaving brush tube, tooth brush tube, razor box, cloth brush, leather covered easel mirror, manicure scissors, comb, and file. Case has sliding strap handle and closes with Snap-Tite lock. Size closed 7x10 inches. Each **$22.50**

MC438. Black cobra grain cowhide, brown trout grain leather lining. Fittings in natural ebony finish consist of military brush, utility box, razor box, tooth brush tube, shaving brush tube, leather covered mirror, manicure scissors, comb and file. Sliding strap handle and Snap-Tite lock. Size closed 6¼x10¾ inches. Each **$19.50**

Mastercraft dressing cases for men made of black shark grain and cobra grain cowhide with ebony fittings available from A.C.Becken in 1932.

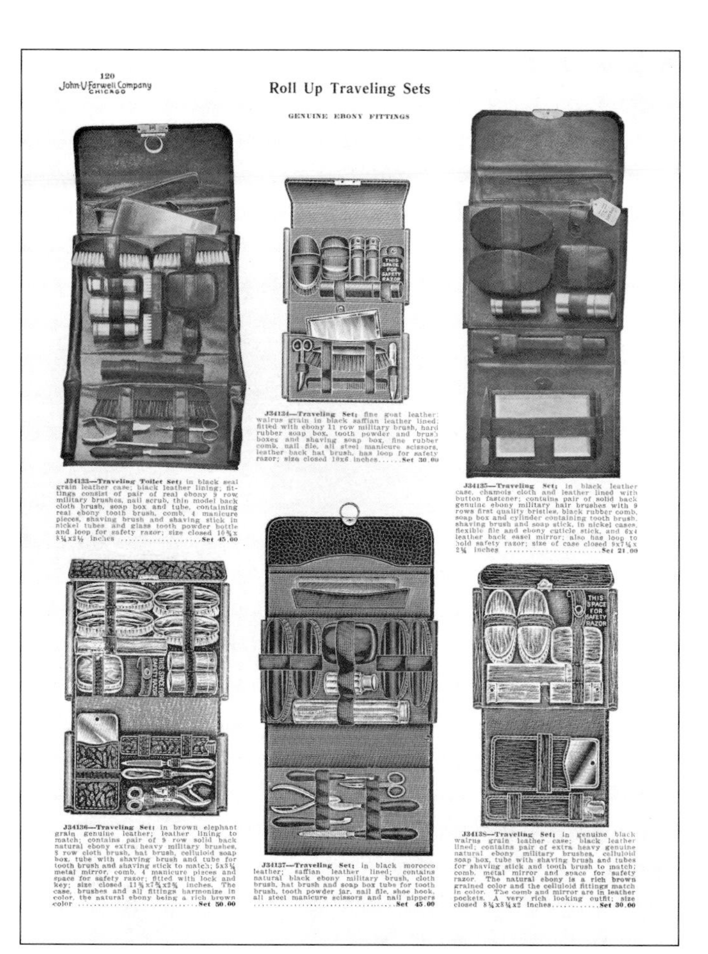

120

John V. Farwell Company Chicago

Roll Up Traveling Sets

GENUINE EBONY FITTINGS

J34133—Traveling Toilet Set; in black seal grain leather case; black leather lining; fittings consist of pair of real ebony 9 row military brushes, nail scrub, thin model back cloth brush, soap box and tube, containing real ebony tooth brush, comb, 4 manicure pieces, shaving brush and shaving stick in nickel tubes and glass tooth powder bottle and loop for safety razor; size closed 10¾x8¼x2½ inches . . . . **Set 45.00**

J34134—Traveling Set; fine goat leather; walrus grain in black saffian leather lined; fitted with ebony 11 row military brush, hard rubber soap box, tooth powder and brush boxes and shaving soap box, fine rubber comb, nail file, all steel manicure scissors, leather back hat brush, has loop for safety razor; size closed 10x6 inches. . . . . . **Set 30.00**

J34135—Traveling Set; in black leather case, chamois cloth and leather lined with button fastener; contains pair of solid back genuine ebony military hair brushes with 9 rows first quality bristles, black rubber comb, soap box and cylinder containing tooth brush, shaving brush and soap stick, in nickel cases, flexible file and ebony cuticle stick, and 6x4 leather back easel mirror; also has loop to hold safety razor; size of case closed 9x7¼x2¼ inches . . . . **Set 21.00**

J34136—Traveling Set; in brown elephant grain genuine leather; leather lining to match; contains pair of 9 row solid back natural ebony extra heavy military brushes, 8 row cloth brush, hat brush, celluloid soap box, tube with shaving brush and tube for tooth brush and shaving stick to match; 5x3¾ metal mirror, comb, 4 manicure pieces and space for safety razor; fitted with lock and key; size closed 11¾x7¾x2¾ inches. The case, brushes and all fittings harmonize in color, the natural ebony being a rich brown color . . . . **Set 50.00**

J34137—Traveling Set; in black morocco leather; saffian leather lined; contains natural black ebony military brush, cloth brush, hat brush and soap box tube for tooth brush, tooth powder jar, nail file, shoe hook, all steel manicure scissors and nail nippers . . . . **Set 45.00**

J34138—Traveling Set; in genuine black walrus grain leather case; black leather lined; contains pair of extra heavy genuine natural ebony military brushes, celluloid soap box, tube with shaving brush and tubes for shaving stick and tooth brush to match; comb, metal mirror and space for safety razor. The natural ebony is a rich brown grained color and the celluloid fittings match in color. The comb and mirror are in leather pockets. A very rich looking outfit; size closed 8¾x8¼x2 inches . . . . **Set 30.00**

Roll up traveling sets equipped with genuine ebony appointments were offered for sale in 1920. A brown elephant grain genuine leather roll up with a leather lining and fourteen ebony fittings sold for $50.00.

Besides the popular roll-up cases common in the 1920s, cases with Talon hook-less fasteners became the rage in the 1930s. Fittings were made with a natural ebony finish and the cases were constructed of exotic grains of leather. Cobra grain, shark grain and buffalo grain were some of the exterior leathers used while linings were made of tan monkey grain and brown trout grain. These were very unusual combinations and this is what created consumer interest and kept them popular for many years. Page after page in wholesale catalogs featured these sets for men.

In 1932, novel and compact travel kits were described as having the finest quality leathers, workmanship and fittings. Tan Russian boar, brown jaguar grain cowhide and black real shrunken pig leather were popular choices and fitted with celluloid fittings.

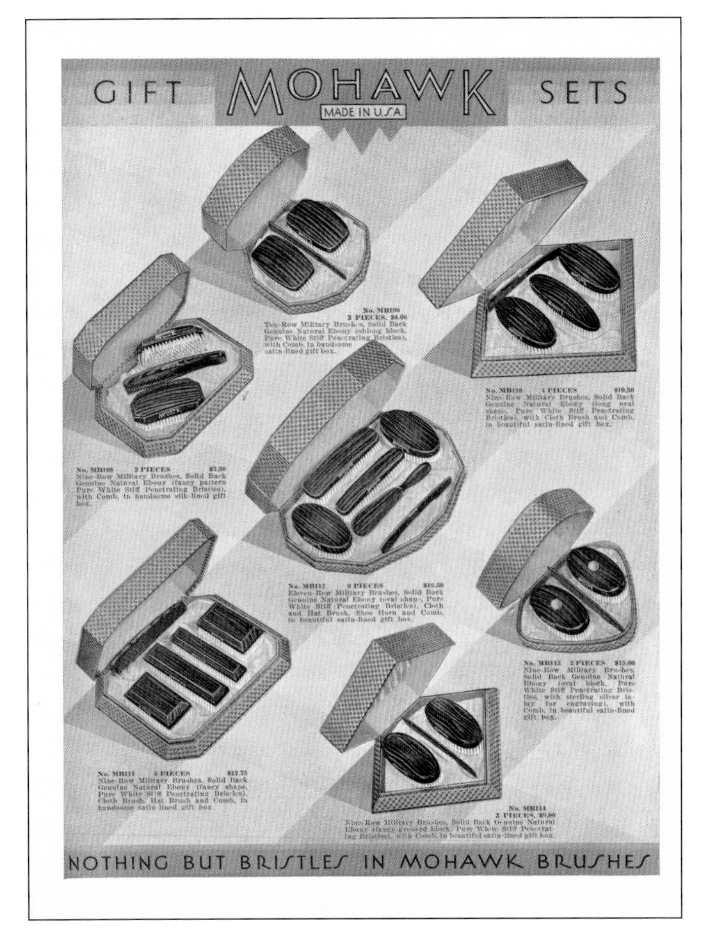

Popular styles in men's satin-lined gift sets by *Mohawk* offered in 1932. The brushes were described as having *pure white stiff penetrating bristles.*

Four-piece men's toilet kit made of jet Pyralin. The set contains a razor box, soap box, shaving brush tube and toothbrush holder. $60-85 set.

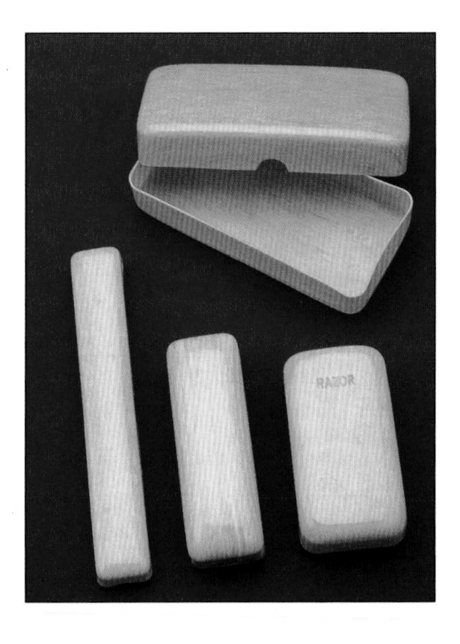

Another four-piece toilet kit for men made of wood-grained celluloid. $60-85 set.

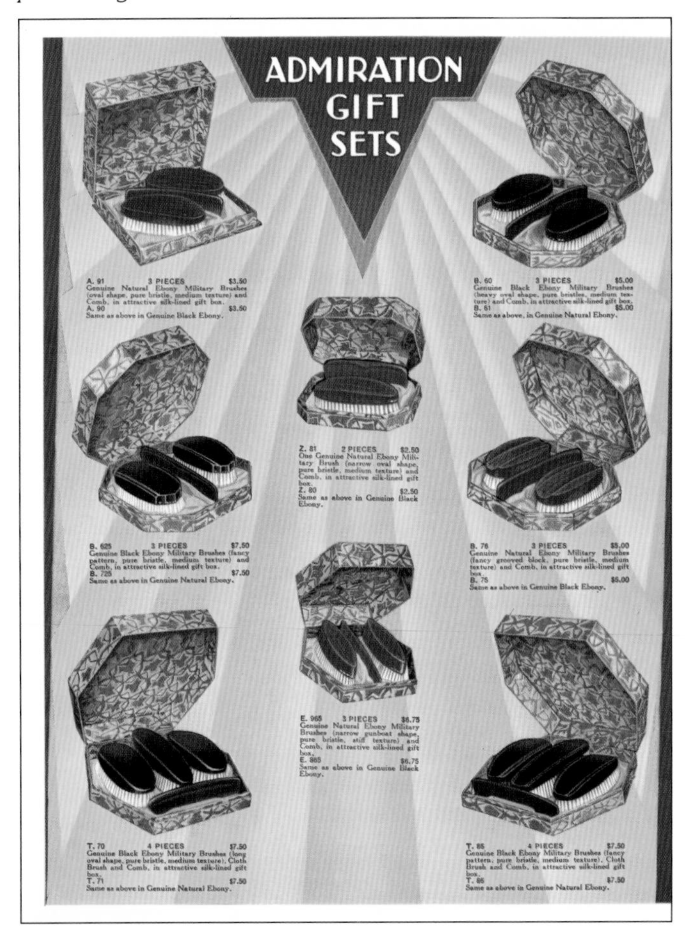

Admiration Gift Sets for men fitted consisted of military brushes, cloth brushes and a comb made of genuine black and natural ebony.

Military gift sets were also popular for men in the early 20th century. Consisting of usually two military hairbrushes, a cloth brush and a comb, these sets were advertised as being *mannishly designed*. Early on, the fittings were celluloid, either in a tan color or black. Then genuine ebony and ebony-finished fittings were offered. By the mid 1930s, chrome and enamel military sets were in vogue.

Fancier gift sets, made of genuine ebony, were decorated with sterling silver shields and advertised as *attractive modern designs*. Some brush sets consisted of six to eight pieces. Besides the two military hairbrushes and the cloth brush, sets would include a hat brush and a genuine badger shaving brush. The leather cases were often monogrammed with gold lettering. Theses sets became a perfect gift item for a man for any occasion.

KENTLEYCRAFT
Talon "Zipper" fastenings, soft, flexible type; finest leathers, highest quality fittings.
DRESSING CASES
MC536. Pillow type zipper case, opens on two sides with Talon fastener. Brown walrus grain leather. Rubberized lining with extra pocket. Fittings in black ebony finish include: military brush, comb, tooth brush tube, shaving brush tube, and razor box. Size 10½"x8". Each . $8.50
MC542. New type zipper case. Brown African buffalo grain cowhide. Tan leather lining. Fittings in natural ebony finish, military brush, tooth paste tube, utility box, shaving brush tube, razor box, comb, file, and mirror. Size closed 9¼"x7½". Each $19.50
MC544. Pillow type zipper case opening on three sides with Talon fastener. Black shark grain cowhide with brown trout grain leather lining. Leather snap flap compartment for paste tubes. Fittings in black ebony finish include: military brush, shaving brush tube, tooth brush tube, lotion bottle, soap box, razor box, comb, cloth brush, file, nail scissors, and shoe horn. Case has leather handle. Size closed 13"x8½". Each $22.50
MC539. New type zipper case. Walrus grain genuine leather, rubberized lining. Fittings in black ebony finish, military brush, razor box, shaving and tooth brush tubes, paste tube, comb, nail file. Size closed 9"x6¾". Each $10.00
MC541. New type zipper case with wide gussets. Opens on three sides with Talon fastener. Brown shark grained cowhide. Trout grain leather lining. Fittings in black ebony finish include: military brush, razor box, tooth brush tube, shaving brush tube, utility box, cloth brush, comb, and file. Size closed 11"x6½". Each $15.00
MC534. Brown shark grain genuine leather envelope type case, fitted on one side with Talon zipper, rubberized lining. Fittings in black ebony finish consist of: handle hair brush, cloth brush, comb, tooth brush tube, shaving brush tube, and razor box. Size 10"x6". Each $5.00
MC541
MC534

Kentleycraft dressing cases for men were made of exotic grains of cowhide. Brown African buffalo, shark grain and walrus grain were a few that were offered in 1933.

Kentleycraft offered folding style dressing cases for men in 1933. Advertised as *luxurious travel necessities*, these sets became the perfect, practical and appreciated gift. Common designs were the black shark-grain cowhide with the tan monkey grain interiors. Others were made with tan trout grain leather linings. Fittings were usually natural ebony. Prices ranged from $7.50 for a six-piece set to $31.50 for a fifteen-piece set.

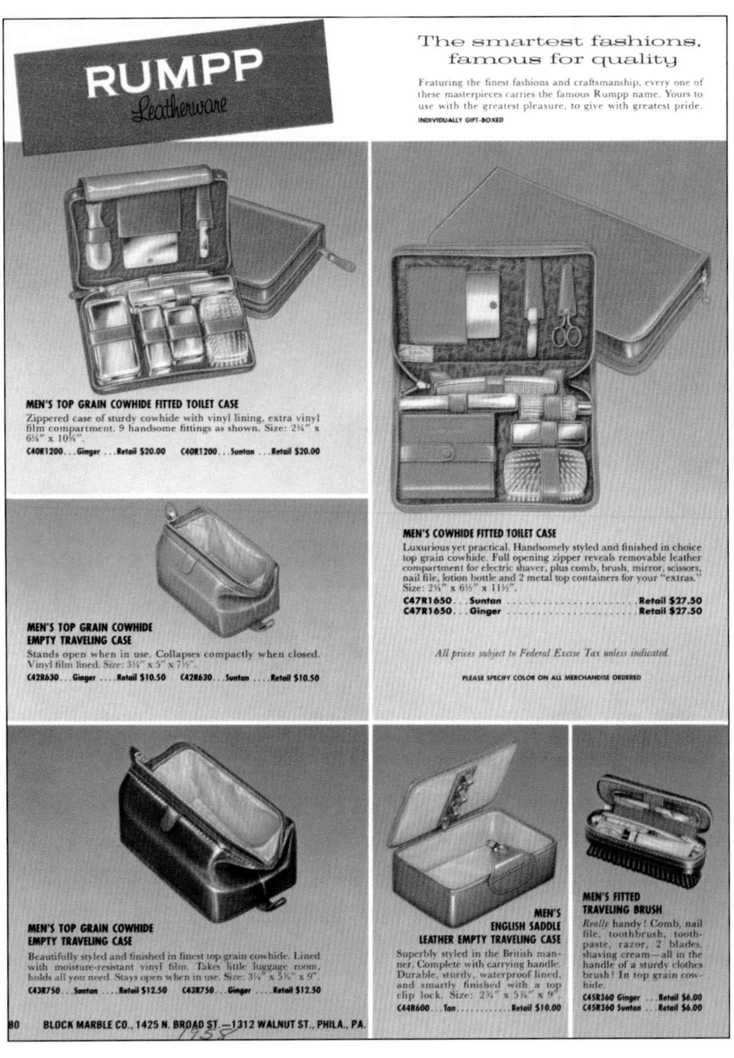

Men's top grain cowhide fitted toilet case with chromium-plated fittings by *Rumpp* offered for sale in 1958.

Men's zipper toilet sets or fitted dressing cases were available for purchase in 1942 by Empire Brand Luggage. A popular model was the Suntan Russet Case. Made of genuine top grain cowhide leather in the new "Suntan Shade," the case had a suede cloth lining, an oiled silk utility pocket and Talon zipper fastener around the sides. The kit contained two military brushes, clothes brush, utility box, razor box, toothbrush holder, comb, easel-backed mirror, shoehorn, scissors, file and combination nail pick and extractor. The set retailed for $16.95.The variety was endless and the most useful travel companion for men on the go.

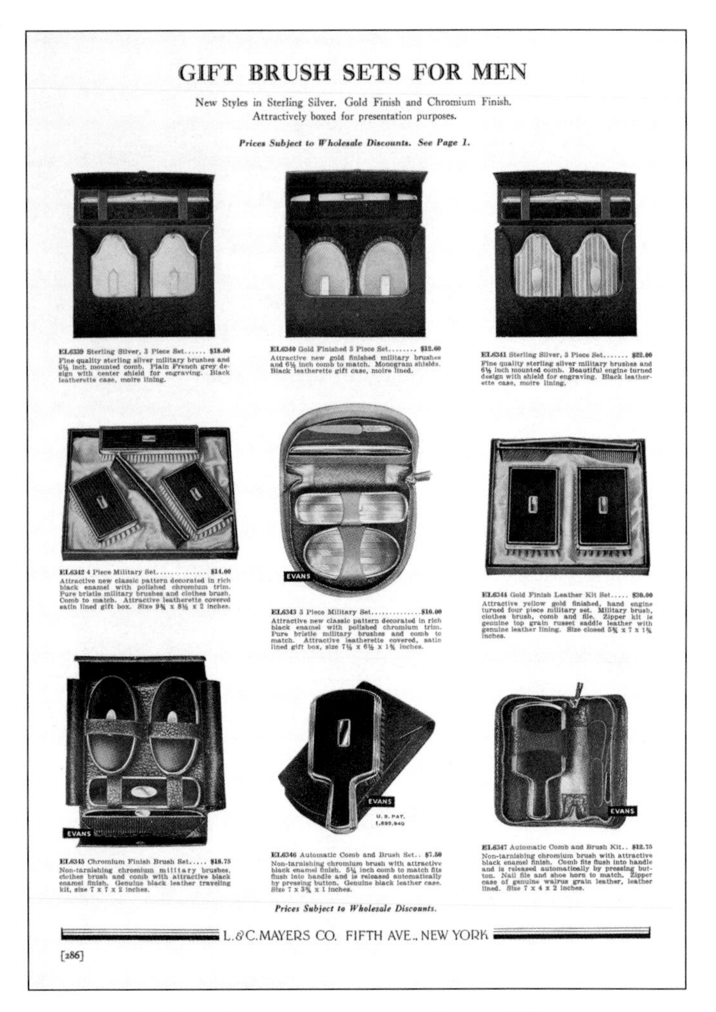

In the mid 1930s, gift sets for men were also available with sterling silver fittings in addition to gold finished and non-tarnishing chromium finished sets. They were packaged attractively for gift-giving.

Chapter Seven

Manicure Sets

In the 1890s, the finest sterling silver and quadruple plate was used in the manufacture of manicure implements. Exquisite sets were offered for sale with heavily embossed, engraved and chased designs. Manicure sets could be as simple or as complex as one wanted them to be. A simple set could consist of a pair of manicure scissors, a nail file and a shoe buttoner in a plush-lined box. A ten-piece set, including a tray, would also include scissors, files, cuticle knife, nail buffer, cream boxes and button hook. But an even more elaborate set would contain items such as tweezers, tooth brushes, hair curlers, nail brushes, hair curlers, corn knives and glove buttoners. A fancy sterling silver seven-piece manicure set in a silk-lined kid leather case listed for $19.00 in 1896. That was quite a bit of money at that time for a manicure set.

In 1902, Sears, Roebuck & Company, known as the "Cheapest Supply House on Earth," offered a large assortment of sterling silver novelties and manicure implements in their mail-order catalog. These goods became so popular that Sears stated... *the demand having increased to such a wonderful extent that necessitated our pulling in a much larger line of these goods.* Other companies followed suit and page after page in mail order and wholesale catalogs were filled with sterling silver manicure implements to purchase for personal use or gift giving. All of the novelties, as they were sometimes referred to, were advertised as being 925/1000 fine and all were hand engraved or hand chased.

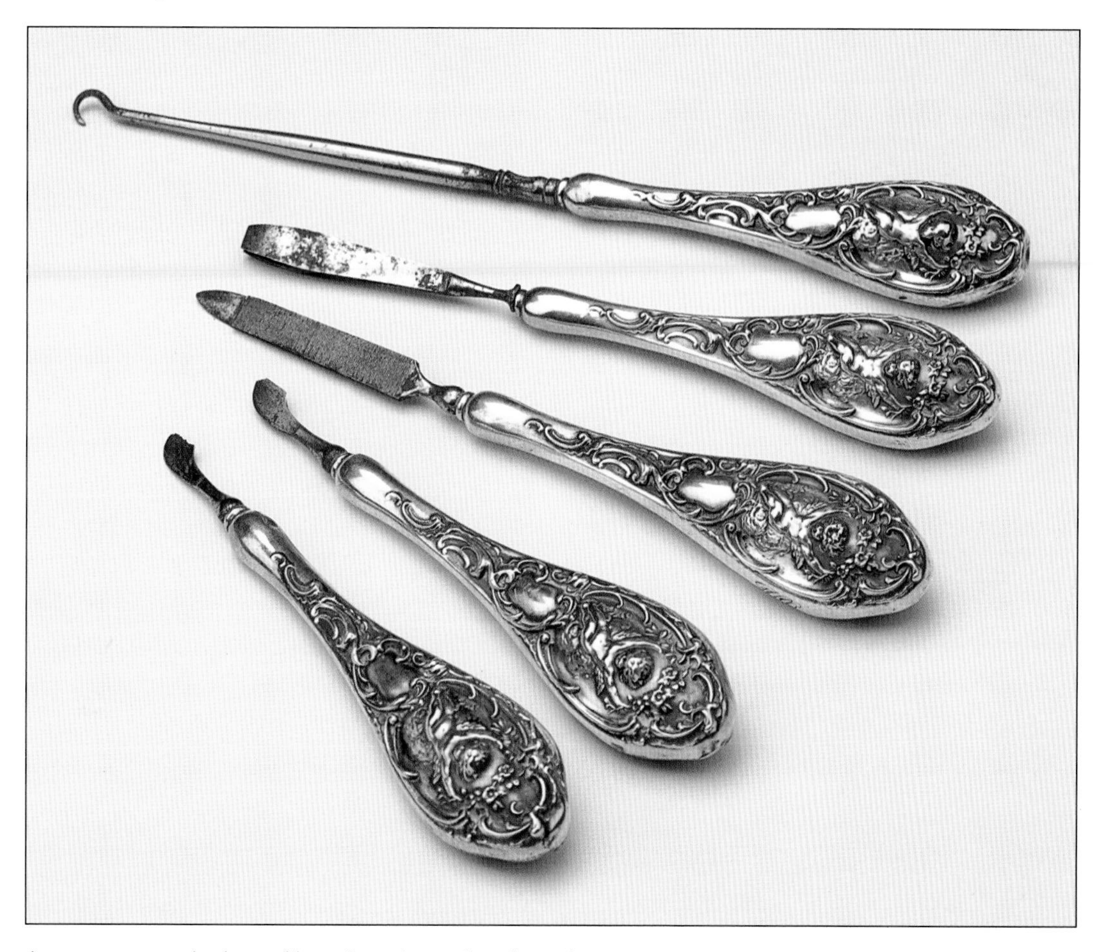

A gorgeous matched set of heavily embossed sterling silver manicure implements with cupid motifs. $350-400 set.

Cuticle knife and nail files all made with heavy embossed sterling silver handles in Victorian and Art Nouveau designs. Cuticle knife $75-100; nail files $95-125 each.

Ebony, sterling silver and ormolu nail buffers measuring three to six inches long. $75-135 each.

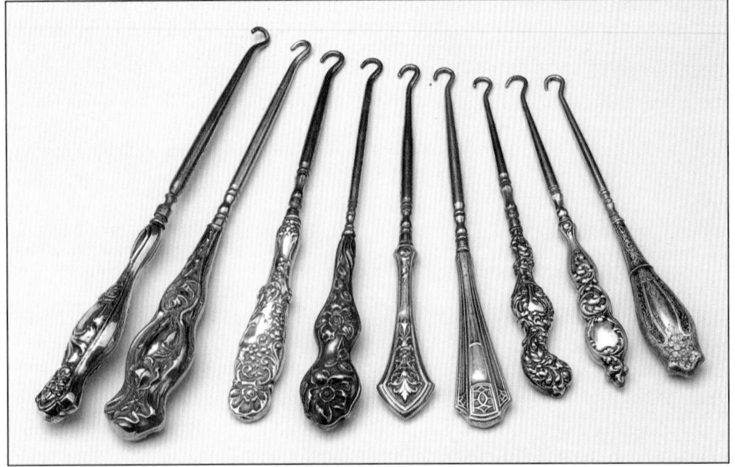

Exceptional embossing and chasing are evident in these period shoe buttoners made of sterling silver from the late 19th and early 20th centuries. $95-145 each.

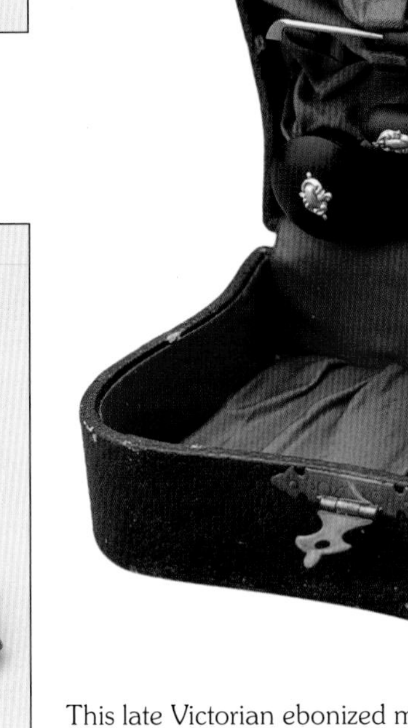

This late Victorian ebonized manicure set with ornate silver trim fits nicely into a purple silk-lined case. The set consists of a nail buffer, nail file, shoe buttoner, cuticle knife, cuticle stick and cream jar. This was a very popular style in toilet accessories around 1895. $165-195 set.

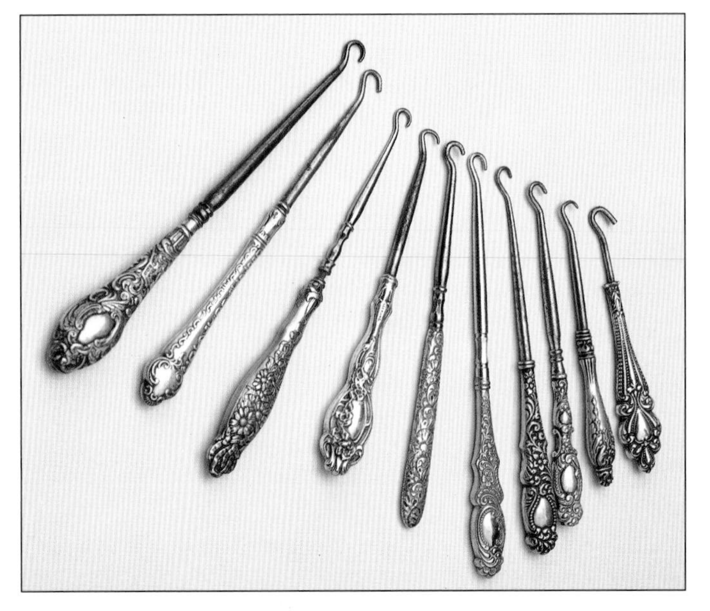

Ten different Victorian shoe and glove buttoners made of fancy embossed sterling silver. $75-135 each.

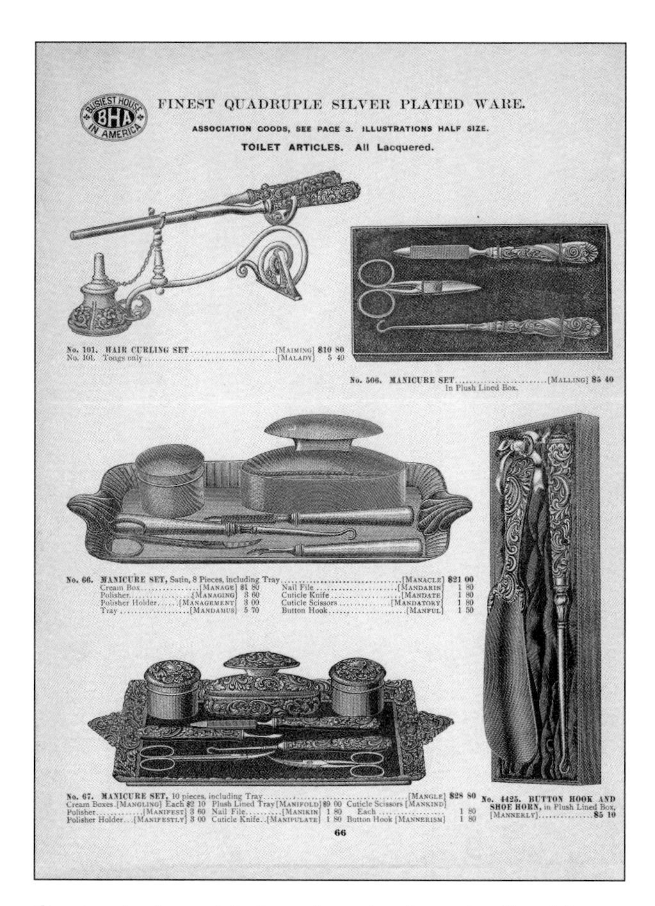

Quadruple silver plated manicure implements offered for sale in 1895 from the BHA (Busiest House in America) Illustrated Catalog. All of the plated items were lacquered to prevent tarnishing.

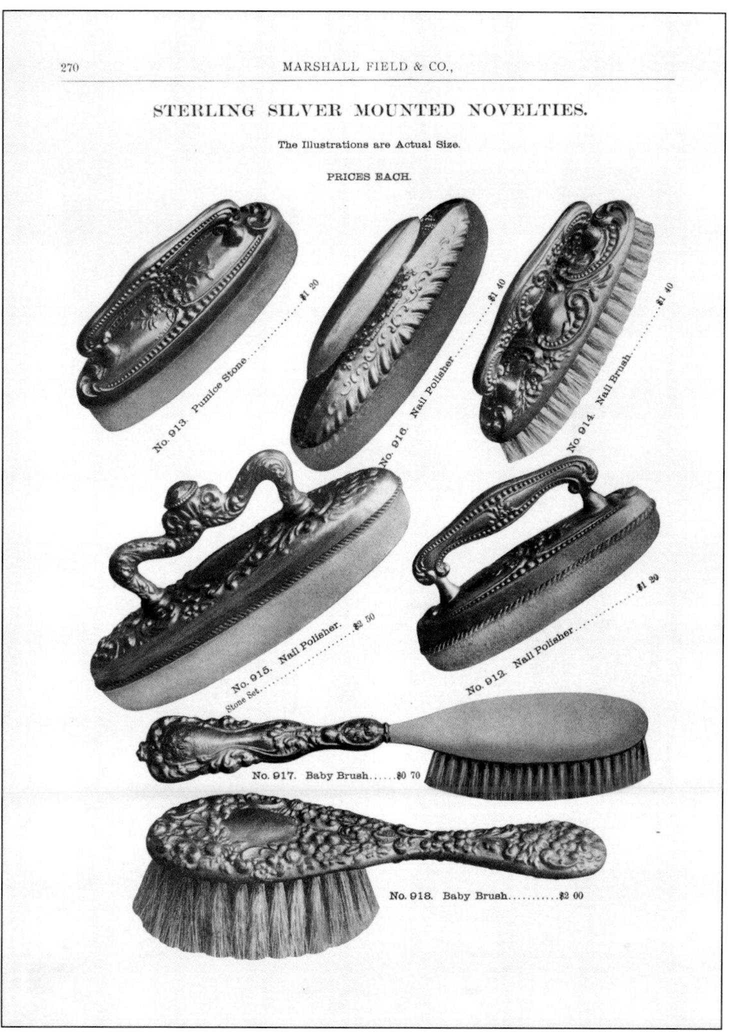

Pumice stones, nail polishers and baby items mounted with embossed sterling silver and offered for sale from Marshall Field in 1900.

Elaborate sterling silver mounted manicure implements and miscellaneous toilet goods pictured here in a popular pattern for 1900.

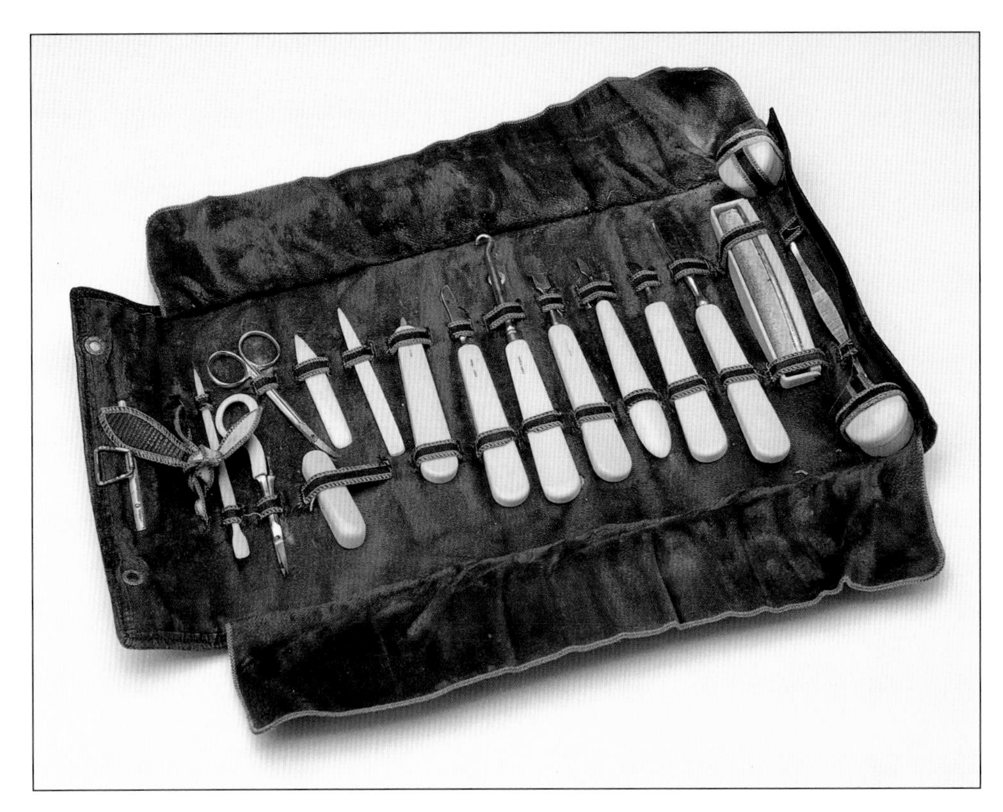

Seventeen-piece French Ivory manicure set in green velvet lined leather roll case. $225-265 set.

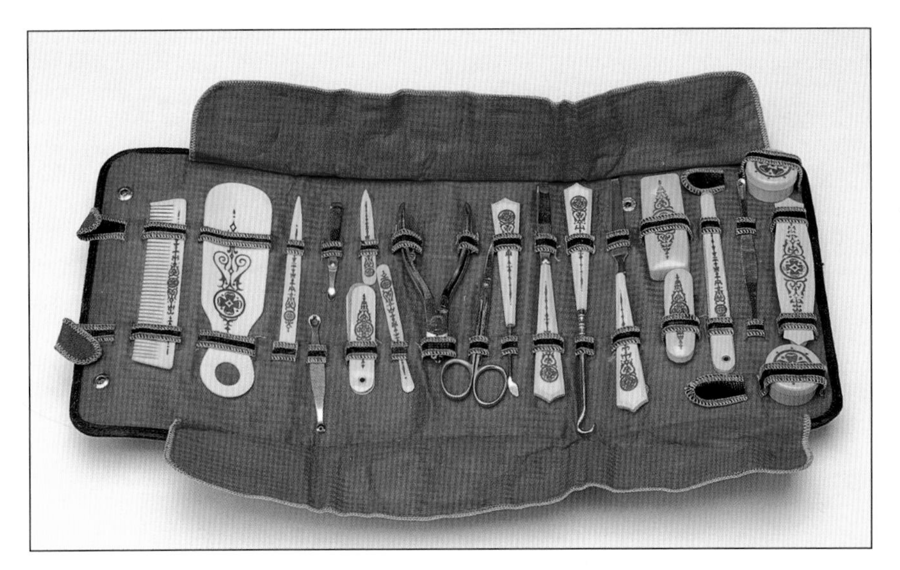

This is a complete twenty-one piece manicure set made of celluloid accented with inlaid and colored designs. Each piece fits into a silk-lined leather case. It is very rare to find a set this complete. $250-300 set.

This fancy figured genuine leather snap case with embossed and colored cameo top houses an early 20th century manicure set.

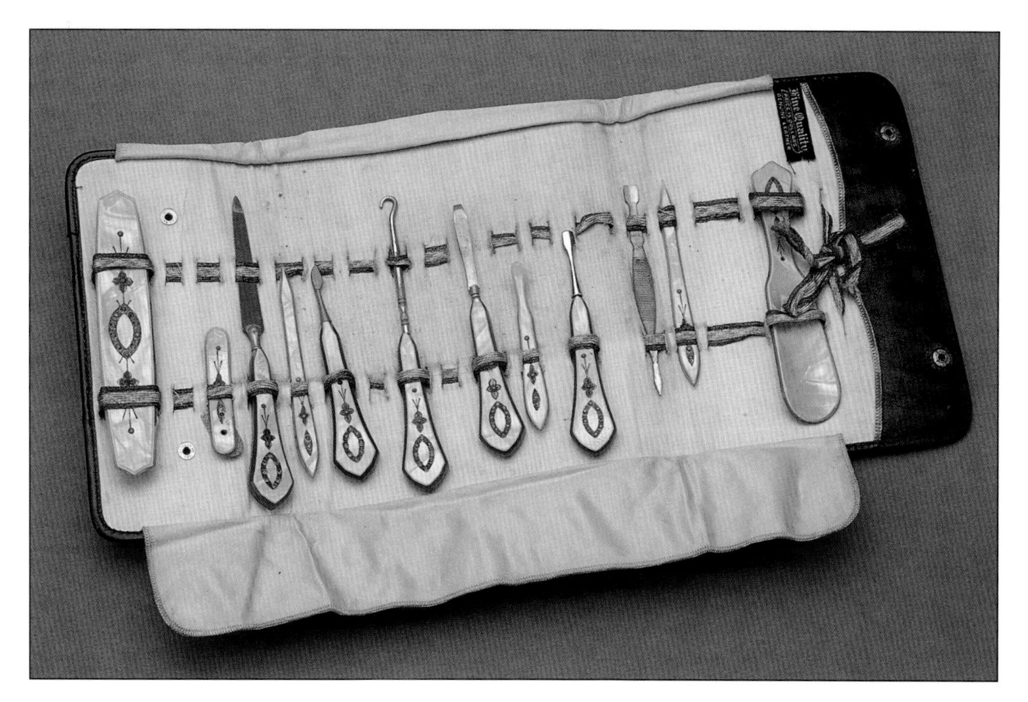

Same leather case shown open to reveal the jeweled rose pearl-on-amber pyralin manicure set consisting of twelve of the original seventeen pieces. $145-185 set.

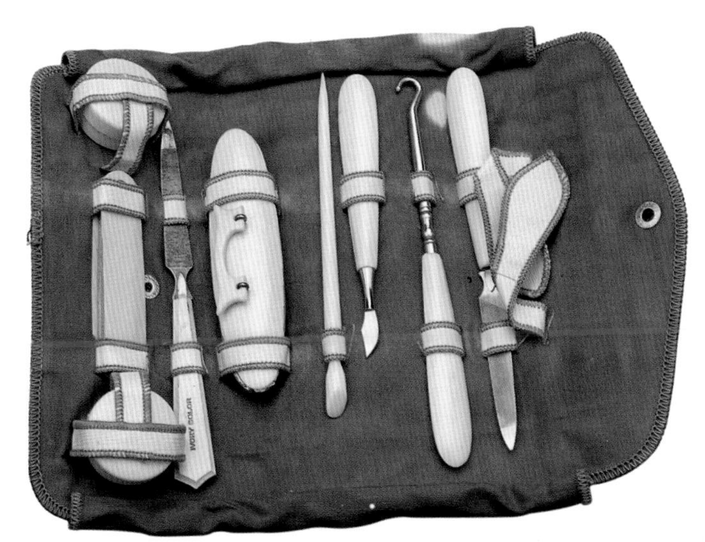

Early 20th century manicure set fitted with nine French Ivory implements in a genuine leather roll case. $135-175 set.

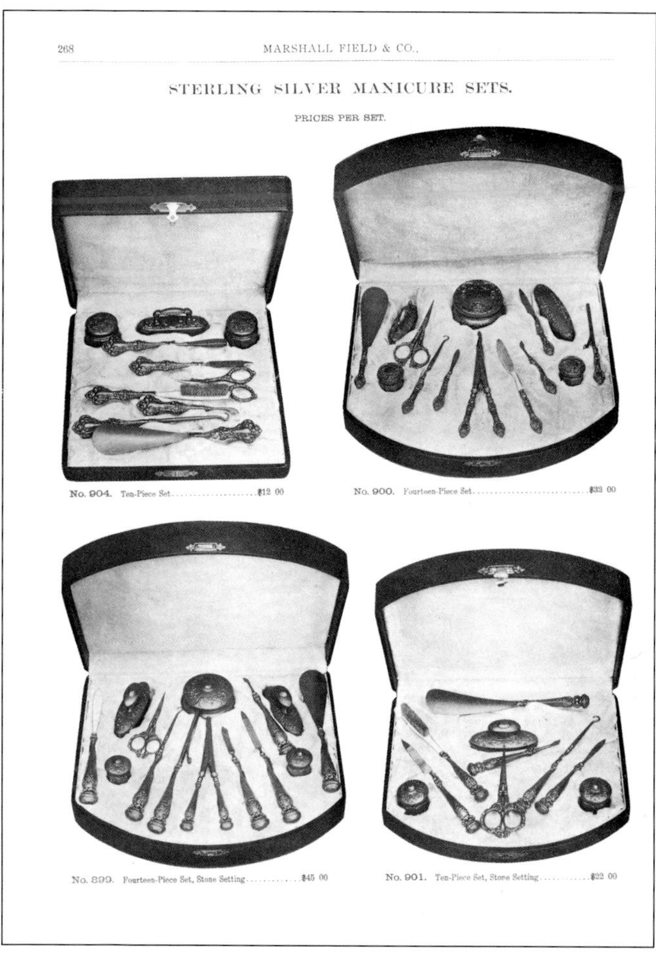

268 MARSHALL FIELD & CO.,

STERLING SILVER MANICURE SETS.

PRICES PER SET.

No. 904. Ten-Piece Set....................$12 00

No. 900. Fourteen-Piece Set..........................$33 00

No. 899. Fourteen-Piece Set, Stone Setting............$45 00

No. 901. Ten-Piece Set, Stone Setting............$22 00

Ten to fourteen-piece sterling silver manicure sets from 1900 ranging in price from $12.00 to $45.00 per set.

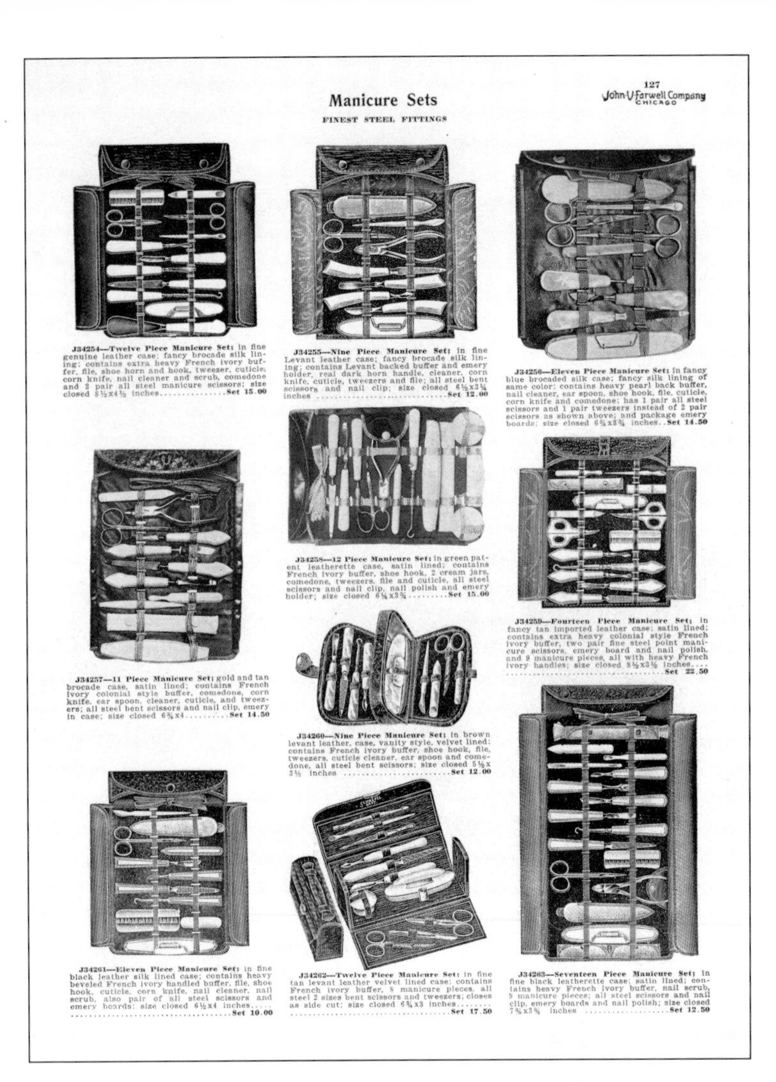

127

Manicure Sets

John V. Farwell Company CHICAGO

FINEST STEEL FITTINGS

J34254—Twelve Piece Manicure Set; in fine genuine leather case; fancy brocade silk lining; contains extra heavy French ivory buffer, file, shoe horn and hook, tweezer, cuticle, corn knife, nail cleaner and scrub, comedone and 2 pair all steel manicure scissors; size closed 8½x4½ inches................**Set 15.00**

J34255—Nine Piece Manicure Set; in fine Levant leather case; fancy brocade silk lining; contains Levant backed buffer and emery holder, real dark horn handle, cleaner, corn knife, cuticle, tweezers and file; all steel bent scissors and nail clip; size closed 6½x3¼ inches ....**Set 12.00**

J34256—Eleven Piece Manicure Set; in fancy blue brocaded silk case; fancy silk lining of same color; contains heavy pearl back buffer, nail cleaner, ear spoon, shoe hook, file, cuticle, corn knife and comedone; has 1 pair all steel scissors and 1 pair tweezers instead of 2 pair scissors as shown above; and package emery boards; size closed 6¾x3¾ inches..**Set 14.50**

J34257—11 Piece Manicure Set; gold and tan brocade case, satin lined; contains French ivory colonial style buffer, comedone, corn knife, ear spoon, cleaner, cuticle, and tweezers; all steel bent scissors and nail clip, emery in case; size closed 6¾x4..........**Set 14.50**

J34258—12 Piece Manicure Set; in green patent leatherette case, satin lined; contains French ivory buffer, shoe hook, 2 cream jars, comedone, tweezers, file and cuticle, all steel scissors and nail clip, nail polish and emery holder; size closed 6¼x3¾..........**Set 15.00**

J34259—Fourteen Piece Manicure Set; in fancy tan imported leather case; satin lined; contains extra heavy colonial style French ivory buffer, two pair fine steel point manicure scissors, emery board and nail polish, and 9 manicure pieces, all with heavy French ivory handles; size closed 8½x3½ inches.... ....**Set 22.50**

J34260—Nine Piece Manicure Set; in brown levant leather, case, vanity style, velvet lined; contains French ivory buffer, shoe hook, file, tweezers, cuticle cleaner, ear spoon and comedone, all steel bent scissors; size closed 5½x3½ inches ....**Set 12.00**

J34261—Eleven Piece Manicure Set; in fine black leather silk lined case; contains heavy beveled French ivory handled buffer, file, shoe hook, cuticle, corn knife, nail cleaner, nail scrub, also pair of all steel scissors and emery boards; size closed 6½x4 inches..... ....**Set 10.00**

J34262—Twelve Piece Manicure Set; in fine tan levant leather velvet lined case; contains French ivory buffer, 8 manicure pieces, all steel 2 sizes bent scissors and tweezers; closes as side cut; size closed 6¾x3 inches........ ....**Set 17.50**

J34263—Seventeen Piece Manicure Set; in fine black leatherette case; satin lined; contains heavy French ivory buffer, nail scrub, 9 manicure pieces; all steel scissors and nail clip, emery boards and nail polish; size closed 7¾x2¾ inches ....**Set 12.50**

Manicure sets offered for sale in 1920 with French Ivory handles and the finest steel fittings presented in genuine leather and leatherette cases.

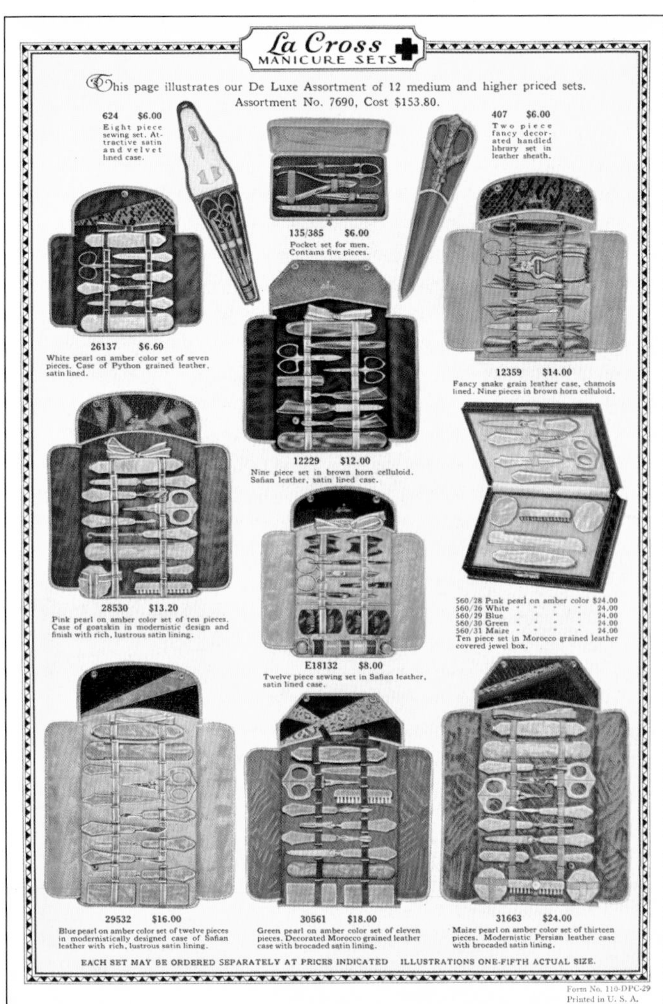

La Cross
MANICURE SETS

This page illustrates our De Luxe Assortment of 12 medium and higher priced sets.
Assortment No. 7690, Cost $153.80.

624 $6.00
Eight piece sewing set. Attractive satin and velvet lined case.

135/385 $6.00
Pocket set for men. Contains five pieces.

407 $6.00
Two piece fancy decorated handled library set in leather sheath.

26137 $6.60
White pearl on amber color set of seven pieces. Case of Python grained leather, satin lined.

12229 $12.00
Nine piece set in brown horn celluloid. Safian leather, satin lined case.

12359 $14.00
Fancy snake grain leather case, chamois lined. Nine pieces in brown horn celluloid.

28530 $13.20
Pink pearl on amber color set of ten pieces. Case of goatskin in modernistic design and finish with rich, lustrous satin lining.

E18132 $8.00
Twelve piece sewing set in Safian leather, satin lined case.

560/28 Pink pearl on amber color $24.00
560/26 White " " " " 24.00
560/29 Blue " " " " 24.00
560/30 Green " " " " 24.00
560/31 Maize " " " " 24.00
Ten piece set in Morocco grained leather covered jewel box.

29532 $16.00
Blue pearl on amber color set of twelve pieces in modernistically designed case of Safian leather with rich, lustrous satin lining.

30561 $18.00
Green pearl on amber color set of eleven pieces. Decorated Morocco grained leather case with brocaded satin lining.

31663 $24.00
Maize pearl on amber color set of thirteen pieces. Modernistic Persian leather case with brocaded satin lining.

EACH SET MAY BE ORDERED SEPARATELY AT PRICES INDICATED ILLUSTRATIONS ONE-FIFTH ACTUAL SIZE.

Form No. 110-DPC-29
Printed in U. S. A.
BACK

Colorful pearl-on-amber Pyralin and brown horn celluloid manicure sets by *La Cross* and available for sale from May & Malone in 1930. The cases were made of goatskin, Safian leather, Morocco grained leather and Python grained leather with colorful satin and brocaded linings.

By the 1920s, manicure sets were becoming extremely large and often complex. Sets consisted of over twenty-five pieces and occasionally it was hard to figure out what each piece was actually designed for. Large sets were usually housed in either genuine leather or leatherette roll cases. The fittings were made of the finest steel and the handles were French Ivory, tortoise, or mother of pearl. Two-toned composites like amber and Ivory Pyralin were common as well as handles made of jeweled Pyralin in different colors. Unusual implements such as corn knives, nail scrubs, blackhead removers and hoof sticks were found in large sets.

LaCross manicure sets were popular in 1930. They were offered in a colorful assortment of blue, maize, pink or white pearl-on amber Pyralin. A more subdued set was available in brown horn celluloid. Seven to thirteen-piece sets were housed in cases made of Morocco grained leather, safian leather or fancy snake grained leather. Prices ranged from $6.60 to $24.00 per set. LaCross sets were advertised as *modernistic and keeping in step with the trend of the times.* The implements were described as having *smart Sheffield finishes.*

Art Deco manicure set consisting of ten manicure implements, eight of which are made of blue Premalite. The inside of the box is fitted with a mirror in the lid surrounded by silk fabric creating a very elegant look. $165-195 set.

Same box shown closed to reveal colorful Art Deco designs on the lid.

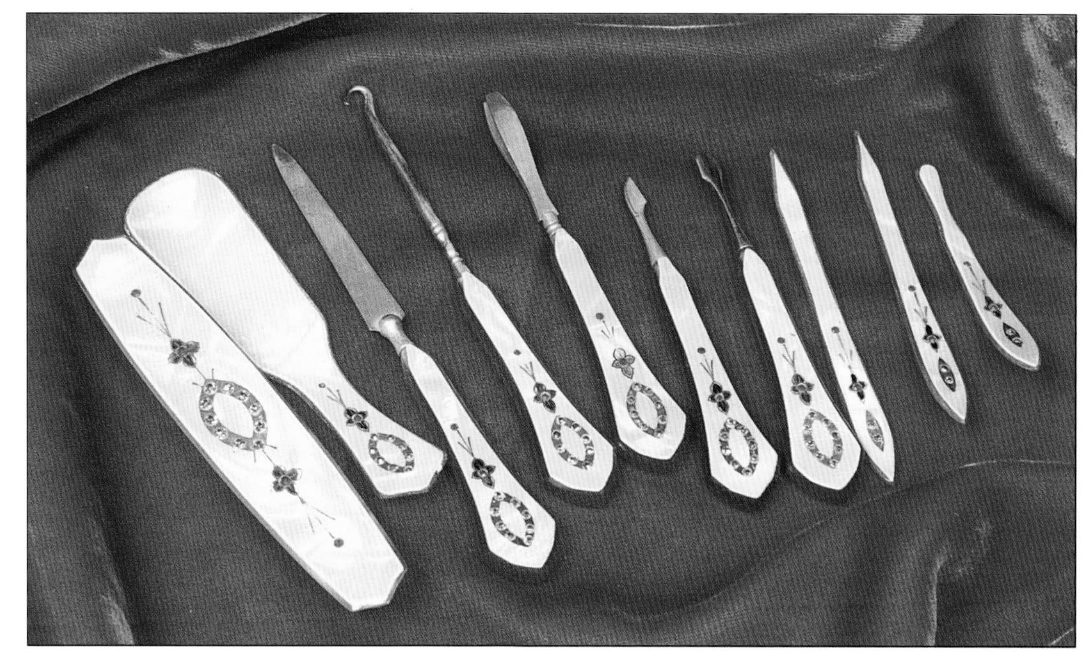

Ten-piece manicure set made of rose pearl-on-amber Pyralin. The set was further enhanced with applied decorations and paste stones. $120-145 set.

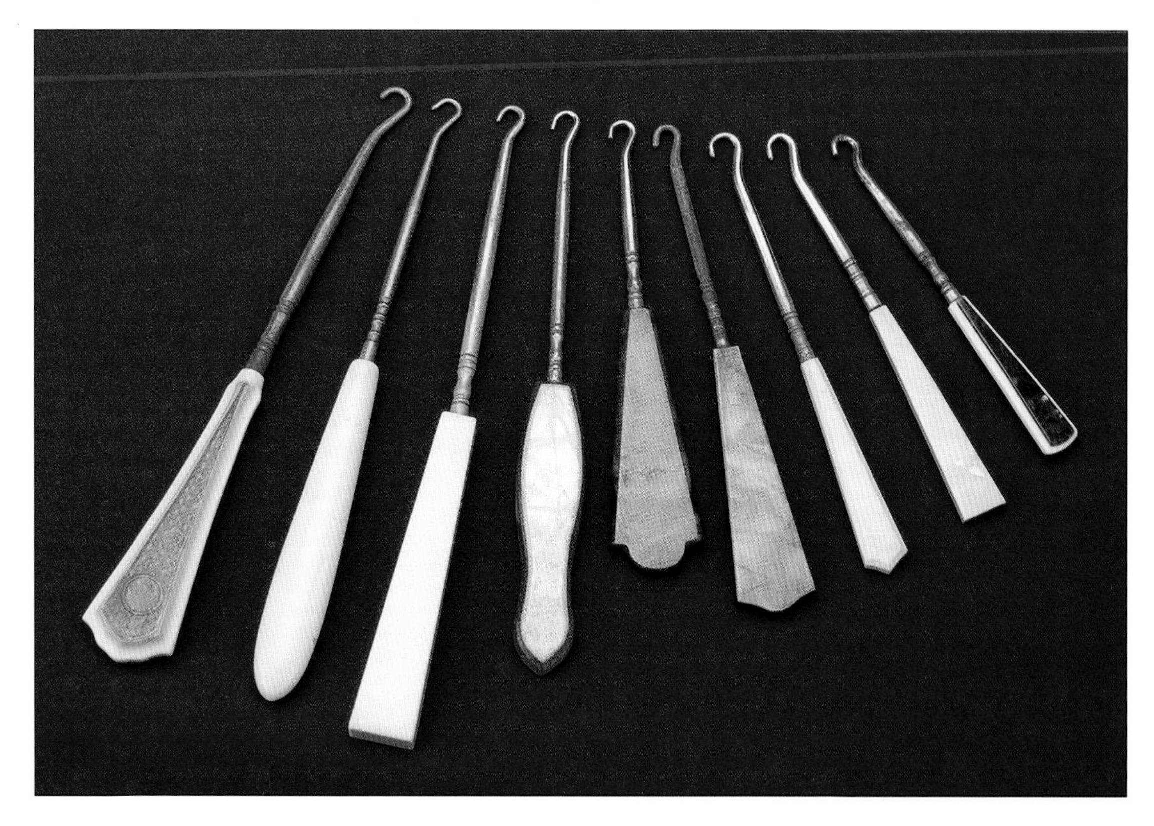

Assortment of nine shoe buttoners made of French Ivory, Lucite and Pyralin in a variety of shapes and sizes. $25-50 each.

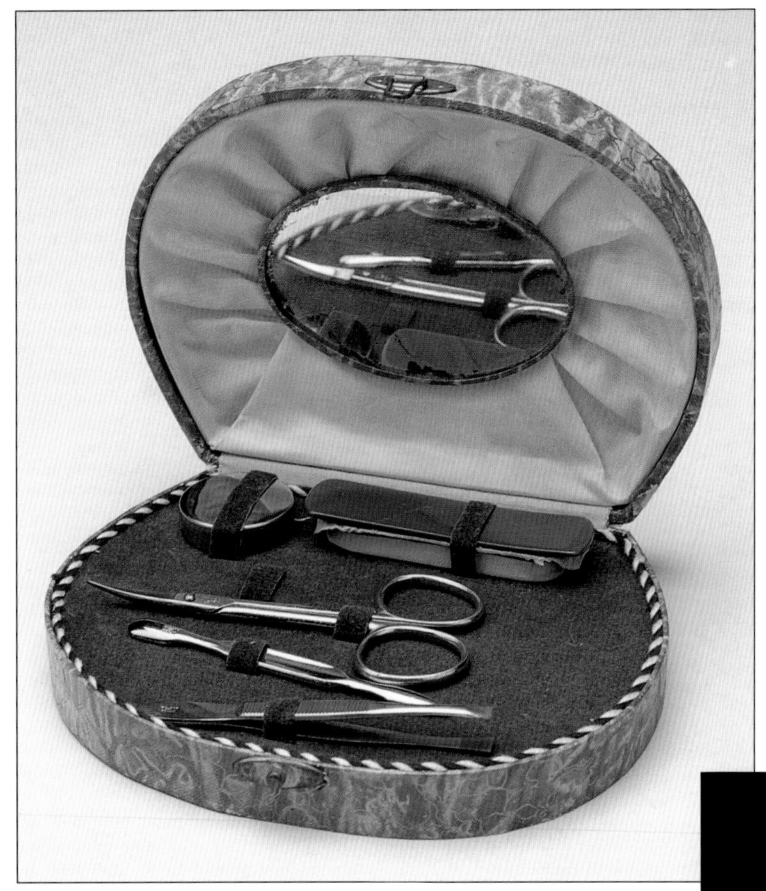

1940s manicure set equipped with five manicure implements in a charming cardboard and fabric display box. $70-90.

The box is all that remains of an early 20th century manicure set. Obviously the original owner thought enough of this wonderful old box to cherish it long after the implements had disappeared. $35-50.

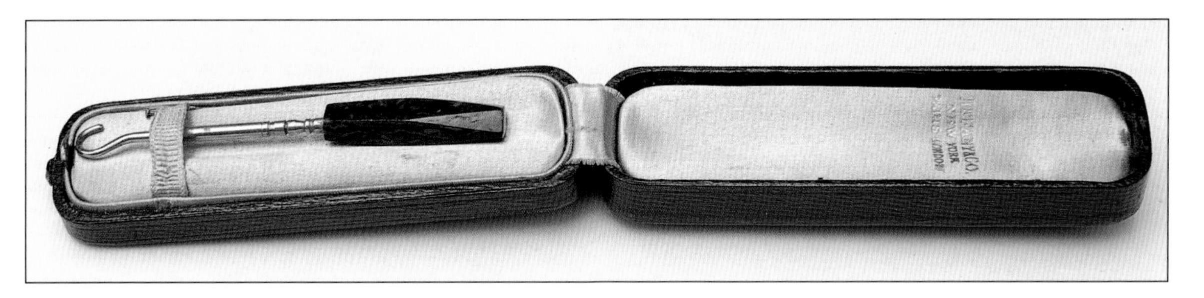

Exquisite *Tiffany & Co.* glove buttoner made of lapis pyralin and still nestled in its original silk-lined leather presentation box. $185-235.

In 1932, Premier Manicure Ensembles were made of a new material called Premalite. Similar to celluloid, premalite was described as a *non-inflammable creation rivaling the exquisite beauty of glittering jewels –of jade, of coral, cornelian and a myriad of others.* Since celluloid was known for being highly flammable, premalite became a desirable alternative. Premier manicure ensembles had a modern appearance and they were packaged in colorful Art Deco presentation boxes or satin-lined leather cases. They were wonderful gift items.

As the years went on, the overall appearance of the manicure set changed. In the late 19th and early 20th centuries, sterling silver was the material used most often. In the 1920s and 1930s, early plastics were common. By the 1940s, 18kt gold plate was used in addition to chrome and nickel. In 1932, manicure sets made with Galilith handles and genuine steel were offered for sale. Prices ranged from $1.50 for a six-piece set to $6.35 for an eleven-piece set. Sets were styled for utility and designed not only for home use but travel as well. In the late 1930s, the overall look of a traditional manicure set was changing. Certain items were removed from the kit while others were added. Cuticle creams, lotions and polishes were now part of a manicure ensemble.

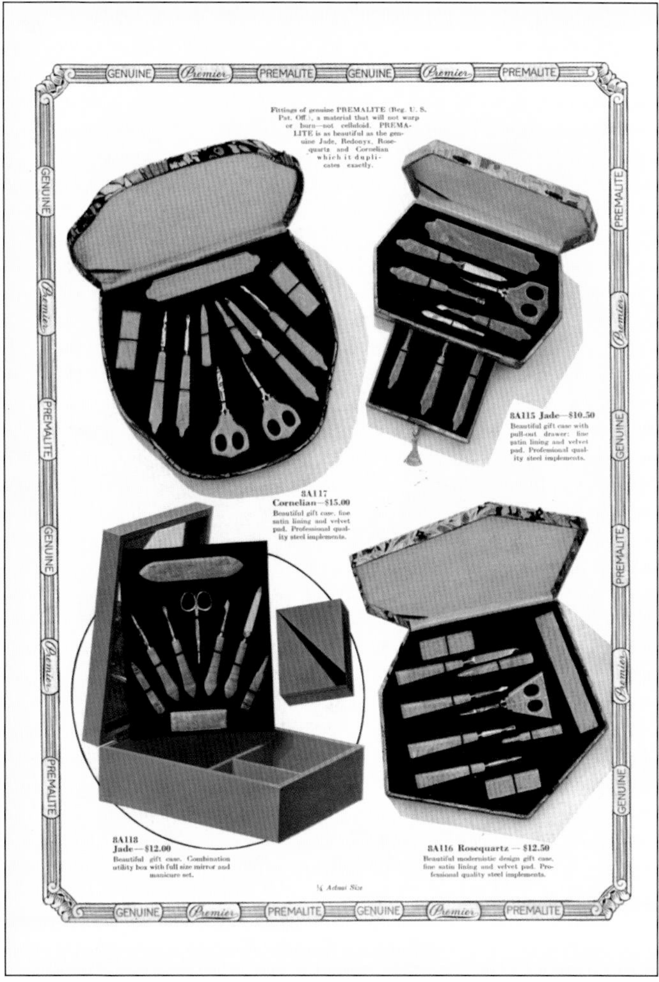

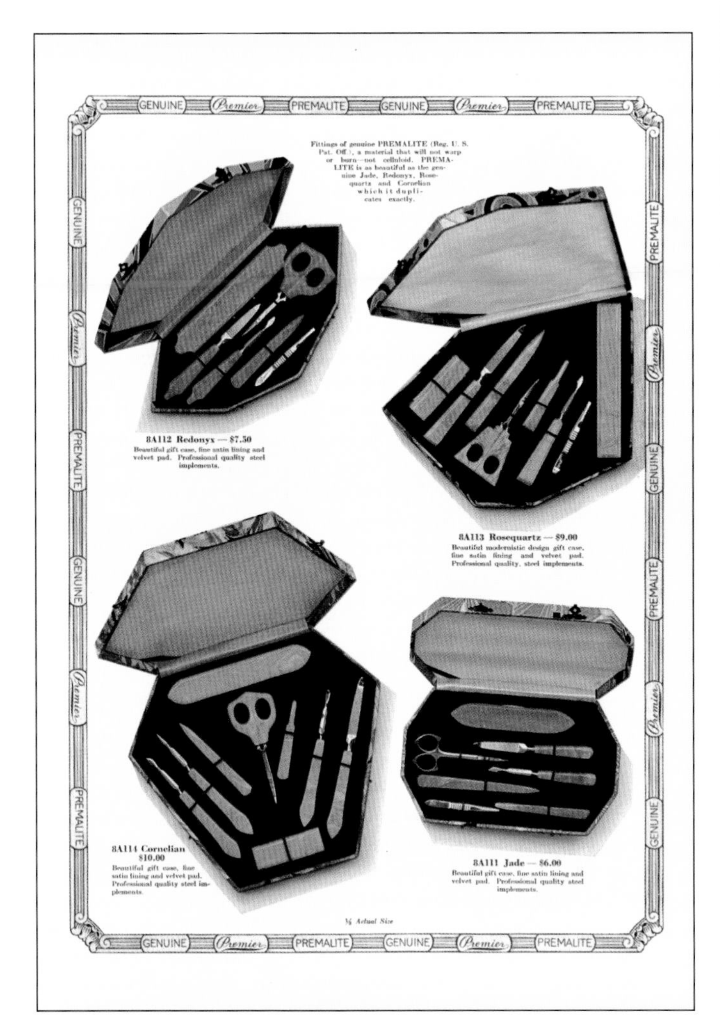

Presentation boxes for *Premalite* manicure sets from 1932 were colorful as well as geometric and modernistic exemplifying the Art Deco movement of the time.

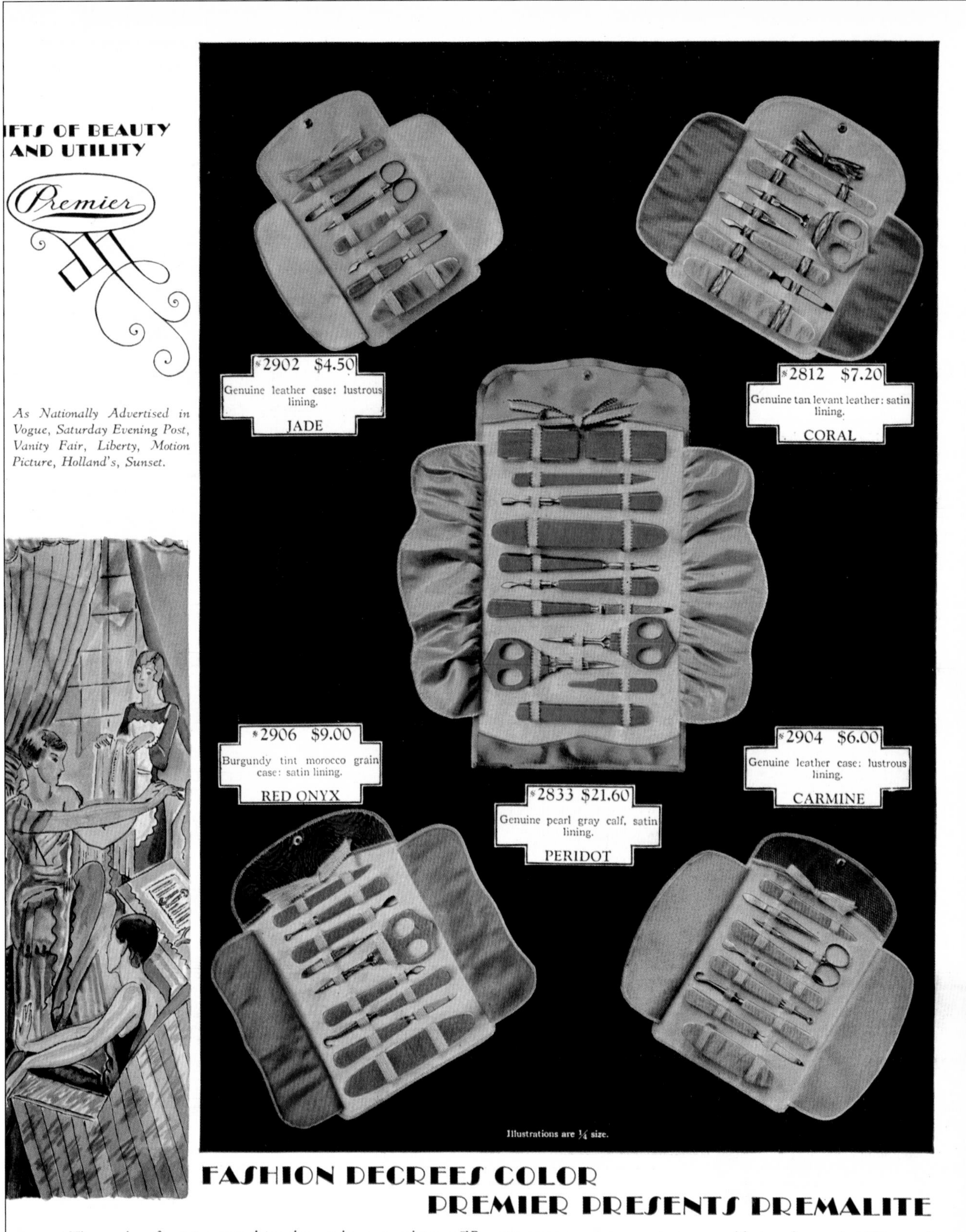

IFTS OF BEAUTY
AND UTILITY
Premier
As Nationally Advertised in Vogue, Saturday Evening Post, Vanity Fair, Liberty, Motion Picture, Holland's, Sunset.
*2902 $4.50
Genuine leather case: lustrous lining.
JADE
*2812 $7.20
Genuine tan levant leather: satin lining.
CORAL
*2906 $9.00
Burgundy tint morocco grain case: satin lining.
RED ONYX
*2904 $6.00
Genuine leather case: lustrous lining.
CARMINE
*2833 $21.60
Genuine pearl gray calf, satin lining.
PERIDOT
Illustrations are ¼ size.
FASHION DECREES COLOR
PREMIER PRESENTS PREMALITE
The modern feminine appeal is color—color everywhere. Premier presents a new manicure ensemble in colors to her liking and in harmony with her own intimate surroundings. Beautiful Premalite—an exclusive creation by Premier, comes in a variety of hues simulating the color and exclusiveness of jade, cornelian, rose quartz and other semi-precious stones. Here indeed, is a gift of utility and beauty—smart to the 'nth degree!

Genuine exotic leather cases with sumptuous linings were backdrops for the even more exquisite and colorful *Premalite* manicure ensembles by *Premier* in 1932. Premalite manicure accessories were offered in different hues resembling jade, carnelian, rose quartz, red onyx, peridot and coral.

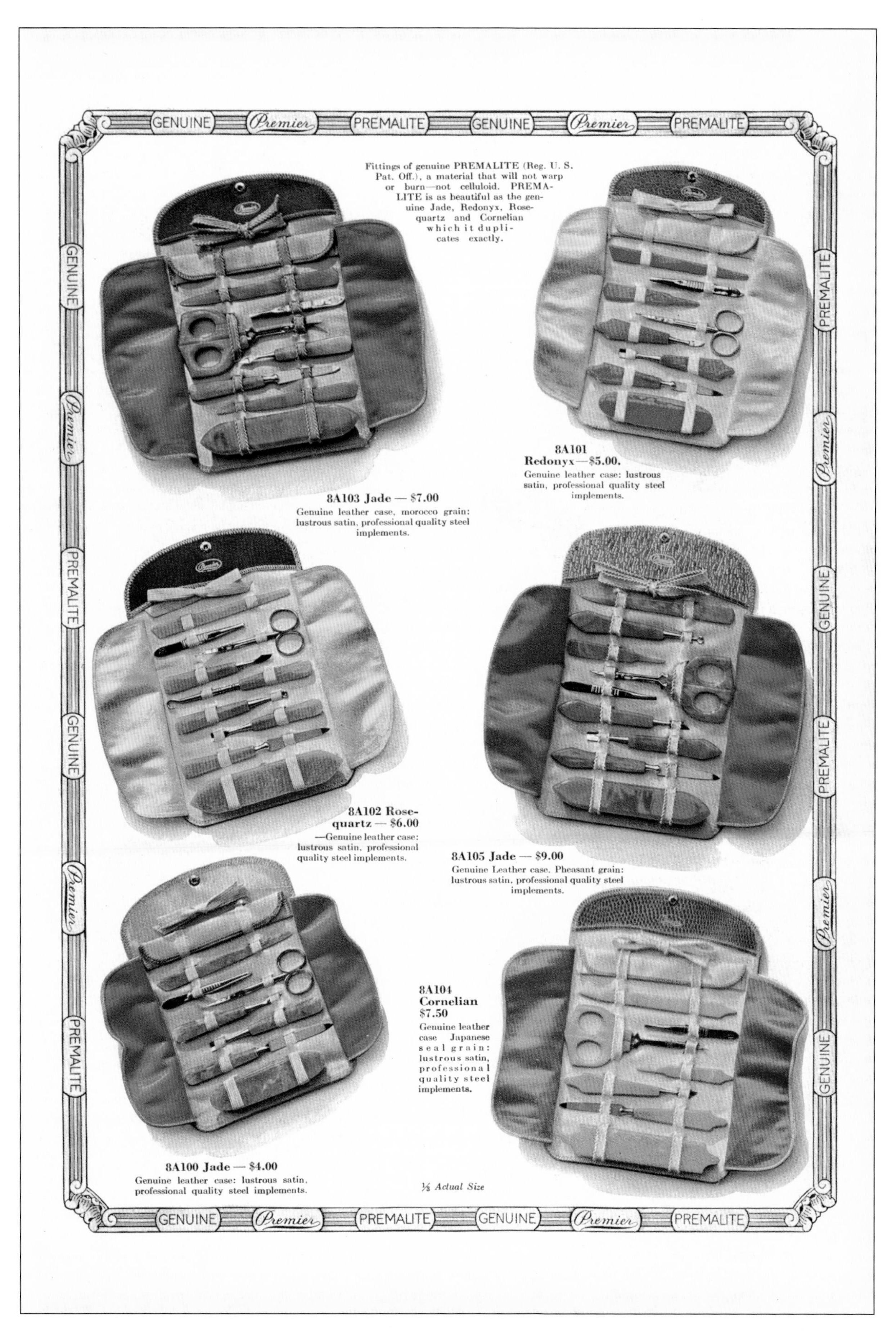

GENUINE
Premier
PREMALITE
Fittings of genuine PREMALITE (Reg. U. S. Pat. Off.), a material that will not warp or burn—not celluloid. PREMALITE is as beautiful as the genuine Jade, Redonyx, Rose-quartz and Cornelian which it duplicates exactly.
8A101
Redonyx—$5.00.
Genuine leather case: lustrous satin, professional quality steel implements.
8A103 Jade — $7.00
Genuine leather case, morocco grain: lustrous satin, professional quality steel implements.
8A102 Rose-quartz — $6.00
—Genuine leather case: lustrous satin, professional quality steel implements.
8A105 Jade — $9.00
Genuine Leather case. Pheasant grain: lustrous satin, professional quality steel implements.
8A104
Cornelian
$7.50
Genuine leather case Japanese seal grain: lustrous satin, professional quality steel implements.
8A100 Jade — $4.00
Genuine leather case: lustrous satin, professional quality steel implements.
½ Actual Size

Roll-up manicure cases by *Premier* were constructed of exotic grains of leather. Pheasant grain, Japanese seal grain and Morocco grain were some examples found in the 1932 models offered for sale.

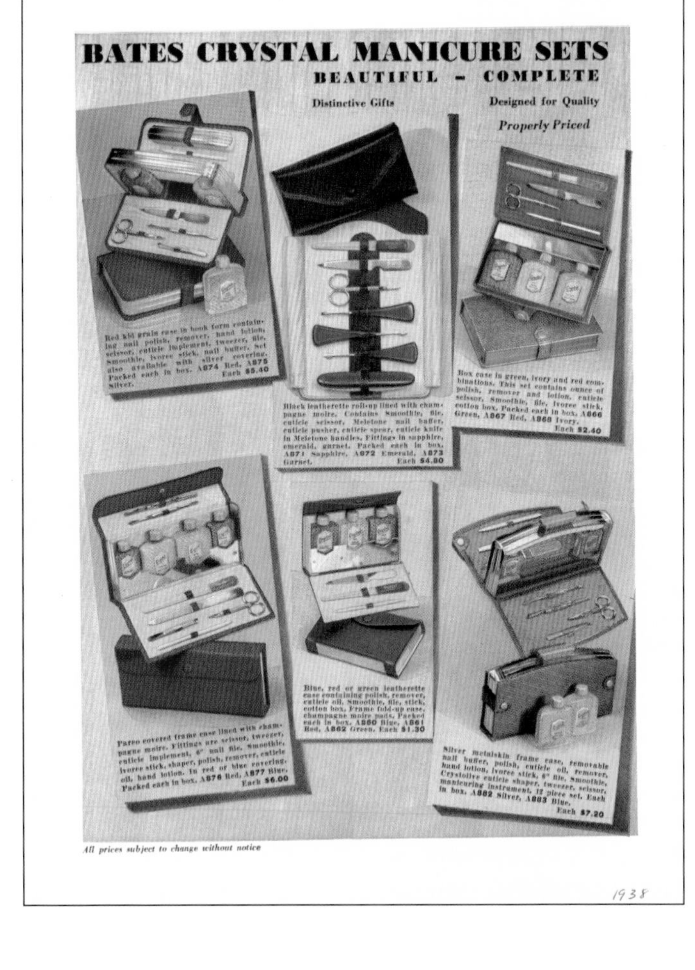

BATES CRYSTAL MANICURE SETS

BEAUTIFUL - COMPLETE

Distinctive Gifts

Designed for Quality

Properly Priced

Red kid grain case in book form containing nail polish, remover, hand lotion, scissor, cuticle implement, tweezer, file, Smoothie, Ivoree stick, nail buffer. Set also available with silver covering. Packed each in box. A874 Red, A875 Silver. Each $5.40

Black leatherette roll-up lined with champagne moire. Contains Smoothie, file, cuticle scissor, Meletone nail buffer, cuticle pusher, cuticle spear, cuticle knife in Meletone handles. Fittings in sapphire, emerald, garnet. Packed each in box. A871 Sapphire, A872 Emerald, A873 Garnet. Each $4.80

Box case in green, ivory and red combinations. This set contains ounce of polish, remover and lotion, cuticle scissor, Smoothie, file, Ivoree stick, cotton box. Packed each in box. A866 Green, A867 Red, A868 Ivory. Each $2.40

Pareo covered frame case lined with champagne moire. Fittings are scissor, tweezer, cuticle implement, 6" nail file, Smoothie, Ivoree stick, shaper, polish, remover, cuticle oil, hand lotion. In red or blue covering. Packed each in box. A876 Red, A877 Blue. Each $6.00

Blue, red or green leatherette case containing polish, remover, cuticle oil, Smoothie, file, stick, cotton box. Frame fold-up case, champagne moire pads. Packed each in box. A860 Blue, A861 Red, A862 Green. Each $1.30

Silver metalskin frame case, removable nail buffer, polish, cuticle oil, remover, hand lotion, Ivoree stick, 6" file, Smoothie, Crystolite cuticle shaper, tweezer, scissor, manicuring instrument, 12 piece set. Each in box. A882 Silver, A883 Blue. Each $7.20

All prices subject to change without notice

1938

Manicure sets by *Bates* in 1938 were made with steel fittings, some with *Meletone* handles, and assorted bottles of polish, remover and lotions. The traditional look of the manicure set was changing.

Solingen import, cuticle, nail, sewing scissors, cuticle nippers. Cowhide case, suede lining, flap protector. Shocking Pink, Apple Green.
56070................Retail **$15.75**
plus .45 Fed. Tax

Combination pocket book-manicure set. Removable pad—22 karat gold plated implements. Confetti brocade covering. Compact, lipstick holder and perfume bottle.
56031................Retail **$12.50**
plus .50 Fed. Tax

Imported essential set with cuticle nippers. Solingen best quality. Cowhide case. Suede lined. Lipstick Red or Turquoise Blue.
56056............Retail **$7.95**
plus .25 Fed. Tax

Griffon **LADIES' SETS**

Finest in Implements. Made in Solingen—world famous city for master craftsmen. Forged Steel, sharp and practical instruments

Best in Leather. Finest imported genuine top grain cowhides—rich colors. Tailored and stitched for beauty and long wear.

Quality Cutlery Ever Since 1888

These Solingen implements make the job easy. Featuring two nippers and three scissors. Cowhide case. Carrying strap. Suede lining and flap protector. Lipstick Red or Shocking Pink.
56072......Retail **$20.00**
plus .50 Fed. Tax

Useful set from Solingen with two scissors. Genuine leather, pin seal grain, silk lined. Red or Green.
56052........Retail **$5.00**
plus .12 Fed. Tax

Solingen import, top quality implements, including two scissors and cuticle nippers. Cowhide case. Lining and flap protector of soft suede. Turquoise Blue or Parchment White.
56062...........Retail **$12.50** *plus .30 Fed. Tax*

6 Piece Pinking Shears-Dressmaker set includes all sewing essentials. Genuine leather case—Silk lined—Red or Green.
56212..................Retail **$16.50**

Griffon offered ladies' and gents' manicure sets in 1957 with top quality Soligen steel implements in fine genuine leather cases.

In the early 1950s, a company called Clauss, noted as a cutlery maker since 1877, began manufacturing manicure sets. Slim pocket-sized sets with the finest professional implements were housed in zippered cowhide cases. Larger sets that were advertised as distinguished and custom-made were lined with natural pigskin. Wafer-thin sets, also made of top grain smooth cowhide were lined with genuine suede. Five professional implements were fitted inside. The sets listed for $10.45 in 1954. Three years later, Jack Kellmer Company offered Griffon manicure sets for ladies and men in their wholesale catalog. Advertised as *the finest available anywhere...deluxe models of real leather with famous precision "Soligen" implements.* Sets ranged from $5.00 to $20.00.

Rumpp Leatherware Company manufactured combination toilet and manicure sets for men and women in the late 1950s. A large variety of travel sets, empty traveling cases and fitted toilet sets were made of fine leathers.

Another cutlery company called Regent Sheffield began manufacturing manicure sets in 1960. The emphasis was now placed on the precision steel implements whereas in the earlier decades, the material that the handles were made of and the exotic grains of leather that the case was made out of seemed to be top priority. In 1966, travel kits by Lion Brand now offered imported manicure sets *attractively designed with practical manicuring and pedicuring instruments.*

Jr. Miss Vanity Case made of black and gold vinyl by *Hassenfeld Bros., Inc., Central Falls, R.I.*

Open view of the child's vanity case showing the cutest vanity items for a little girl made just like Mom's! $85-100.

Chapter Eight
Boxes

Celluloid

Celluloid collar and cuff box with hunting dog scene on the top panel, a farm scene on the front panel and lovely floral decorations everywhere else. The top of the box opens up as storage for the collars and the bottom is home for the cuffs. $475-550.

Interior view of celluloid collar and cuff box. It is lined with blue silk.

Celluloid toilet and manicure case featuring a lovely young lady and cupid on the lid surrounded by a purple and green floral motif. The photo is surrounded by fancy gilded scroll work, which frames the picture. $450-525.

Two female Edwardian tennis players adorn the lid of this celluloid toilet box which originally housed a brush, mirror and comb set. $425-500.

Celluloid handkerchief box decorated with a well-dressed woman lounging in a formal setting. The rest of the box is floral decorations. The celluloid is embossed and gilded with fancy scrollwork surrounding the center photo. $395-450.

Celluloid necktie box with a wonderful scenic lid featuring mountains, streams, trees and even a fisherman with gold embossing framing the picture from both sides. A floral motif covers the rest of the box. $295-345.

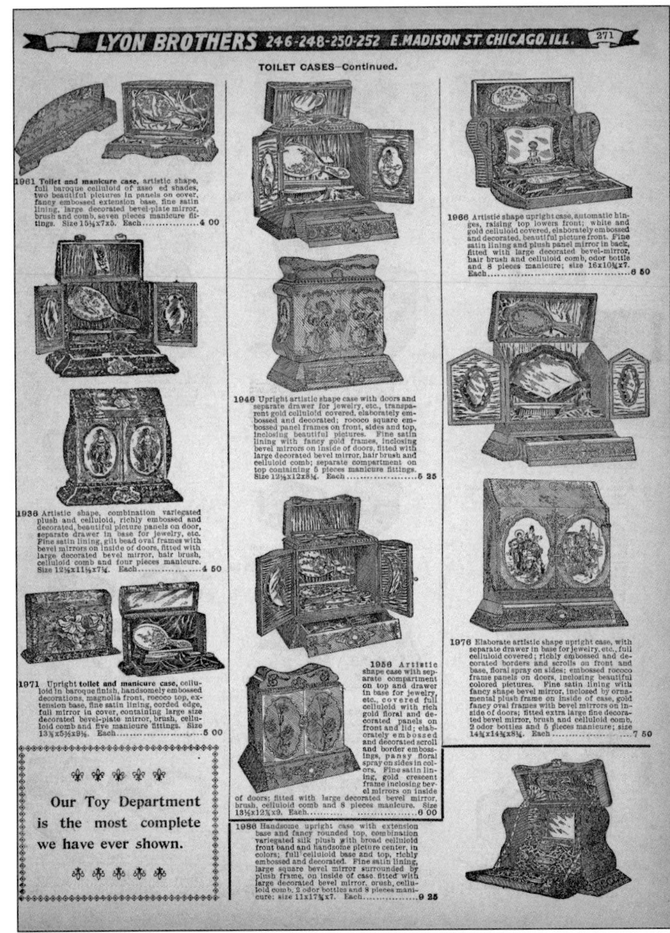

LYON BROTHERS 246-248-250-252 E. MADISON ST. CHICAGO, ILL. 271

TOILET CASES—Continued.

1961 **Toilet and manicure case,** artistic shape, full baroque celluloid of asso ed shades, two beautiful pictures in panels on cover, fancy embossed extension base, fine satin lining, large decorated bevel-plate mirror, brush and comb, seven pieces manicure fittings. Size 15½x7x5. Each 4 00

1938 Artistic shape, combination variegated plush and celluloid, richly embossed and decorated, beautiful picture panels on door, separate drawer in base for jewelry, etc. Fine satin lining, gilt bead oval frames with bevel mirrors on inside of doors, fitted with large decorated bevel mirror, hair brush, celluloid comb and four pieces manicure. Size 12½x11½x7¼. Each 4 50

1971 Upright **toilet and manicure case,** celluloid in baroque finish, handsomely embossed decorations, magnolia front, rococo top, extension base, fine satin lining, corded edge, full mirror in cover, containing large size decorated bevel-plate mirror, brush, celluloid comb and five manicure fittings. Size 13¾x5½x9½. Each 5 00

Our Toy Department is the most complete we have ever shown.

1946 Upright artistic shape case with doors and separate drawer for jewelry, etc., transparent gold celluloid covered, elaborately embossed and decorated; rococo square embossed panel frames on front, sides and top, inclosing beautiful pictures. Fine satin lining with fancy gold frames, inclosing bevel mirrors on inside of doors, fitted with large decorated bevel mirror, hair brush and celluloid comb; separate compartment on top containing 5 pieces manicure fittings. Size 12½x12x8¾. Each 5 25

1956 Artistic shape case with separate compartment on top and drawer in base for jewelry, etc., covered full celluloid with rich gold floral and decorated panels on front and lid; elaborately embossed and decorated scroll and border embossings, pansy floral spray on sides in colors. Fine satin lining, gold crescent frame inclosing bevel mirrors on inside of doors; fitted with large decorated bevel mirror, brush, celluloid comb and 8 pieces manicure. Size 13½x12¾x9. Each 6 00

1986 Handsome upright case with extension base and fancy rounded top, combination variegated silk plush with broad celluloid front band and handsome picture center, in colors; full celluloid base and top, richly embossed and decorated. Fine satin lining, large square bevel mirror surrounded by plush frame, on inside of case, fitted with large decorated bevel mirror, brush, celluloid comb, 2 odor bottles and 8 pieces manicure; size 11x17¾x7. Each 9 25

1966 Artistic shape upright case, automatic hinges, raising top lowers front; white and gold celluloid covered, elaborately embossed and decorated, beautiful picture front. Fine satin lining and plush panel mirror in back, fitted with large decorated bevel-mirror, hair brush and celluloid comb, odor bottle and 8 pieces manicure; size 16x10¾x7. Each 6 50

1976 Elaborate artistic shape upright case, with separate drawer in base for jewelry, etc., full celluloid covered; richly embossed and decorated borders and scrolls on front and base, floral spray on sides; embossed rococo frame panels on doors, inclosing beautiful colored pictures. Fine satin lining with fancy shape bevel mirror, inclosed by ornamental plush frame on inside of case, gold fancy oval frames with bevel mirrors on inside of doors; fitted extra large fine decorated bevel mirror, brush and celluloid comb, 2 odor bottles and 5 pieces manicure; size 14¾x14¾x8¾. Each 7 50

Artistic and elaborate celluloid fitted toilet cases offered for sale in 1899.

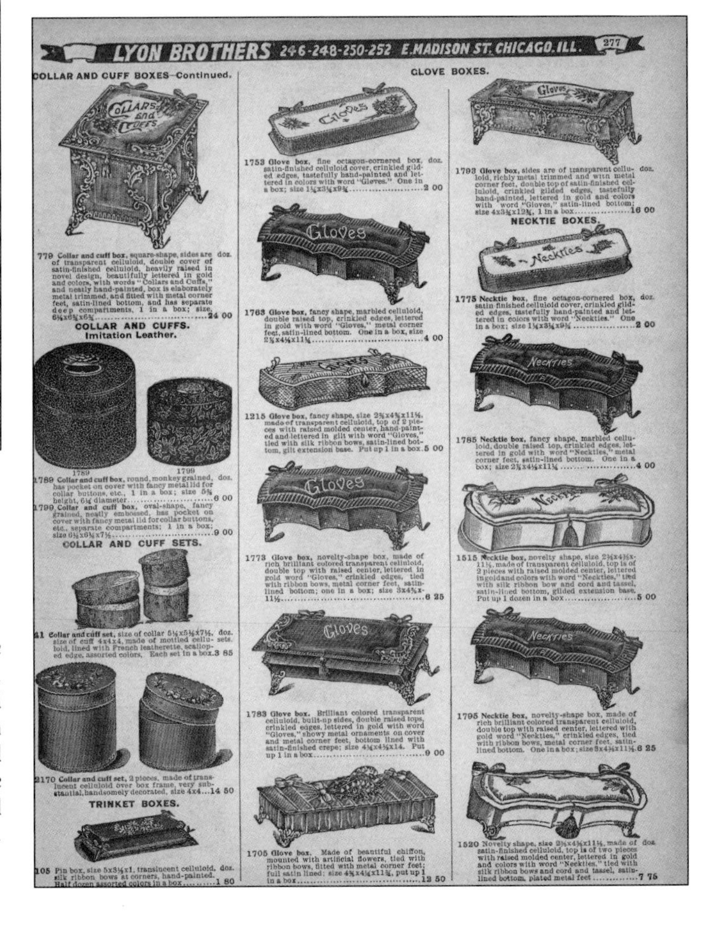

LYON BROTHERS 246-248-250-252 E. MADISON ST. CHICAGO, ILL. 277

COLLAR AND CUFF BOXES—Continued.

779 **Collar and cuff box,** square-shape, sides are of transparent celluloid, double cover of satin-finished celluloid, heavily raised in novel design, beautifully lettered in gold and colors, with words "Collars and Cuffs," and neatly hand-painted, box is elaborately metal trimmed, and fitted with metal corner feet, satin-lined bottom, and has separate deep compartments, 1 in a box; size, 6¼x6⅝x6⅝. doz. 24 00

COLLAR AND CUFFS. Imitation Leather.

1789 **Collar and cuff box,** round, monkey grained, has pocket on cover with fancy metal lid for collar buttons, etc., 1 in a box; size 5⅝ height, 6¼ diameter. doz. 6 00

1799 **Collar and cuff box,** oval-shape, fancy grained, neatly embossed, has pocket on cover with fancy metal lid for collar buttons, etc., separate compartments; 1 in a box; size 6½x6¾x7½ 9 00

COLLAR AND CUFF SETS.

21 **Collar and cuff set,** size of collar 5¼x5½x7½, size of cuff 4x4x4, made of mottled celluloid, lined with French leatherette, scalloped edge, assorted colors. Each set in a box. doz. sets. 3 85

2170 **Collar and cuff set,** 2 pieces, made of translucent celluloid over box frame, very substantial, handsomely decorated, size 4x4...14 50

TRINKET BOXES.

105 Pin box, size 5x3½x1, translucent celluloid, silk ribbon bows at corners, hand-painted. Half dozen assorted colors in a box. doz. 1 80

GLOVE BOXES.

1753 **Glove box,** fine octagon-cornered box, satin-finished celluloid cover, crinkled gilded edges, tastefully hand-painted and lettered in colors with word "Gloves." One in a box; size 1¼x3¼x9¾. doz. 2 00

1763 **Glove box,** fancy shape, marbled celluloid, double raised top, crinkled edges, lettered in gold with word "Gloves," metal corner feet, satin-lined bottom. One in a box, size 2⅞x4½x11¼ 4 00

1215 Glove box, fancy shape, size 2⅝x4⅝x11¼, made of transparent celluloid, top of 2 pieces with raised molded center, hand-painted and lettered in gilt with word "Gloves," tied with silk ribbon bows, satin-lined bottom, gilt extension base. Put up 1 in a box. 5 00

1773 **Glove box,** novelty-shape box, made of rich brilliant colored transparent celluloid, double top with raised center, lettered in gold word "Gloves," crinkled edges, tied with ribbon bows, metal corner feet, satin-lined bottom; one in a box; size 3x4¾x11½ 6 25

1783 **Glove box.** Brilliant colored transparent celluloid, built-up sides, double raised tops, crinkled edges, lettered in gold with word "Gloves," showy metal ornaments on cover and metal corner feet, bottom lined with satin-finished crepe; size 4¼x4½x14. Put up 1 in a box 9 00

1705 **Glove box.** Made of beautiful chiffon, mounted with artificial flowers, tied with ribbon bows, fitted with metal corner feet; full satin lined; size 4¾x4¼x11¾, put up 1 in a box 13 50

1793 **Glove box,** sides are of transparent celluloid, richly metal trimmed and with metal corner feet, double top of satin-finished celluloid, crinkled gilded edges, tastefully hand-painted, lettered in gold and colors with word "Gloves," satin-lined bottom; size 4x3¾x12¾, 1 in a box. doz. 16 00

NECKTIE BOXES.

1775 **Necktie box,** fine octagon-cornered box, satin-finished celluloid cover, crinkled gilded edges, tastefully hand-painted and lettered in colors with word "Neckties." One in a box; size 1¼x3¼x9¾. doz. 2 00

1785 **Necktie box,** fancy shape, marbled celluloid, double raised top, crinkled edges, lettered in gold with word "Neckties," metal corner feet, satin-lined bottom. One in a box; size 2⅞x4½x11¼ 4 00

1515 **Necktie box,** novelty shape, size 2½x4½x11¼, made of transparent celluloid, top is of 2 pieces with raised molded center, lettered in gold and colors with word "Neckties," tied with silk ribbon bow and cord and tassel, satin-lined bottom, gilded extension base. Put up 1 dozen in a box 5 00

1795 **Necktie box,** novelty-shape box, made of rich brilliant colored transparent celluloid, double top with raised center, lettered with gold word "Neckties," crinkled edges, tied with ribbon bows, metal corner feet, satin-lined bottom. One in a box; size 3x4¾x11¼. 6 25

1520 Novelty shape, size 2¾x4½x11¼, made of satin-finished celluloid, top is of two pieces with raised molded center, lettered in gold and colors with word "Neckties," tied with silk ribbon bows and cord and tassel, satin-lined bottom, plated metal feet. doz. 7 75

Transparent celluloid was used to make boxes for gloves, neckties, collars and cuffs. To embellish the boxes, some were hand painted while others were decorated with applied silk ribbons, cords, tassels and artificial flowers. Fancier models were accented with ornate metal lettering, and decorative metal corner feet.

Beautiful decorative and functional celluloid boxes from the late 19th and early 20th centuries have become a very desirable collectible in the 21st century. These celluloid treasures were offered for sale in abundance from mail-order and wholesale catalogs in the 1890s. Boxes designed to hold toiletware, jewelry, gloves, collars, cuffs, handkerchiefs, shaving kits, neckties, sewing items and other trinkets were lavishly made in large numbers.

Thin sheets of celluloid covered wooden or cardboard frames in a patented process beginning in the early 1890s. The overall appearance of the boxes was described as being shaped artistically while others were square or rectangular. Many boxes were decorated with colorful lithography creating exquisite works of art. Celluloid boxes and photo albums were sometimes backdrops for displaying the works of famous artists' of the period. Rich embossing was also found on many of the boxes, sometimes around the borders and other times creating the name of what the box was designed for. For example, a long slender rectangular box designed to hold neckties would have the word *Neckties* embossed on the lid. The interior of the boxes would be lined with silk, satin and velvet. Some catalogs described the linings as being *puffed moiré mercerized lining* while others just stated that the boxes were lined. Beveled mirrors were fitted into the lids. Gilded trim added the finishing touches to certain boxes while gilt locks, hinges and fancy corners were added. Toilet cases designed to hold brush, mirror and comb sets and occasionally manicure implements as well were made in many unusual shapes. Some were upright cases with an extra drawer at the bottom to hold jewels and a separate compartment at the top. Other cases were flat, some were round and on rare occasion, a heart-shaped case was offered. An "exceptionally attractive" upright case offered for sale in 1900 was described as:

> *A number that not only contains good, useful articles but is very pretty and an exceptional value at our very low price of $3.79. This combination upright toilet case and manicure outfit has a very artistic shape, covered with fancy combination celluloid and plush; two pretty pictures on swinging doors in gilt frame; has a very beautiful decorated top; extension base covered with celluloid. This case is fitted with good quality highly decorated brush with bevel edge mirror to match, also a good quality white celluloid comb, with bone handled file, buttonhook, manicure scissors and chamois buffer. On inside of doors are two beveled mirrors in gilt frames. Satin lined throughout.*

This description tells it all.

Occasionally, artistically shaped, upright celluloid cases would be made with automatic hinges. This meant that when the top was opened, the front was lowered automatically. Odd shapes were available. A box called "Napoleon's Tomb" was made with a curved top and sides, and an extension base. It had full tinted celluloid, embossing, and hand painting. The box was lined with satin and fitted with a bevel-plate mirror, brush, comb and two manicure implements. The case listed for $27 in 1899.

Celluloid handkerchief box depicting a beautiful lady in Edwardian attire adorned with furs. The rest of the box is a garden of tulips. $300-395.

Celluloid collar box decorated with a pretty young girl wearing a large hat. The rest of the box is floral decoration. $275-325.

Ivory-colored celluloid box with the word *Neckties* embossed on the lid. The rest of the box displays more embossed decorations. $175-225.

Each box was uniquely decorated with colorful celluloid, rich embossing, hand painting and decorated borders. Images of beautiful women in beautiful clothing, gorgeous floral themes and darling children graced the lids of the boxes. Boxes with animals or scenes were also popular. Some boxes were made with a combination of variegated silk plush with a broad celluloid front band and handsome picture center. Less expensive models were made with celluloid tops and printed paper sides. Celluloid was sometimes plain and colored like the French Ivory toiletware. Embossing was the only distinguishing feature with the exception of the fancy hardware on each box. Transparent celluloid with crinkled edges, gilt metal trimmings and hand painted decorations were also used to make glove, collar, cuff, necktie, hankies and trinket boxes.

Metal

Boxes designed to hold powder in addition to those manufactured to hold jewelry, gloves, handkerchiefs, hairpins, collars, cuffs, neckties and many other trinkets were lavishly made of assorted metals. In the 1890s, department stores like *Marshall Field & Company* offered jewel and trinket boxes in a variety of shapes and sizes made of quadruple silver plate. Oval, oblong, square and heart-shaped boxes were designed with heavy embossing, bright finishes and satin linings. Some boxes were designed with decorative feet; some had ball feet, while others had flat bottoms with no feet at all. The most elaborate box offered for sale in their 1896 catalog was one which was designed to hold handkerchiefs and listed for $15.00.

Rare figural Art Metal jewel casket with a silver-plated finish. The jewel box looks like a lovely lady in fancy attire sitting on a bench. $195-245.

Square brass jewel casket accented with sapphire blue glass cabochons and applied decorations. $200-250.

Four *Art Metal Jewel Boxes* or jewel caskets, as they were sometimes called, with hinged tops and ormolu gold finishes. They were all designed with ornate embossed decorations. Three of the boxes have floral themes while the fourth box has an unusual grape motif. They are all silk-lined. Small box $75-95; medium box $95-145; large box $150-200.

Three lovely Art Nouveau jewel caskets made of Art Metal with silver and ormolu finishes. All are embossed with wonderful floral decorations and each box is silk-lined $95-150 each.

FINEST QUADRUPLE SILVER PLATED WARE.

ASSOCIATION GOODS, SEE PAGE 3.

LADIES' TRINKET OR JEWEL BOXES. All Lacquered.

ILLUSTRATIONS TOP ROW ONE-HALF SIZE.

No. 196. TRINKET BOX, $4 20
[LIGAMENT.]

No. 172. TRINKET BOX..............[LIMITARY] **$4 80**

No. 165. TRINKET BOX......**$3 60**
[LIMITABLE.]

Above Boxes are all nicely lined with Satin or Plush.

ILLLUSTRATIONS BELOW ALL ONE-THIRD SIZE.

No. 150. JEWEL CASKET, [LIMPING] **$11 40**
Fine Satin Lining.

No. 174. JEWEL CASKET.....**$13 20**
[LIMPNESS.]
Fine Satin Lining.

No. 163. JEWEL CASKET.....[LIMPER] **$10 80**
Fine Satin Lining.

FANCY GILT.

No. 16. JEWEL CASKET.....[LIMPID] **$11 40**
Opens and closes by moving handle backward and forward. Satin Lined.

No. 68. JEWEL CASKET.[LINGER] **$14 40**
Opens and closes by pushing down or raising handle. Satin Lined.

No. 65. JEWEL CASKET.[LINGERING] **$12 00**
Opens and closes by moving handle.
Satin Lined.

58

Extremely fancy jewel caskets and trinket boxes made of the finest quadruple silver plate that was available in 1895.

JEWEL BOXES.

QUADRUPLE PLATE.

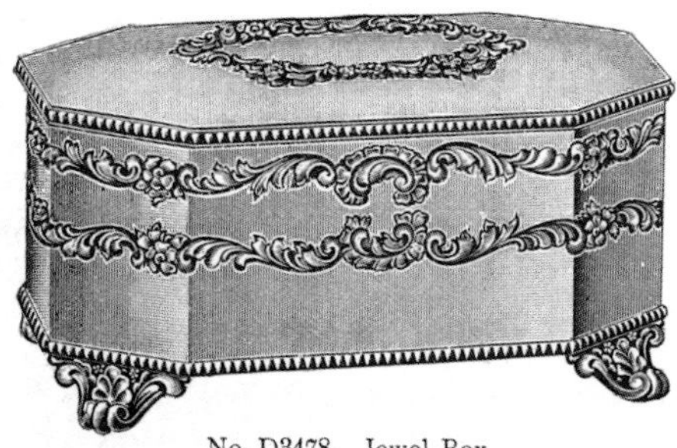

No. D3478. Jewel Box.
Bright Silver, Satin Lined........................$9 80

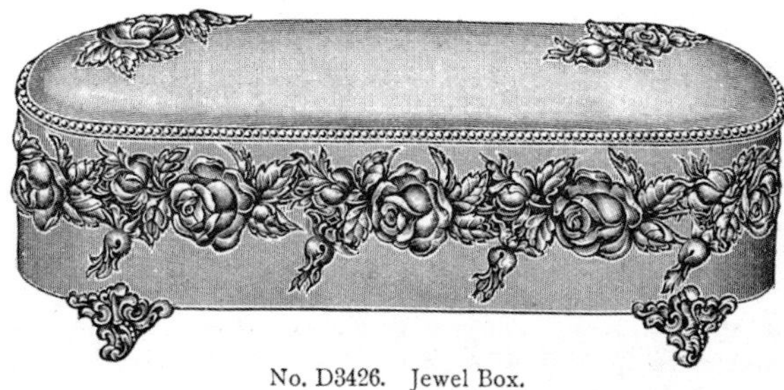

No. D3426. Jewel Box.
Bright Silver, La France Rose Decoration, Gray Silver Finish....$7 75
Satin Lined.

No. D3410. Jewel Box.
Satin, Gold Inlaid, Satin Lined...................$9 80
Hand Painted Medallion Top.

No. B9. Jewel Box.
Bright Embossed, Satin Lined..................................$6 25
Length 6¼ inches.

No. D3430. Jewel Box.
Bright Silver, Satin Lined.............$8 75

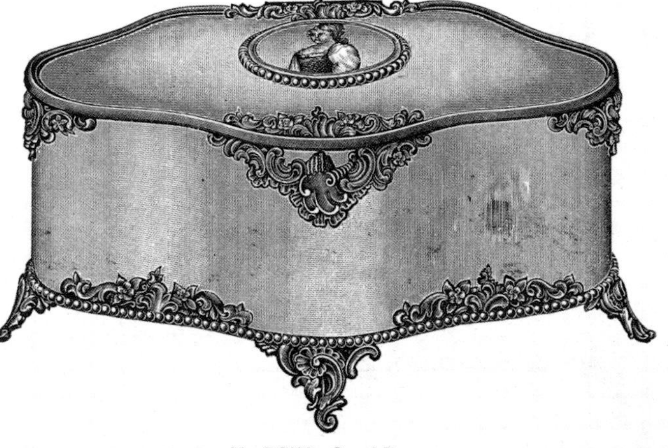

No. D3406. Jewel Box.
Satin, Gold Inlaid, Satin Lined..............................$11 90

In 1900, jewel boxes made of quadruple silver plate was available in bright and satin finishes with embossed and engraved designs. More expensive models were inlaid with gold. Some had hand painted medallion tops. Prices ranged from $6.25 to $11.90.

A mail-order company called the Busiest House in America offered quadruple plated boxes with lovely embossed designs. Other boxes sported embossed lettering on the lid, which described the exact purpose of the box. For example, one box was embossed to read "Rings & Things" while another read "Trinkets". Still others had cute phrases like "Ladies Friend" or a "Friend in Need" which described boxes specifically designed to hold hairpins. Besides being embossed, boxes were burnished, hammered, chased, engraved and engine-turned.

Metal jewel boxes were also referred to as "Jewel Caskets". In the last decade of the 19th century, jewel caskets were highly ornamented and very elaborate. Mostly made of fancy gilt and quadruple plate, some jewel caskets were designed with a large handle, that when pushed in either direction, the box would automatically open. Ornate finials were perched on top of box and underneath the handle. Some finials were made like cupids, birds or other animals. A more extreme example was one made in the shape of a perfume bottle with a cut crystal stopper. The boxes were heavily engraved and very appealing. They definitely retained the ostentatious nature of the Grand Victorian style.

In 1910, Chicago's Boston Store catalog advertised the *newest designs in jewel cases with an Ormula gold finish.* The spelling was slightly different but the effect was basically the same.

Jewel Cases Ormolu Gold Finish

1120 1098

1097

J. C. 1120, J. C. 9098, J. C. 1097. Ass styles and designs, Jewel Boxes, Ormolu Gold Finish, Silk lined Raised floral decorations, Average sizes, 1⅝x2, 1½x2 about 1¼x2.
Dozen 90c. Gross $10.50

J. C. 1059. Jewel case, with silk lining, fancy floral design, ormolu finish. Size 2½x2½.
Dozen $1.75

J. C. 1023. Ormolu gold plated fancy design jewel box. 3½x2½ inches, handsomely decorated on top and sides, good silk lining.
Dozen $3.00

J. C. 1057. Jewel box, gilt ormolu finish, embossed with floral designs, silk lined. 3x1¾.
Dozen $1.75

J. C. 1004. Fuchsia design, ormolu gold plated, fine silk lining.
Dozen $3.50

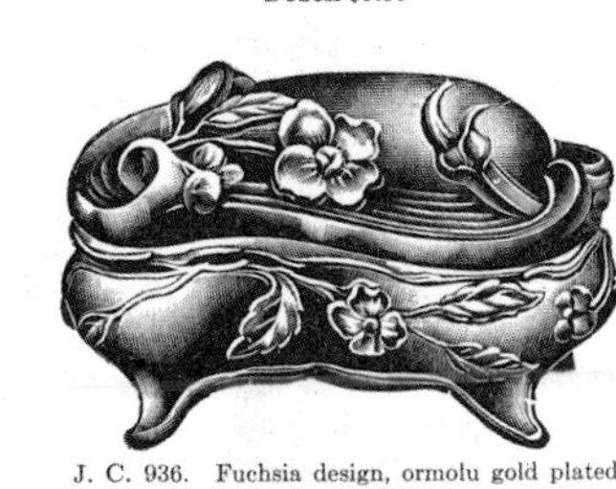

J. C. 1073. Jewel case, with fine silk lining, gold ormolu finish, wild rose design.
Dozen $3.50

J. C. 1053 ass. Same as above; assorted styles and designs, ormolu finish; each in a box; about 6x5½ inches.
Each $1.00

J. C. 1058. Jewel Case, 2½x1¾ inches, ormolu gold finish, silk lined.
Dozen $1.75

J. C. 1022. Fancy embossed top and body, ormolu gold plated finish, silk lining. Size 3x3.
Dozen $3.50

J. C. 936. Fuchsia design, ormolu gold plated, with fine silk lining, handsomely decorated on top and sides.
Dozen $3.50

J. C. 926 ass. Same as above, assorted designs and styles, ormolu gold finish, about 5½x5½ inches; each in a box.
Each $1.00

J. C. 1018. Embossed top and sides, with beautiful floral designs, silk lined, about 3x2½ inches.
Dozen $3.50

J. C. 1017 ass. Same as above, but extra large, about 6x5½ inches; very showy and attractive ass, designs; each in a box. **Each $1.00**

J. C. 922. Jewel box, silk lined elaborate embossed decorations, ormolu gold plated extra large size, 6 inches long 4½ inches wide, 4½ inches high.
Each 65c.

J. C. 1136. 4 inches long 3½ inches wide 4 inches high, ormolu gold plated Jewel Box, exceptionally good value. **Each 65c.**

The limited space does not allow us to display our entire variety of Jewel Boxes.

In 1899, lovely Art Nouveau designs and raised floral decorations were common on these jewel cases with ormolu gold finishes.

Jewel cases or caskets were also designed in the Art Nouveau style made of ormolu or gilt metal finishes. Wonderful designs in all shapes and sizes were handsomely embossed on top and along the sides with raised decorations. Each box was silk or satin lined. This style of box was popular from the late 1890s until the early 1930s. Some companies referred to them as art metal boxes with hinged tops. Many of them had raised floral decorations, especially large roses. Some footed jewel cases had French Gray finishes. In 1910, one oblong box with a raised pattern, footed base and silk lining in a French Gray finish listed for $2.48. A slightly larger example, finished in bright and dull ormolu listed for $3.98. A small ring box with a handle on top was $.48.

Two large and two small fluted glass jars with fancy embossed 24kt gold-plated lids. All four dresser jars fit nicely on a tray with a lace insert under glass. $195-235 set.

Crown-shaped musical jewel box made of gold-plated metal and bright red enamel. The inside of the box is fitted with a mirror and a velvet lining. $115-145.

Advertised in 1960 as a *Travel-Minded*, this *Phinney Walker* travel alarm made of brocaded fabric with rhinestone trim is also a velvet-lined jewel box. It is still accompanied by its original box. $75-115.

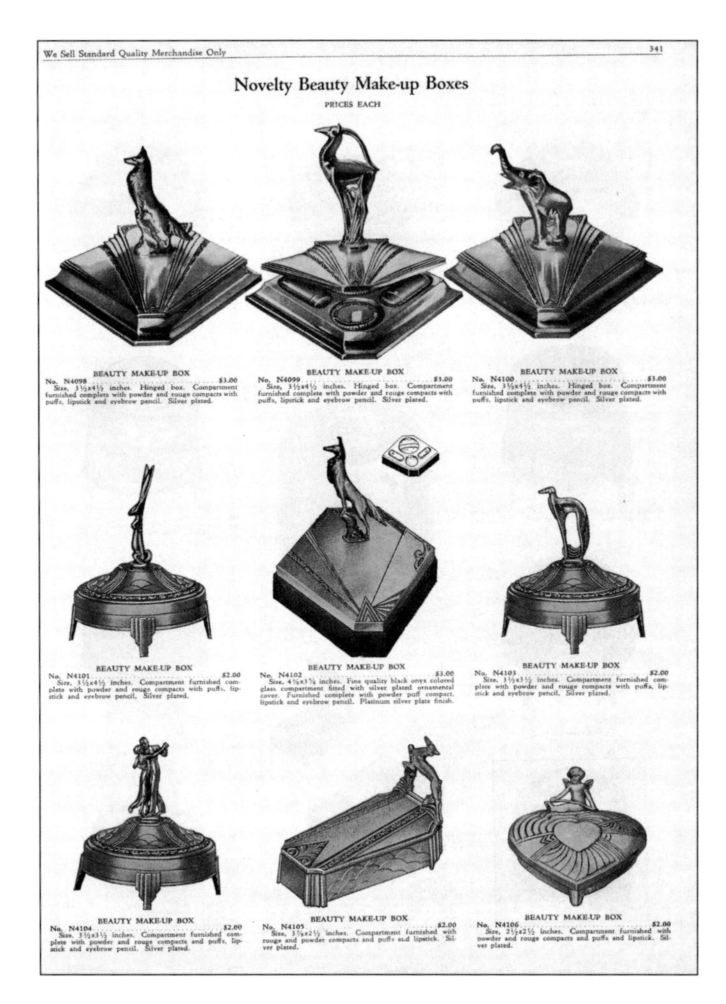

We Sell Standard Quality Merchandise Only 341

Novelty Beauty Make-up Boxes

PRICES EACH

BEAUTY MAKE-UP BOX
No. N4098 .... $3.00
Size, 3½x4½ inches. Hinged box. Compartment furnished complete with powder and rouge compacts with puffs, lipstick and eyebrow pencil. Silver plated.

BEAUTY MAKE-UP BOX
No. N4099 .... $3.00
Size, 3½x4½ inches. Hinged box. Compartment furnished complete with powder and rouge compacts with puffs, lipstick and eyebrow pencil. Silver plated.

BEAUTY MAKE-UP BOX
No. N4100 .... $3.00
Size, 3½x4½ inches. Hinged box. Compartment furnished complete with powder and rouge compacts with puffs, lipstick and eyebrow pencil. Silver plated.

BEAUTY MAKE-UP BOX
No. N4101 .... $2.00
Size, 3½x4½ inches. Compartment furnished complete with powder and rouge compacts with puffs, lipstick and eyebrow pencil. Silver plated.

BEAUTY MAKE-UP BOX
No. N4102 .... $3.00
Size, 4⅝x3⅝ inches. Fine quality black onyx colored glass compartment fitted with silver plated ornamental cover. Furnished complete with powder puff compact, lipstick and eyebrow pencil. Platinum silver plate finish.

BEAUTY MAKE-UP BOX
No. N4103 .... $2.00
Size, 3½x3½ inches. Compartment furnished complete with powder and rouge compacts with puffs, lipstick and eyebrow pencil. Silver plated.

BEAUTY MAKE-UP BOX
No. N4104 .... $2.00
Size, 3½x3½ inches. Compartment furnished complete with powder and rouge compacts and puffs, lipstick and eyebrow pencil. Silver plated.

BEAUTY MAKE-UP BOX
No. N4105 .... $2.00
Size, 3⅞x2½ inches. Compartment furnished with rouge and powder compacts and puffs and lipstick. Silver plated.

BEAUTY MAKE-UP BOX
No. N4106 .... $2.00
Size, 2½x2½ inches. Compartment furnished with powder and rouge compacts and puffs and lipstick. Silver plated.

Silver-plated make-up and powder boxes offered for sale in 1932 from A.C.Becken were streamlined and modernistic. Large finials depicting animals were very popular at that time.

214

Jewel, Candy, Powder and Cigarette Boxes

EXCELLENT GIFT ARTICLES

No. 701747....5¾ •
CANDY JAR
Candy jar. Minutely reproduced from a high priced import. Cast metal, electro-plated in antique silver or gold non-tarnishable finish, with removable cobalt blue glass liner. Height 8¼", width 4¼".

No. 701748....3¾ •
Mirror tray, 14½" long and 8½" wide. One-piece lightweight cast metal frame fashioned in a dainty floral design. High quality clear sparkling mirror, silvered with a lasting finish. Suedine base to prevent surface scratches on furniture. Frame is plated and lacquered in antique gold finish.

No. 701749....4½ •
Jewel box. Dainty cast jewel box. Velveteen lining and velveteen base. Removable lid with a Watteau group miniature in the center against a background of either ivory or black. Size 5½" long, 3½" wide, and 1½" high. Antique gold finish.

No. 701750....5¼ •
Candy box. Metal box in a new chased design inspired by an expensive import. Metal base holds a 2-compartment liner. The cover has an embossed Pastoral scene finished in old ivory. Size 6½" wide and 2½" high. Antique gold finish.

No. 701751....3¾ •
Powder box. Reproduced from a high priced French import. This round daintily designed metal box is 4½" in diameter and 1¾" high. The lid is removable and has a 3" miniature insert of famous ladies of French history (4 different subjects), and a clear crystal mirror inside to aid in powdering. Antique gold finish.

No. 701752....5¼ •
Candy box. 6½" in diameter and 2½" high. Beautifully cast one-piece metal frame with 3 compartment crystal glass liner. Ivory, black, or green covers with medallion miniature in center. Antique gold finish.

No. 701753....3¾ •
Powder box. Same specifications as No. 701751, but in a new rounded shape with dome top. 4½" in diameter and 3" high with a 2" miniature slightly raised so that lid may be easily lifted.

No. 701754....6 •
Cigarette box. Two compartment metal cigarette box. Holds 50 cigarettes. Lined with California redwood. Antique gold finish.

No. 701755....6 •
Candy box. Same specifications as No. 701752, but with bas-relief plaque in center of cover. 3-compartment crystal glass liner. Antique gold finish.

• THIS IS NOT A PRICE BUT A CODE SYMBOL FROM WHICH THE PURCHASER FINDS HIS COST. SEE PAGE 1

In 1950, powder boxes, jewel boxes, cigarette boxes and even candy boxes were attractively made of embossed metal with antique gold finishes and accented with miniature portrait medallions.

By the early 1930s, a modern look was becoming evident in jewel boxes and make-up and powder boxes due to the changes taking place because of the Art Deco movement. The smooth flowing lines and ornate decorations of the previous decades were being replaced by geometric lines and streamlined appearances. Hinged boxes made of silver-plated metal and natural gold finishes were designed with rouge and powder compartments, lipstick and eyebrow pencil. The boxes were designed with geometric shapes and topped with stylized animal finials.

Fancy musical powder boxes advertised in 1932 by the A.C.Becken Company were made of what they referred to as *Britannia* metal with gold-plated interiors. The tops were fitted with porcelain plaques which were *reproductions of famous European masters.* All of the music boxes had French Empire feet. Prices ranged from $4.50 to $10.50 each. By 1950, musical powder boxes were now made with Jeweler's Bronze. Other boxes were silver-plated or enameled in pretty pastel colors. The tops were porcelain cloisonné or novelty tops like butterflies under glass. Other powder or jewel boxes *reproduced from high priced French imports* were round in shape and made of embossed metal. The lids had miniature inserts of *famous ladies of French History or Pastoral scenes.* The impetus on French art had a huge impact on the manufacture of American dresser accessories in the 1950s.

Wood

In 1895, handsome boxes made of selected quartered oak were made in many shapes and many sizes. Oblong boxes were designed to hold women's gloves or men's neckties. A smaller but wider rectangular box was made to house handkerchiefs. Deeper boxes were meant to hold jewelry. Large square boxes were perfect for collars and cuffs. The boxes were highly polished, satin lined and accented with silver-plated trimmings placed on the corners or each box. Some boxes were even perfumed and advertised as a *rich and tasteful present for either a Lady or Gentleman.*

Replicas of quartered-oak boxes were made with the finest ebony and mahogany finishes . They were described as being made of "thoroughly seasoned hardwood." Instead of the silver-plated trimmings on the oak boxes, ebony and mahogany boxes often had gold-plated trim. The prices for all these were almost identical.

FINEST EBONY AND MAHOGANY FINISH BOXES.

Made of Thoroughly Seasoned Hardwood. Surfaces Highly Polished, Lined with Finest Satin (except Cigar Box). Perfumed. Gold Plated, Lacquered Trimmings. Rich and Tasteful Present for Lady or Gentleman.

Ebony Finish. Gilt Trimmings.
No. 1620E. Glove Box..................[RANGE] $8 00
13½x4x3 inches. Lock and Key.

Ebony Finish. Gilt Trimmings.
No. 1780E. Handkerchief Box.......$9 00 [RANGING.]
9¼x7½x3 inches. Lock and Key.

Ebony Finish. Gilt Trimmings.
No. 1520E. Jewel Box..[RANKLE] $8 00
Partition Tray.
8x5¼x5 inches. Lock and Key.

Ebony Finish. Gilt Trimmings.
No. 2013E. Collar and Cuff..[RANSACK] $5 00
7½x7½x5½ inches. With Catch.

Ebony Finish. Gilt Trimmings.
No. 1713E. Handkerchief Box..........$4 40 [RAPTURE.]
7½x7½x3 inches. With Catch.

Ebony Finish. Gilt Trimmings.
No. 1513E. Jewel Box........[RARITY] $5 00
8x5¼x5 inches. Lock and Key.

Mahogany Finish. Gilt Trimmings.
No. 1615M. Glove Box.........[RASHLY] $5 50
13½x4x3 inches. Lock and Key.

Mahogany Finish. Gilt Trimmings.
No. 1715M. Handkerchief Box $5 50 [RATABLE.]
7½x7½x3 inches. Lock and Key.

Mahogany Finish. Gilt Trimmings.
No. 2115M. Necktie Box..........[RATIFY] $5 50
13½x4x3 inches. With Catch.

Mahogany Finish. Gilt Trimmings.
No. 2015M. Collars and Cuffs [RATLIN] $6 00
7½x7½x5½ inches. With Catch.

Mahogany Finish. Gilt Trimmings.
No. 1515M. Jewel Box........[RAVINE] $6 00
Partition Tray.
8x5¼x5 inches. Lock and Key.

Mahogany Finish. Gilt Trimmings.
No. 2215M. Cigar Box........[REACH] $6 00
Capacity, 50 Cigars.
10x6½x4 inches. Lock and Key.

137

Ebony and mahogany finished hardwood boxes were designed for jewels, handkerchiefs, gloves, collars and cuffs to name a few. They were perfumed, lined with satin and accented with gold and silver trimmings. They were advertised as wonderful gifts for men and women in the 1890s.

In the 1920s and 1930s, wooden boxes made with reproduction prints under glass were popular and were used as jewel boxes. Small cedar boxes were plentiful in the 1940s and 1950s.

Wooden jewel box with a print of a pretty lady displayed under a panel of glass. $95-135.

Same box shown open to reveal a large mirror and three sections designed to store jewelry.

Rectangular wooden box with a *Godey's* print under a clear glass panel. The inside is fitted with a large mirror and three sections for jewelry storage. $95-130.

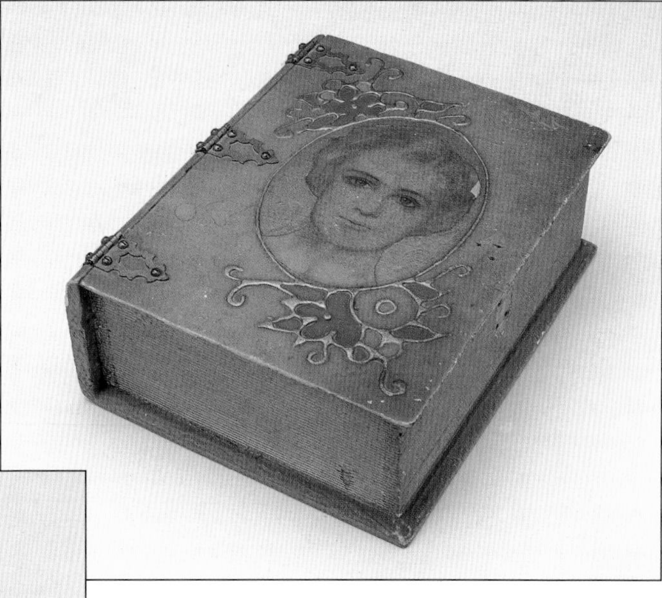

Fancy hinged wooden jewel box made in the shape of a book with a woman's face in an oval medallion surrounded by applied decorations. $75-90.

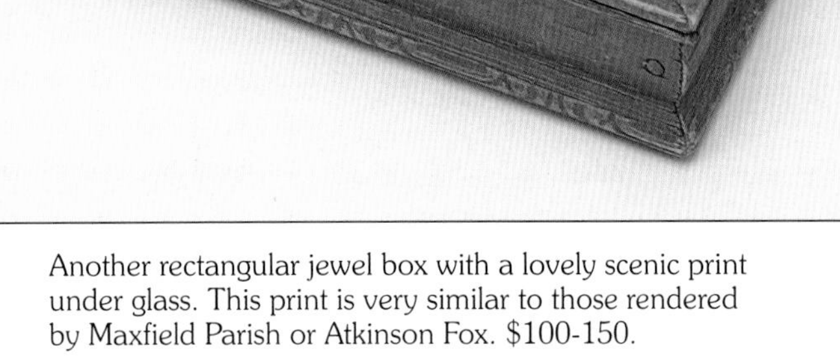

Another rectangular jewel box with a lovely scenic print under glass. This print is very similar to those rendered by Maxfield Parish or Atkinson Fox. $100-150.

Wooden jewel box made in the shape of a dresser with five drawers. $115-140.

Bibliography

Books

Cerwinske, Laura. *Russian Imperial Style.* New York, New York:Wings Books, 1990.

Dunn, Shirley. *Celluloid Collectibles: Identification & Value Guide.* Paducah, Kentucky: Collector Books, 1996.

Hayhurst, Jeanette (Consultant). *Miller's Glass Buyer's Guide.* Great Britian: Octopus Publishing Group, 2001.

Lauer, Keith and Robinson, Julie. *Celluloid: Collector's Reference and Value Guide.* Paducah, Kentucky: Collector Books, 1999.

North, Jacquelyne Y. *Czechoslovakia Perfume Bottles and Boudoir Accessories.* Marietta, Ohio: Antique Publications, 1990.

Patten, Joan Van and Williams, Peggy. *Celluloid Treasures of the Victorian Era: Identification & Values.* Paducah, Kentucky: Collector Books, 1999.

Sloan, Jean. *Perfume & Scent Bottle Collecting.* Radnor, Pennsylvania: Wallace-Homestead Book Company, 1989.

Snowman, A. Kenneth. *Carl Faberge: Goldsmith to the Imperial Court of Russia.* New York, New York: Crown Publishers, Inc., 1983.

Truitt, Robert & Deborah. *Collectible Bohemian Glass 1880-1940.* Kensington, Maryland: B&D Glass, 1995.

Whitmyer, Margaret & Kenn. *Bedroom & Bathroom Glassware of the Depression Years.* Paducah, Kentucky: Collector Books, 1990.

Catalogs

ACB Jewelers Wholesale Price List, Chicago, 1920.

Benjamin Allen & Co., Chicago, 1935 and 1952.

BHA Illustrated Catalog, Schwenksville, Pennsylvania, 1895.

Bloomingdales Illustrated Catalog, 1886, Dover Reprint, 1988.

Boston Store, Chicago, Fall & Winter 1910/1911.

Carson Pirie Scott & Company, Chicago, 1942.

Ft. Dearborn Watch & Clock Company, Chicago, 1923/1924 and 1936.

Geo. T. Brodnax Co., Memphis, Tennessee, 1915.

H.M.Manheim & Company, Catalogue # 72, New York, 1933.

Hagn's Holiday Gift Flyer, 1934.

Jason Weiler & Sons, Boston, Mass., 1927.

John V. Farwell Company, Chicago, 1920/1921.

John Wanamaker Store & Home Catalogue, Philadelphia, 1913.

L & C Mayers, New York, 1936 and 1938.

Lyon Bros. Catalog #258, Chicago, 1899/1900.

May & Malone Company, Chicago, 1930 and Red Book 1942.

Marshall Field & Company, Chicago, 1896, 1900 and 1933/1934.

M.Gerber Wholesale Company, Philadelphia, 1899/1900.

National Cloak & Suit Company, 1925.

New York Jeweler Illustrated Catalogue No.39, 1895.

N.Shure Company, Chicago, 1932.

Oskamp Nolting Company, 1950.

R.T.& Company Wholesale Supply House, Chicago, 1900.

The Becken Book, Chicago, 1938.

W.H.Sims Catalog, 1940.

Magazines

Ehrich's Fashion Quarterly (Summer 1879 and 1880).

Keystone (1919 and September 1929).

Index